www.brookscole.com

www.brookscole.com is the World Wide Web site for Brooks/Cole and is your direct source to dozens of online resources.

At *www.brookscole.com* you can find out about supplements, demonstration software, and student resources. You can also send email to many of our authors and preview new publications and exciting new technologies.

www.brookscole.com
Changing the way the world learns®

Mathematical Modeling
and Computer Simulation

Mathematical Modeling and Computer Simulation

Daniel P. Maki
Indiana University

Maynard Thompson
Indiana University

Australia • Canada • Mexico • Singapore • Spain
United Kingdom • United States

THOMSON
BROOKS/COLE

Publisher: *Bob Pirtle*

Assistant Editor: *Stacy Green*

Editorial Assistant: *Katherine Cook*

Technology Project Manager: *Earl Perry*

Marketing Manager: *Tom Ziolkowski*

Marketing Assistant: *Erin Mitchell*

Advertising Project Manager: *Bryan Vann*

Project Manager, Editorial Production:
Cheryll Linthicum

Art Director: *Vernon T. Boes*

Print/Media Buyer: *Lisa Claudeanos*

Permissions Editor: *Kiely Sisk*

Production Service: *Matrix Productions Inc.*

Text Designer: *Roy R. Neuhaus*

Copy Editor: *Connie Day*

Illustrator: *Interactive Composition Corporation*

Cover Designer: *Roy R. Neuhaus*

Cover Image: *Getty Images: Maciei Frolow/Brand X
Pictures, Comstock Images, Digital Vision, Digital
Vision/Adam Gault*

Cover Printer: *Quebecor World/Taunton*

Compositor: *Interactive Composition Corporation*

Printer: *Quebecor World/Taunton*

For more information about our products,
contact us at:
**Thomson Learning Academic
Resource Center
1-800-423-0563**

For permission to use material from this text
or product, submit a request online at
http://www.thomsonrights.com.
Any additional questions about permissions
can be submitted by email to
thomsonrights@thomson.com.

Library of Congress Control Number: 2004109840

ISBN 0-534-38478-1

Thomson Higher Education
10 Davis Drive
Belmont, CA 94002-3098
USA

Asia (including India)
Thomson Learning
5 Shenton Way
#01-01 UIC Building
Singapore 068808

Australia/New Zealand
Thomson Learning Australia
102 Dodds Street
Southbank, Victoria 3006
Australia

Canada
Thomson Nelson
1120 Birchmount Road
Toronto, Ontario M1K 5G4
Canada

UK/Europe/Middle East/Africa
Thomson Learning
High Holborn House
50/51 Bedford Row
London WC1R 4LR
United Kingdom

Latin America
Thomson Learning
Seneca, 53
Colonia Polanco
11560 Mexico
D.F. Mexico

Spain (includes Portugal)
Thomson Paraninfo
Calle Magallanes, 25
28015 Madrid, Spain

Brief Contents

Contents

CHAPTER **2**

Model Building: Selected Case Studies 25

CHAPTER **5**

Linear Programming Models   212

APPENDIX

Addendum for Students and Teachers on Projects and Presentations   **266**

Preface

The use of the concepts and methods of mathematics as an aid to understanding problems arising in the life and social sciences and in business is now a well-established approach to gaining knowledge in these areas. One of the most important contributions that mathematics makes in the study of such situations and problems is through the construction and analysis of models. This text is intended for upper-division undergraduate students in the mathematical, life, and social sciences and in business, and for graduate students with interest in the applications of mathematics in these areas. It is designed as an introduction to the use of mathematical and computer models in fields such as biology, ecology, finance, political science, and psychology. Students who have taken basic calculus, linear algebra, and a beginning course in probability should find the book accessible.

It is our experience that students learn mathematical modeling by doing it. However, most students are unprepared to simply "do it" without background knowledge of the process, examples of the process, and some practice. Consequently, the early chapters of the text are devoted to developing a useful framework in which to view the modeling process (Chapter 1) and to discussing several examples that illustrate the basic ideas in specific situations (Chapter 2). Chapter 2 includes examples that arise in a number of application areas and involve a range of mathematical concepts and methods.

The last section of Chapter 2, Section 2.8, contains a brief discussion of the important concepts of parameter estimation and model validation. In many situations, parameter estimation is an essential step in comparing predictions based on a model to data collected in observations. Model validation, also discussed in Chapter 4, is concerned with the question of whether a model is an accurate representation of the situation under study. Although both of these topics are worthy of substantial discussion, they involve ideas and techniques that lie outside the focus of the book, and they are not pursued in depth.

Much of the remainder of the book expands and develops important ideas that are drawn from the mathematical sciences and have proved their worth in model building. The topics chosen (the problems or situations and the mathematical structures) provide good examples of the process and of models that have wide and successful applicability, but there are other topics that would have served equally well. Chapter 3 illustrates the use of stochastic processes to model several situations arising in biology, ecology, psychology, and similar fields. Markov chains are singled out for more detailed study, and several basic results are developed. Some of the mathematical details are collected in Chapter 3's Appendix: Mathematical Details.

Although several topics related to computer modeling are included in Chapters 1 and 2, in Chapter 4 we focus in a systematic way on the development of computer simulation models. In some respects, Chapter 4 provides, for computer simulation, a discussion analogous to that of Chapters 1 and 2 for mathematical modeling. The chapter includes a section with more complex applications and material on model validation.

A class of models that are widely used in economic and financial decision making is introduced in Chapter 5. The mathematical setting and a variety of basic results are discussed in order to provide an overview of linear programming models. The chapter continues with a more detailed consideration of several special situations.

Student projects and presentations are a significant part of the modeling courses we teach, and we have included an Appendix on sources of projects—designed to be useful to students and to instructors—in which we discuss the use of projects as a vehicle for students to participate actively in the model-building process. Effective communication, at the beginning to be sure the goals of the project are understood and at the end to describe results, is an important part of project activity. The Appendix provides an overview and several examples.

Almost every section of the book includes exercises. Some exercises are fairly direct applications of the concepts discussed in the book. Others involve situations that are quite similar to those used as a context for model building in the examples but differ from the examples in some respects. A few are concerned with more complex versions of the models developed in the text or require a fair amount of creative work to extend an idea discussed in the examples. Some are feasible only if mathematical software is used to assist with the calculations and other numerical work.

Throughout the book, we rely on the use of mathematical software as a tool to facilitate the study and evaluation of models. The use of such software enables the reader to construct models for situations that are more realistic, and therefore more complex, than those that can be studied analytically or with less sophisticated computational aids. In order to make the discussion specific, we work with Maple and Excel, although we recognize that several alternatives would serve the same purpose. In our courses we do not dwell on the basics of the software. However, from time to time we find it useful to spend a few minutes in class discussion of specific aspects of a package that enables us to accomplish a goal.

While teaching modeling courses and writing this book, we have benefitted greatly from discussions with colleagues and students, including many graduate students majoring in fields other than mathematics. Our ideas on content and presentation have evolved over many years, and earlier versions of several of the topics addressed here were included in our book *Mathematical Models and Applications* (Prentice-Hall, 1973). We appreciate the comments of many colleagues on that book, and we hope the discussions here have benefitted from those comments. For the present book, we are grateful to Ted Hodgson for a careful reading of a preliminary version of the manuscript and for many helpful suggestions, and to the reviewers: Ethan Berkove, Lafayette College; William Gearhart, California State University Fullerton; and Christopher Hee, Eastern Michigan University. We also appreciate the contributions of the students who provided very useful assistance in preparing figures and simulation code and in working exercises. In particular, Lynn Winebarger and Mike Blandford contributed significantly to Section 4.5. Help with proofreading and the exercises was provided by Lynn, Mike, Frederic Picard, and Kynthia Brunette. Ms. Stela Adam provided outstanding and invaluable help with manuscript preparation.

Daniel P. Maki
Maynard Thompson

CHAPTER 1

Basic Principles

1.0   Overview of the Uses of the Term *Model*

What are models and why are people interested in them? Our goal in this book is to discuss mathematical models: to show what they are, how they are constructed, and how they are used. But before moving on to that task, it will be useful to provide a little background on various uses of the term *model*. In everyday conversation the term frequently refers to a display version of something (a fashion model or a model house) or to a miniaturization of something (a model boat or an architect's model of a building). Although such uses are standard, models of this sort, which are also known as physical models, are not the focus of this book. Other uses of the term—uses that we will discuss in more detail later—include real models, theoretical models, logical models, computational models, simulation models, and of course mathematical models. It is likely that you have encountered some of these uses previously, and we will "set the stage" by commenting briefly on each of them here.

Physical models, as noted above, include miniature versions of real objects (a familiar example is a model plane) and idealized versions of real people (a good example is a model teacher). They are intended for study, experimentation, display, and sometimes emulation. You might test the aerodynamics of a new airplane design by experimenting with a model in a wind tunnel, you might study the effects of new pedagogical techniques by observing an experienced teacher in carefully controlled situations, and you might learn the public's reactions to a new style of clothing by using a fashion model and videotapes in focus groups. A map or a globe is a physical model of the earth; it is a miniature version of a real object. Today, in many situations, physical models are being replaced by the creative use of computer-assisted design and testing, but the use of the term remains common.

Theoretical models are commonly associated with science, although they also occur in other areas. They are used in an attempt to explain or account for observed phenomena by creating a conceptual or hypothetical mechanism or process. Well-known examples include the Niels Bohr "planetary motion" model of an atom and Gregor Mendel's model of genetics. In many cases there is more than one theoretical model for a set of observations. Such models have played a very important role in most of the physical and life sciences.

Logical models, in contrast to physical models, are a part of the abstract side of mathematics. Abstract mathematics is a formal subject in which one begins with undefined terms, definitions, axioms, and standard rules of inference and then, using these rules, deduces theorems. The system of definitions, axioms, and theorems forms a mathematical theory. For example, if the axioms are those of Euclidean geometry, then they yield the theory of Euclidean geometry. A logical model is a concrete representation of an abstract system. As we shall see later in this chapter, finding a logical model for an axiom system is important in determining some useful characteristics of the system.

Mathematical models are not physical models, and they are not logical models. But a mathematical model may be closely related to a theoretical model, and it may be associated with one or more logical models. In all cases, mathematical models attempt to represent reality, or aspects of reality, by using mathematical concepts, symbols, and relations. The construction of a mathematical model involves representing real objects and processes with mathematical objects and processes. For a specific real situation, the first step in this representation is usually a simplification of the real situation, assuming that certain aspects can be neglected (at least for the moment) and that other aspects can be assumed to be simpler than they actually are. This simplification enables us to deal with fewer objects and processes than would be necessary if we retained all the detail of the real situation. The step of observing, simplifying, and eliminating detail is a very important aspect of the model-building process. It is an attempt to identify the most important parts of the setting—the parts we must incorporate into our study if we are to succeed in understanding the observations and making predictions. The result of this process of simplification and idealization is called a *real model,* and developing a real model is a crucial step in the task of constructing a mathematical model.

Mathematical models consist of symbols, assumptions about symbols, assumptions about relations among the symbols, and a connection between the real model and these symbols and relations. Mathematical models are often expressed in terms of equations, such as the famous $f = ma$, but mathematical models need not consist of equations. For example, a model for group decision making (a group of people trying to select among a set of alternatives) might consist of a set of assumptions about how the group should choose to make a collective decision—assumptions based, in part, on how individuals make their decision. Proposals for decision-making processes might involve diagrams and tables, but not equations. Examples of such processes are discussed in Section 2.3.

Mathematical models that involve equations can be studied using the same tools that we apply to equations whatever their origin—namely, analytical, geometrical, and computational tools. Engineers and physical scientists usually begin their study of a real situation with models involving equations that can be solved analytically, often by using calculus. However, models leading to predictions that agree closely with observations usually involve equations that cannot be solved analytically but instead require computational tools. Once the equations of a mathematical model are coded into a computer program, it becomes a *computational model,* and it is studied numerically and graphically. The equations are solved numerically, the results are displayed in either numerical or graphical form, and the results are examined to determine how well they correspond to observations.

Mathematical and computational models have been used with great success, but they also have limitations. How should models be evaluated and what determines whether a model is successful? To some extent, the answer depends on the purpose behind the creation of the model. Some models are constructed in an attempt to provide an *explanation* of certain observations, other models are constructed with a goal of making *predictions* or facilitating *decisions,* and still others are constructed for a combination of reasons. Mendel's model for inheritance can be viewed as an example of a model constructed to explain a set of observations. Newton's model for classical mechanics, which can be used to describe the motions of a set of interacting particles subject to gravitational forces, was applied to explain the observed motions of the planets around the sun. Also, with the help of careful observations of otherwise unexplained motions of known planets, it led to the discovery of new planets.

Although the earliest uses of mathematical models were in the sciences, it is now common for the concept and techniques to be applied in other fields. For instance, mathematical models based on assumptions about production are commonly used in business and industry to help managers make decisions about the allocation of resources. Also, sophisticated mathematical models of various financial instruments are now commonly used on Wall Street. In general, a model is judged by how well it accomplishes the task for which it was intended. If it was designed to explain, then it should provide an acceptable description of the observed phenomena. If it was designed to predict, then it is judged by the accuracy of the predictions based on it. If it was designed to facilitate decision making, then it is judged by the efficiency and accuracy of decisions based on it, compared with decisions based on other criteria.

In addition to explaining, predicting, and providing a basis for decision making, models are sometimes created with a goal of *control* and *guidance*. Such models include components for intervention and control through processes external to the system. A biomedical model of this type might include the administration of drugs intended to combat a disease; an economic model of this type might include the control of interest rates (as the Federal Reserve does with the U.S. economy). Situations as complex as human disease and the U.S. economy are known to be difficult to control, but there are great benefits to be derived merely from making progress. For instance, the U.S. economy itself is extraordinarily complex, and historical data provide only limited evidence of responses to various controls. In addition, the U.S. economy is interconnected with the economies of many other nations—which are themselves complex—in ways that are not completely understood. The decisions of individuals, individual businesses, specific sectors of the economy, governments, and similar parts of other economies all affect the economic consequences of efforts at influence and control, and it may be very difficult to assess the consequences of various actions. Nevertheless, the study of models that include mechanisms for control may provide useful information about the ways in which complex systems respond to external influences.

When traffic planners consider changing traffic flow in a city (for example, through the introduction of one-way streets, traffic lights, and bus lanes), it is difficult to use systems of equations to predict the effects. There are many individual drivers who make decisions using different criteria, and some may even make choices in a random manner. In settings such as this, it is natural to use a special type of mathematical model called a *simulation model* to predict the consequences of various proposals for modifying traffic flow. In a simulation model of traffic flow, a computer is used to create an artificial traffic flow system in which various assumptions about individual behaviors can be tested. The goal is to gather information about the consequences of the proposals to use in evaluating their merits.

This brief overview has touched on the various uses of the term *model,* and we will return to many of them in the sections that follow. Next we turn to our main topic: the creation and study of mathematical models. Where do such models come from? How are they created? Of what value are they?

1.1 The Process of Constructing Mathematical Models

An examination of the origins of any scientific field, be it astronomy or anatomy, physics or psychology, indicates that the discipline began with a set of observations and experiments. It is natural, then, that the first steps in quantifying the subject should involve the collection, presentation, and treatment of data. Consequently, the study of statistics has played a dominant role in the mathematical preparation of students working in the quantitative areas of the social and life sciences. A statistical treatment of data may be quite elementary, involving little more than listing, sorting, and a few straightforward computations. It may also be quite sophisticated, involving substantial mathematical ideas and delicate problems of experimental design. Once enough data have been collected and adequately analyzed, the researcher tries to imagine a process that accounts for these results. It is this activity, the mental or pencil-and-paper creation of a theoretical system, that is the topic of this book. In the scientific literature, this activity is commonly known as theory construction and analysis. In those cases where the system is expressed as a mathematical model, we shall refer to the process as the construction, development, and study of mathematical models.

The original problem nearly always arises in the real world (Figure 1.1), sometimes in the relatively controlled conditions of a laboratory and sometimes in the much less completely understood environment of everyday life. For example, a psychologist observes certain types of behavior in rats running in a maze, a geneticist notes the results of a plant hybridization experiment, or an economist records the volume of international trade under a specific tariff policy, and then each makes conjectures, or proposes reasons, for what she or he has observed. These conjectures may be based completely on intuition, but more often they are the result of detailed study and the recognition of some similarities to other situations that are better understood. This close study of a system, which for the experimenter usually precedes the forming of conjectures, is really the first step in model building. Much of this initial work must be done by a researcher who is familiar with the origin of the problem and with the basic biology, psychology, or whatever else is involved.

The next step (after initial study) is an attempt to make the problem as simple and precise as possible. This includes arriving at a clear and definite understanding of the words and concepts to be used. This process typically involves making simplifications, idealizations, and approximations. One important aspect of this step is the attempt to identify and select those concepts that are to be considered basic in the study. The purpose here is to eliminate unnecessary information and to simplify that which is retained as much as possible. For example, with regard to a psychologist studying rats in a maze, the experimenter may decide that it makes no difference that all the rats are gray or that the maze has 17 compartments. On the other hand, she may consider it significant that all the rats are siblings or that one portion of the maze is illuminated more brightly than another. This step of identification, approximation, and idealization will be referred to as constructing a **real model.** This terminology is intended to reflect the fact that the context is still that of real things (animals, apparatus, and the like) but that the situation may no longer be completely realistic. Returning to the maze, the psychologist may construct a real model that contains rats and compartments, but with the restriction that a rat is always in exactly one compartment. This restriction involves the idealization that rats move instantaneously from compartment to

Figure 1.1

compartment and are never half in one compartment and half in another. The psychologist might also construct a model in such a way that the rat moves from one compartment to another regularly in time, an approximation that may or may not be appropriate, depending on just what behavior is to be investigated. The model-building process up to this point is shown in Figure 1.2.

The third step (after study and formation of a real model) is usually much less well defined and frequently involves a high degree of creativity. One looks at the real model and attempts to identify the operative processes at work. The goal is to express the entire situation in symbolic terms. That is, the real model evolves into a mathematical model in which the real quantities and processes are represented by symbols and mathematical operations. Much of the value of the study hinges on this step, because an inappropriate identification between the real world and the mathematical world is unlikely to lead to useful results. The result of this process of model construction is usually not unique, and there may be several mathematical models for the same real situation. In such circumstances, it may happen that one of the models can be shown to be distinctly better than any of the others as a means of accounting for observations. In fact, it often happens that an elaborate experiment is designed for the purpose of showing that one model is truly better than others. Naturally, if this can be shown, then one generally chooses to use the best model. However, it may also happen that each of a number of models proves to be useful in the study—each model contributing to the understanding of some aspects of the situation, but no one model adequately accounting for all facets of the problem under consideration. Thus there may not be a best model, and which to use will depend on the precise questions to be studied. The process at this stage is shown in Figure 1.3.

After the problem has been transformed into symbolic terms, the resulting mathematical system is studied using appropriate mathematical ideas and techniques. The results of the mathematical study are theorems, from a mathematical point of view, and predictions, from the empirical point of view. The motivation for the mathematical study is not to produce new mathematics (new abstract ideas or new theorems), although this may happen, but to produce new information about the situation being studied. In fact, in many situations such information can be obtained using well-known mathematical concepts and techniques. The important contribution of the study may be recognition of the relationship between known mathematical results and the situation being studied. Figure 1.4 shows the steps of the model-building process that we have described so far. There is one important step remaining before the process is completed.

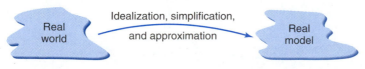

Figure 1.2

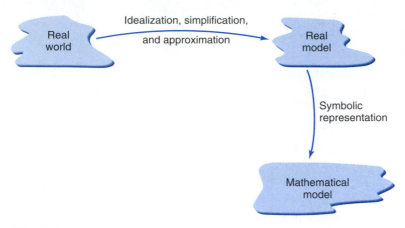

Figure 1.3

The final step in the model-building process is comparison of the results predicted on the basis of the mathematical work with the real world. The most desirable situation is that everything actually observed is accounted for in the conclusions of the mathematical study and that other predictions are subsequently verified by experiment. Frequently, however, such agreement is not observed, at least not on the first attempt. A much more typical situation would be that the set of conclusions of the mathematical theory contains some that seem to agree and some that seem to disagree with the outcomes of experiments. In such a case, one has to examine every step of the process again. Has there been a significant omission in the step from the real world to the real model? Does the mathematical model reflect all the important aspects of the real model, and does it avoid introducing extraneous behavior not observed in the real world? Is the mathematical work free from error? It usually happens that the model-building process proceeds through several iterations, each a refinement of the preceding one, until finally an acceptable model is found. Pictorially, we can represent this complete cycle of the model-building process as shown in Figure 1.5.

The solid lines in Figure 1.5 indicate the process of building, developing, and testing a mathematical model as we have outlined it. The dashed line is used to indicate an abbreviated

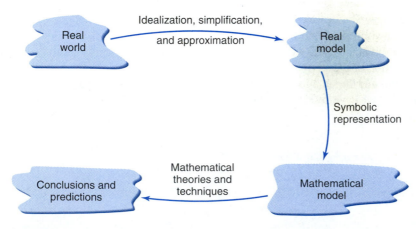

Figure 1.4

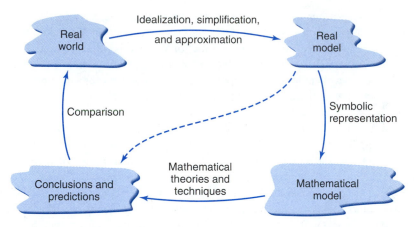

Figure 1.5   The Complete Model-Building Process

version often used in practice. The shortened version is particularly common in the social and life sciences, where "mathematization" of the concepts may be difficult. In either case, the steps in this process may be exceedingly complex, and there may be complicated interactions between them. However, for the purpose of studying the model-building process, such an oversimplification is quite useful.

We also note that this distinction between real models and mathematical models is somewhat artificial. It is a convenient way to represent a basic part of the process, but in many cases it may be difficult to decide where the real model ends and the mathematical model begins. In general, research workers often do not worry about making such a distinction. Hence one frequently finds that predictions and conclusions are based on a sort of hybrid model, part real and part mathematical, with no clear distinction between the two parts. There is, however, some danger in this practice. Although it may well be appropriate to work with the real model in some cases and with the mathematical model in others, one should always keep in mind the setting that is being used. At best, a failure to distinguish between a real model and a mathematical model is confusing; at worst, it may lead to incorrect conclusions. Complications may arise because problems in the social, biological, and behavioral sciences often involve concepts, issues, and conditions that are very difficult to quantify. Thus essential aspects of the problem may be lost in the transition from the real model to the mathematical model. In such cases, conclusions based on the mathematical model may not be conclusions about the real world or the real model. Thus there are circumstances in which it is crucial to identify, and keep clearly in mind, the model to which a conclusion refers.

1.2   Types of Mathematical Models and Some Practical Aspects of Model Building

Many skills are needed for successful model building. Suppose you have a situation arising in the real world and some questions about that system—for example, a population of an endangered species and a question about the long-range size of the population. It is essential to be able to identify those aspects of the population that are most relevant to answering the question. For instance, it may be very important to know the current population size and the reproductive characteristics. Whether the population is located in Maryland or in Florida may be less crucial. Many times, the determination of what is important and what is not must be made by someone who is knowledgeable about the situation. In our example of an endangered species, we would normally look to biologists and ecologists to provide information and advice on which aspects should be included in our study if the predictions are to be reliable.

Another valuable skill is the ability to recognize general structures and relations among components of models. The recognition of patterns and general characteristics is a useful skill in the study of pure mathematics, and it is equally valuable in building and studying mathematical models. For example, very similar models might be expected to be useful in studying the servicing of automobiles at tollbooths and the servicing of customers in a bank. However, it is less obvious that, from a mathematical point of view, these two situations have a great deal in common with certain models for the propagation of a rumor or the transmission of a contagious disease.

■ Deterministic versus Stochastic Models

There are several decisions a model builder must make, and one of the most important is the extent to which probabilistic aspects of the situation are to be incorporated into a model. If such aspects are included, then the model is termed *probabilistic*. It is called *deterministic* if such aspects are not included. These are broad classifications, and many models are hybrids—that is, they include some probabilistic and some deterministic aspects. A model in which we study the evolution of a system through time is said to be *stochastic* if probabilistic characteristics of the situation are incorporated into the model. The predictions are necessarily in probabilistic terms. That is, a **stochastic model** predicts the probabilities of events and the expected values of numerical outcomes. No matter how much we know about a situation, we cannot make a specific prediction; the uncertainty is inherent. In contrast, **deterministic models** have the property that if the parameters of the situation are known, then predictions can be made in specific terms; there is no uncertainty about the results. If we are studying a situation evolving through time (for example, the size of the population of white-tail deer in a certain state), then a deterministic model will lead to predictions of an exact population size, and a stochastic model will lead to predictions of an expected population size or to a prediction that the population will have a certain size with a certain probability.

As another example, this one drawn from the area of personal finance, in most cases a deterministic model is appropriate for the value of a certificate of deposit. If we know the amount invested, the term, the interest rate, and the schedule for compounding interest, then we can determine the exact value in the account at any time between purchase and maturity of the certificate. On the other hand, consider a money market savings account, an account in which the interest rate is determined by economic events that take place after the account is opened. In this case a stochastic model is usually more appropriate. Typically, we use estimates of anticipated interest rates, and the likelihoods of those rates occurring, to determine an expected value for the amount in the account. Because there is uncertainty in the anticipated interest rates, there is uncertainty in the predicted value of the account.

As yet another example, suppose we are concerned with a concept acquisition experiment in which an unbiased subject is provided stimuli intended to convey, in some explicitly defined way, the concepts of redness, roundness, and fourness. Then, we might choose to construct a model in which one uses the knowledge of how quickly the subjects learned the concepts of redness and roundness to predict how quickly the subjects learned the concept of fourness. In most models, it is not possible to predict exactly how quickly the concept of fourness will be acquired, and the strongest statement that can be made is a probabilistic one such as "The subject will learn the concept of fourness upon ten or fewer presentations of a stimulus with probability 0.8." Many of the most useful models in the life and social sciences are of this type—that is, they are models whose mathematical description involves chance and uncertainty. In some cases one can construct both deterministic and stochastic models for the same situation, and in some cases a comparison of the predictions serves as a check on the validity of both.

   Which type of model should be constructed depends on many factors, and the decision is ultimately just a choice of the investigator. A deterministic model is frequently used as a first approximation in a situation where a stochastic model appears more appropriate but also more complex. For example, when formulated in terms of a real model, a situation involving the growth of a biological population may appear to be best modeled in stochastic terms. However, a stochastic model may present technical difficulties that are either impossible to overcome or prohibitively time-consuming, so it may be desirable to use a deterministic model as a first approximation and to compare the predictions of such a model with observation to see whether it is adequate for the goals of the study. In general, one should not assume that predictions based on one type of model are necessarily better (or worse) than those based on another type. The relative merits of models of different types vary from one situation to another.

■ Implementing a Model

In order for a model to help us better understand a real-world situation, we must have results that can be tested or compared with observations. Consequently, realistic computational schemes or algorithms to generate predictions are an essential aspect of model building. There are some particularly important algorithms associated with the topics of this book, and we shall introduce some of them. Most algorithms of practical significance are implemented on a computer.

   In many situations, a model can be studied and predictions developed only after critical parameters are determined, and the task of parameter estimation is an important part of model building. For instance, a model for learning may involve parameters that represent the likelihood of learning a concept upon presentation of a stimulus; a model for population growth may involve birth and death rates; a model for the spread of disease may involve the rate of contact between a susceptible individual and an individual who has the disease; and a model for voting strategies may involve the likelihood of forming certain coalitions. In order to obtain specific predictions, it is necessary to give numerical values to the parameters in the system. In general, the estimation of parameters is a substantive and delicate problem for which each discipline has special and refined techniques. Because of the special nature and restricted applicability of these ideas, and certainly not because of any lack of importance, we consider only some special cases.

   To conclude this section, we remark that the use of mathematical techniques, in particular the use of mathematical models, is only one method among many that can be applied to questions arising in the sciences. As noted earlier, many important aspects of a situation in the social or life sciences may be very difficult to quantify. In such cases the use of mathematical models may be limited, and it may be better to study the situation in the context of a real model by nonmathematical means. Indeed, one might legitimately ask what basis we have for expecting mathematical methods to be effective. Our hopes rest on the proven effectiveness of mathematics in the physical sciences and on scattered but significant successes in the social and life sciences. One of the most impressive examples—indeed perhaps the most impressive example—of the fruitful use of mathematical models occurs in the study of planetary motion. This is also a fine example of the evolution of a model through several stages. Because this example had such a profound influence on science, it is worthwhile to consider it briefly, and it is the topic of the next section.

1.3   A Classic Example

The creation of a coherent system to explain and predict the apparent motions of the planets and stars as viewed from the earth is certainly a significant triumph of human intellect. That the problem had attracted attention from the most ancient times and that the theory was still undergoing modification in the twentieth century give an indication of the enormous amounts of time and energy that have been devoted to its study.

Early ideas of a fixed and flat earth covered by a spherical celestial dome were studied by the Greeks, who in the fourth century B.C. devised a real model that accounted, at least approximately, for the rough observations then available. The earth was viewed as fixed, with a sphere containing the fixed stars rotating about it. The "seven wanderers" (the sun, the moon, and five planets) moved in between. The Greeks' goal was to construct combinations of uniform circular motions centered on the earth by which the movements of the seven wanderers among the stars could be represented. Each body was moved by a set of interconnecting, rotating spherical shells. This system was adopted by Aristotle, who introduced 55 shells to account for observed motions. This real model based on geometry was capable of reproducing the apparent motions, at least to a degree consistent with the accuracy of the contemporary observations. However, because it kept each planet a fixed distance from the earth, it could not account for the varying brightnesses of the planets as they moved.

■ Ptolemy

This system was modified by Ptolemy, the last great astronomer at the famous observatory at Alexandria, in the second century A.D. In its simplest form, the Ptolemaic system can be described as follows: Each planet moved in a small circle (epicycle) in the period of its actual motion through the sky, while simultaneously the center of this circle moved around the earth on a larger circle. In this model the distance of a planet from the earth is not constant, and the variation of distance can be used to account for the observed variation in brightness. Also, this model can account for other observations—the retrograde motion of Mars, for instance. A planet is said to have retrograde motion when its motion about the sun as viewed from the earth is not always in the same direction. For instance, the basic motion of Mars around the sun is interrupted by periods when it appears to slow down, stop, begin to move in the opposite direction, slow again, stop, and resume motion in the original direction. When properly configured, epicycles can account for such observations. The basic epicycle model could accommodate repeated modifications to account for new observations, and such modifications did take place. As a result of repeated modifications, over time the system became more and more complicated. By the thirteenth century the system was quite complicated, for instance, requiring 40–60 epicycles for each planet, and still it required additional modifications as the accuracy of observations improved.

■ Copernicus and Kepler

By the beginning of the sixteenth century, there was widespread dissatisfaction with the Ptolemaic system. Difficulties resulting from more numerous and more refined observations forced repeated and increasingly elaborate revision of the epicycles on which the Ptolemaic system was based. As early as the third century B.C., certain Greek philosophers had proposed the idea of a moving earth, and as the difficulties with the Ptolemaic point of view increased, this alternative appeared more and more attractive. Thus, in the first part of the sixteenth century, the Polish astronomer Copernicus proposed a heliocentric (sun-centered) theory in which the earth, along with the other planets, revolved about the sun. However, he retained the assumption of uniform circular motion—an assumption with a purely philosophical basis—and consequently he was forced to continue the use of epicycles to account for variation in the apparent velocity and brightness of a planet as viewed from the earth.

Mechanical devices (physical models) were developed both for the Ptolemaic system and for systems that followed it. Figure 1.6 shows a physical model of the solar system that had planets circling the sun. These mechanical models were developed in the early 1700s and are often called orreries after Charles Boyle, the Fourth Earl of Orrery.

The next step, and a very significant one, was taken by Johannes Kepler. During the years 1576–1596 a Swedish astronomer, Tycho Brahe, had collected masses of observational data on the motion of the planets. Kepler inherited Brahe's records and undertook to modify Copernican theory to fit these observations. He was particularly bothered by the orbit of Mars, the large eccentricity of which made Mars very difficult to fit into circular-orbit– epicycle theory. Kepler was eventually led to make a very creative step, a complete break with the circular-orbit hypothesis. He posed, as a model for the motions of the planets, the following three "laws":

1. Each planet revolves around the sun in an elliptical orbit with the sun at one focus (1609).
2. The radius vector from the sun to a planet sweeps out equal areas in equal times (1609).
3. The squares of the periods of revolution of any two planets are in the same ratio as the cubes of their mean distances to the sun (1619).

These laws are simply statements of observed facts, or, more accurately, statements consistent with the observations available to Kepler. Together they form a mathematical model based on elementary concepts from geometry (elliptical curves, areas, distances) and physics (motions). Nevertheless, these laws are perceptive and useful summaries of observations. In addition to formulating these laws, Kepler attempted to identify a physical mechanism for the motion of the planets. He hypothesized a sort of force that emanated from the sun and influenced the motion of the planets. This model described well the accumulated observations and provided a useful starting point for the *next* refinement, a creative jump by Isaac Newton.

Figure 1.6 **The Grand Orrery c. 1750 Signed: Cole, Maker at the Orrery in Fleet Street, London.** This instrument was made by Benjamin Cole (1695–1766). The mechanism is brass, and the orrery is 64 cm high and 80 cm in diameter. Courtesy of the University of St. Andrews.

■ Newton

All models developed up to the middle of the seventeenth century involved geometrical representations with minimal physical interpretation. Newton's fundamental universal law of gravitation provides at once a physical interpretation, a concise and elegant mathematical model for the motion, and a description of the motion of all material particles. The law asserts that every material particle attracts every other material particle with a force directly proportional to the product of the masses and inversely proportional to the square of the distance between them. In this framework, the motion of a planet could be determined by first considering the system consisting only of the planet and the sun. This problem involves only two bodies and is easy to solve, using tools developed by Newton. The resulting predictions, the three laws of Kepler, are good first approximations for the motions of the planets because the sun is the dominant mass in the solar system and the planets are widely separated. However, the law of gravitation asserts that each planet is, in fact, subject to forces due to each of the other planets, and these forces result in perturbations in the predicted elliptical orbits.

The mathematical laws proposed by Newton provide such an accurate mathematical model for planetary motion that they led to the discovery of new planets. One could examine the orbit of a specific planet and take into account the influence of all the other known planets on this orbit. If discrepancies were observed between the predictions and observations, then one could infer that these discrepancies were due to another planet, and estimates could be obtained on its size and location. For instance, the planet Uranus was discovered by William Hershel in 1789, and by the middle of the nineteenth century there were significant discrepancies between the orbit of Uranus predicted on the basis of the known planets and the observed orbit of Uranus. Using these deviations, the location of a hypothetical planet was determined in 1845. When astronomers looked in the predicted location, the planet Neptune was discovered.

Yet even this remarkable model does not account for all the observations made of the planets. Early in the twentieth century, small perturbations in the orbit of Mercury, unexplainable in Newtonian terms, provided some motivation for the development of the theory of relativity. The relativistic modification of Newtonian mechanics apparently accounts for these observations. Nevertheless, one should not view this model as ultimate, but rather as the best available at the present time.

Newton's law and the associated model for planetary motion continue to be useful in predicting astronomical phenomena—for example, the trajectories of planetary probes launched from Earth and the existence of planets in other star systems. Astronomers have inferred the existence of planets around other stars by observing slight, periodic movements of the star back and forth across the line of sight. The extent and timing of the wobbles give information on the size, shape, and period of the conjectured planet's orbit. In late 1999, observations of the decrease in brightness of a dim star in the constellation Pegasus were just as predicted on the basis of a planet—with the characteristics deduced from analyzing the wobbles—crossing the face of the star. This transit was the most direct evidence yet of an extrasolar planet. Since 1999, many examples of such planets have been found.

The laws of Newton, viewed as a mathematical model, have provided the physical sciences with an extremely effective tool. The concepts of force, mass, and velocity can be made quite precise, and the model can be studied from a very abstract point of view. Although the social and life sciences do not yet have their equivalents of Newton's laws, the utility of mathematical models in the physical sciences gives hope that their use may contribute to the development of other sciences as well.

1.4   Axiom Systems and Models

In the preceding sections we stated that we intend to use mathematical models to study situations arising outside mathematics, and we considered one example wherein the use of mathematical models has been particularly rewarding. Before continuing, it is appropriate for us to make the concept of a model more precise. As we have noted, a mathematical model can be viewed as a set of assumptions. Because a set of assumptions is a key component of an axiom system, we begin with this topic and later relate it to the modeling process.

Remark   This section is an unusual one, and at first glance, it may not seem to be closely related to the rest of the chapter. It deals with formal axiom systems and with proofs of theorems, and it appears to be part of "pure" mathematics. However, in our view of the applications of mathematics, the topics of this section are a key factor underlying the power of mathematical modeling and the usefulness of mathematics in other areas of study. In this section we show how (and why) proving theorems in a formal setting, where the assumptions (axioms) are clear, provides results that can be used in many different settings without the need to prove them again. This section also shows that proofs of general theorems must be based on the axioms assumed and not on special circumstances that, although they may hold in specific examples, do not hold in the general setting. We suggest that the reader revisit the ideas of this section after studying several of the examples in Chapter 2.

This discussion of axiom systems and models has been greatly influenced by R. L. Wilder and his book *Introduction to the Foundations of Mathematics,* 2nd ed., (New York: Wiley, 1965).

■ Axioms

Because the use of the word *axiom* has changed over the years, we begin our discussion of axioms by contrasting the current use of this term with an earlier use that is still common outside mathematics. At one time (for example, with Euclid and other Greek mathematicians), the term *axiom* meant a self-evident truth. An axiom was a universal statement obvious to all and not subject to debate. An example of such a statement is the proposition "Equals added to equals yield equals." In addition to axioms, mathematicians of the day were also concerned with postulates. These were statements of a more specific character, and they presumably expressed "true facts" about a particular subject, such as geometry. The statement "Through two distinct points there exists one and only one line" qualifies as a postulate. Thus the postulates of geometry use terms such as *point* and *line,* which are special to geometry and are not used in areas such as arithmetic. This separation of basic truths into axioms and postulates has a long history. Euclid used it in his famous *Elements,* calling the axioms "common notions."

The original use of the words *axiom* and *postulate* remained relatively unchanged for almost 2000 years. In fact, it is really only in mathematics itself that a second meaning has arisen. In day-to-day conversation one still hears "It is axiomatic," by which the speaker means that the statement under discussion is a universal truth. In mathematics, however, a change in the meaning of the term *axiom* (and likewise in that of the related term *postulate*) began during the period that saw the development of non-Euclidean geometries. Without going into detail, we simply point out that a number of geometries were developed in which Euclid's fifth postulate failed to be true. (The fifth postulate essentially said that given a line and a point not on the line, there exists one and only one line through this point and parallel to the given line. Gauss, Lobachevski, and Bolyai all used sets of axioms in which this statement is contradicted, whereas all other Euclidean axioms are retained.) The development of these geometries demonstrated the fallacy of the earlier belief that the fifth postulate was a consequence of the other postulates and hence that its denial would lead to contradictions. The assumption of an axiom that contradicted the fifth postulate did not lead to an inconsistent system; instead, it led to new geometries. This means that there were many geometries and that the postulates in one could contradict the postulates in another. This, in turn, implied that postulates were in some sense a matter of choice and that they were not fundamental truths about geometry that the Greeks had discovered. It also implied that there was now more than one candidate for a mathematical description—a mathematical model—for physical space. The initial reaction was that, although other geometries might be of interest to mathematics as areas for abstract investigations, only Euclidean geometry would be important for providing a description of physical space. This has not been the case, however, because a non-Euclidean geometry plays a prominent role in the theory of relativity.

Along with the change in the meaning of the term *postulate,* the notion of axiom was reevaluated. It soon became apparent that these statements could also be denied without necessarily implying a contradiction. Thus the statement "The whole is greater than the part" is in the same category as the fifth postulate. In one context it may be true (in the sense of following logically from other statements), and in another context it may be false, the critical difference being the meaning of the words *whole, part,* and *greater than* in each case. For example, we note that in set theory this statement is false when *part* means "proper subset" and *greater than* means "has a larger cardinal number." Thus, in mathematics, the distinction between axioms and postulates has essentially disappeared. **Axioms** and **postulates** are both statements that one takes as basic in order to study the logical consequences of such assumptions. They are not, in general, considered to be universal truths, and in fact, they can be changed and modified to suit different purposes.

As a final comment about axioms, we remark that frequently some of the words used in phrasing axioms are left undefined. This is necessary if one is to avoid circular and meaningless definitions. For example, in axiom systems for geometry, the words *point* and *line* are often left undefined. In this way, one avoids having pseudodefinitions such as "A point is an object having no length, width, or height, whereas a line has length, but not width or height."

■ Axiom Systems

An **axiom system** is a collection of undefined terms, together with a set of axioms phrased with the use of these undefined terms. Although it is possible to consider axiom systems with an infinite number of undefined terms and/or an infinite number of axioms, we shall always consider both of these sets to be finite.

Our primary concern with axiom systems is with the logical consequences of the axioms. These logical consequences are called **theorems,** and the collection consisting of all theorems that can be logically deduced from an axiom system Σ is called the **theory determined by the axiom system Σ**.

Regarding this definition of a theory, we note that we are using this term as it is used in mathematics and logic. The concept of a theory is somewhat different in the social and life sciences. In these sciences, a theory is usually a collection of basic assumptions that is studied in an attempt to explain certain observed phenomena. Soon we shall connect these two meanings of the term *theory* by using the concept of a model.

Our notion of axiom system is obviously a very general one, because almost every set of statements now constitutes an axiom system. Naturally, not all these axiom systems have interesting or useful theories, and hence restrictions are imposed to obtain useful systems. First, because it is the theory of an axiom system that interests us, we want to make sure that the axioms do not contradict each other. This is necessary because assuming the truth of two contradictory statements is logically equivalent to assuming the truth of all statements. Hence the theory determined by a system containing contradictions would contain all statements and would be of no value. This condition about the absence of contradiction can be made precise as follows: An axiom system Σ is said to be **consistent** if the theory it determines does not contain contradictory statements.

This definition of consistent is logically sound; however, it is not an especially practical definition. The lack of practicality results from the fact that we usually do not know all the statements in the theory determined by an axiom system Σ. Thus, in general, using the definition we cannot tell whether the system is consistent because we do not know which statements to check for contradictions. Fortunately, this difficulty can be circumvented by the use of models. Therefore, our next task is to introduce the notion of a model of an axiom system. We shall then show how models can be used to verify the consistency of an axiom system.

As a final comment on axiom systems, we note that different branches of mathematics can be characterized by the axiom systems they use. Thus group theory is the study of the consequences of the axioms that define a group, topology is the study of the consequences of the axioms for different topological spaces, and geometry is the study of the different axiom systems used to describe geometrical systems.

■ Models and Formal Model Building

As we saw at the beginning of this chapter, the term *model* is often used in different ways in different contexts. Here we first consider how it is used in the framework of mathematical logic, because it is in that setting that it is connected to the important concept of the consistency of an axiom system. To this end, suppose that we are given an axiom system Σ consisting of certain undefined terms and certain statements phrased with these terms. Also, suppose that we assign a meaning to each of the undefined terms and that, with these meanings, the axioms of the system become statements known to be true. Here, by "assign a meaning" we mean associate an object or action of the real world with each of the undefined terms; by "statements known to be true" we mean that the statement is now a meaningful assertion about the real world and that this assertion is an observable and verifiable fact. If such an assignment can be made, then we say that this assignment of meanings to the undefined terms of Σ constitutes a **logical model** of the system Σ. This discussion expands and makes more precise the brief comment on logical models in Section 1.0. We also note that a single axiom system may have several different logical models, each associated with a certain assignment of meanings to the terms.

Before proceeding, a word of caution to the reader is in order. In our discussion of logical models, we are skirting close to very deep and fundamental matters—matters that lie at the basis of mathematical thought. This brief discussion is certainly an inadequate treatment of these profound concepts. However, because a study of the foundations of mathematics and mathematical logic is clearly beyond the scope of our work, we feel that this short survey is adequate for what follows. For a more comprehensive discussions of these matters, the interested reader is referred to the book by R. L. Wilder noted above.

Before continuing our discussion of logical models and their relation to real and mathematical models, we illustrate the notions of an axiom system and a logical model by an example. Our example uses a relatively simple axiom system that was chosen to avoid confusion about the nature of the axioms. We also develop a little of the theory of this system to motivate our later comments on the important relationship between the theory of an axiom system and the logical models of the system. As noted earlier, the theory of an axiom system is the collection of statements that can be logically deduced from the axioms of the system. The development of such a theory (that is, the discovering of the statements) can proceed in many ways, and it can be expressed in many forms. The most economical form, and at the same time the form least likely to be misinterpreted, is in terms of formal theorems. Theorems are shown to be true—*proved* is the usual mathematical term—by using the traditional methods of logical reasoning.

EXAMPLE 1.1   Let Σ be the axiom system whose undefined terms are *player* and *team* and whose axioms are the following:

A_1: Every team is a set of players containing at least two elements.
A_2: There exist at least three players.
A_3: Given any two players P_1 and P_2, there exists one and only one team containing them.
A_4: Given any team T and a player P not in T, then there exists one and only one team T' containing P and disjoint from T (that is, $T \cap T' = \emptyset$).

First, we note that there are several straightforward logical models of Σ:

Model 1   *Player* will have its usual real-world meaning. A collection of three players will then form a logical model of Σ. The set whose three elements are the three players is defined to be a team. There is only one team.

Model 2   *Player* again has its usual meaning. This time we have a collection of four players which we designate A, B, C, and D. We define six teams in the following way: Each is a set of exactly two players, $T_1 = \{A, B\}, T_2 = \{A, C\}, T_3 = \{A, D\}, T_4 = \{B, C\}$, $T_5 = \{B, D\}, T_6 = \{C, D\}$. With these definitions of *team* and *player,* it is easy to see that the axioms are true. To check this, the reader may find it useful to draw some pictures.

Model 3   *Player* is now defined to be a point in the sense of Euclidean geometry, and *team* is defined to be a line. Then the axioms of Σ follow from the axioms of Euclidean geometry.   ■

Model 3 is somewhat different from models 1 and 2 because it uses another axiom system that itself has undefined terms (the axiom system for Euclidean geometry). Strictly speaking, it is not a model of Σ according to our definition. However, the word *model* is often used in this way, and we introduced this interpretation of Σ to illustrate this common stretching of our definition. Henceforth, we shall stay within the strict framework of our definition and shall use only logical models such as models 1 and 2.

One might be inclined to guess that the theory associated with a system as simple as that of Example 1.1 is trivial and uninteresting. But this is not the case, and we shall provide two examples of theorems that are not *a priori* obvious.

THEOREM 1.1   If there are two distinct teams, then there are three distinct teams.

Remark   Two teams T_1 and T_2 are said to be *distinct* if $T_1 \neq T_2$ in the set-theoretic sense.

Proof.   Let T_1 and T_2 be distinct teams. By Axioms A_1 and A_3 and the meaning of set equality, there exists a player in each one who is not in the other. Thus suppose that $P_1 \in T_1$, $P_1 \notin T_2$, and $P_2 \in T_2$, $P_2 \notin T_1$. Next, by A_3 there is a unique team T_3 containing P_1 and P_2. Clearly, $T_1 \neq T_3$ since $P_2 \notin T_1$, and $T_2 \neq T_3$ since $P_1 \notin T_2$. Therefore, T_3 is a distinct team.   ■

The reader is invited to test his or her understanding of the proof of Theorem 1.1 by providing a proof of the following result (Exercise 2).

THEOREM 1.2   If there are three distinct teams, then there are four distinct teams.   ■

It is natural at this point to consider the relationship between the theory (the set of theorems) of any axiom system and any logical model of that system. This relationship is based on certain general assumptions about models. R. L. Wilder has called these assumptions *principles of applied logic,* and they will be basic to our discussion.

Principle 1.1   All theorems in the theory determined by an axiom system Σ are true in every logical model of Σ.

Principle 1.2   The law of contradiction holds for all meaningful statements about each logicial model of an axiom system Σ.

The law of contradiction says that a statement S and its negation S' cannot both be true, and Principle 1.2 is a precise statement of the heuristic "The real world has no contradictions."

Using Principles 1.1 and 1.2, we are now able to provide a more realistic method for checking the consistency of an axiom system. Recall that an axiom system is consistent if it does not imply contradictory statements. Suppose that Σ is an axiom system with a logical model. By Principle 1.1, if Σ implies two contradictory statements, then these statements will both be true about the model. But Principle 1.2 says that the law of contradiction holds for statements about the model. Hence contradictory statements cannot be true about the model, and the theory of Σ must not include contradictory statements. Summarizing these remarks, we have the following important result:

Every axiom system that has a logical model is a consistent system.

In particular, the axiom system Σ of Example 1.1 is a consistent system.

In addition to the result on consistency, the principles of applied logic also yield a result of central importance in the use of axiom systems and logical models. Once a theorem in the theory determined by an axiom system Σ has been established, then this theorem is a true statement about every logical model of Σ. Thus, in regard to the axiom system Σ of Example 1.1, we see that no logical model of this system can have exactly two teams or exactly three teams. This, of course, is an aid in the search for logical models of the system, because it is clear that any candidate for a model should be rejected if it has either exactly two teams or exactly three teams. A similar situation holds for a general axiom system. In the search for logical models of the system, one can use as a guide the knowledge that each theorem must be a true statement about each model.

As the reader is no doubt aware, the axiom system Σ involving players and teams was created with the intention of forming an axiom system that had certain predetermined logical models. It is common to do this—that is, to begin with a specific real situation and to set down statements that describe this situation as accurately as possible. We discussed this process in Section 1.1, and at that time we called the process model building. This brings us to our primary use of the word *model.*

The use of the term *mathematical model* in regard to the idea that we are about to discuss became common first outside of mathematics. It is likely that if mathematicians and logicians had had the prerogative of introducing terminology, then it would have been somewhat different. However, the usage is now well established, and we shall make no attempt to change it.

When an investigator forms statements that she feels express basic principles in an area of observation and study, it is often said that she has formed a model. Actually she has made a major step in forming an axiom system, and it is her hope that this axiom system has a logical model in her area of study. The process of forming this axiom system is called **model building.** The model builder experiments and observes facts about the real world in her area of specialization. She then tries to explain and describe the phenomena that she is studying. She usually does this by proposing certain statements as the ones that are basic and most important. Frequently these statements contain terms that are undefined and are created to aid in explaining certain observed facts. Together these terms and statements constitute an axiom system. It is this axiom system that we call a mathematical model.

The next step in the model-building process is testing and (possibly) modifying the model. This is done by considering the consequences of the axioms that constitute the model and checking whether these consequences are true statements about that part of the real world under study. If they are not true statements, then the model should be changed. Thus model building is frequently an evolutionary process.

We conclude this section with a brief summary where we review and relate the different uses of the term *model*. First, a *real model* is a collection of statements about real objects that is obtained by a process of observation, idealization, simplification, and approximation. A *mathematical model* is an axiom system consisting of undefined terms and axioms that are obtained by abstracting and making precise the essential ideas of a real model. A *logical model* is an association of real objects with the undefined terms of an axiom system so that the axioms become verifiable statements about the real world. The different uses of the term *model* are represented in Figure 1.7.

In this figure, the solid lines indicate the usual relationships between the different models: The real model is obtained from the real world, the mathematical model is obtained from the real model, and the logical model (if one exists) is again in the real world. The dashed line between the real and logical models indicates that very often, the real-world

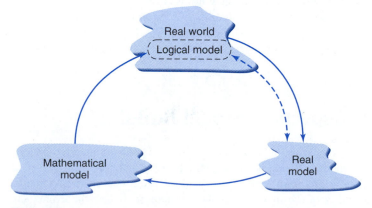

Figure 1.7

setting that led to the real model is itself a logical model. Thus the real-world setting that motivated the study is a natural place to look for a logical model.

1.5   Simulation Models

Whenever we build mathematical models to help us understand a situation, there is continuing tension between incorporating enough of the detail of the situation to yield useful predictions, on the one hand, and having a tractable model—that is, a model for which predictions or results can be obtained—on the other. In general, the more detailed the model, the more likely it is that predictions based on it will agree with observations. However, models that include a large amount of detail, or features very specific to the situation, are likely to be difficult to study using mathematical tools (such as algebra, geometry, calculus, and probability), and it is necessary to look for other alternatives. In many such cases, it may be appropriate to use a simulation model.

The construction of a simulation model begins by using the process described in Section 1.2. However, once the real model and an appropriate mathematical structure have been identified, then—rather than using mathematical tools—we use computers (and appropriate software) to imitate, or *simulate,* the situation being studied. In simulation we use a computer to generate numerical or other output over a set of times (or locations, or relations) of interest. We then evaluate the output (in particular, we assess how representative it is), and ultimately we use the results to draw conclusions about the situation or to make a decision.

For example, suppose that the habitat of an endangered species is to be modified to try to increase the population size, and there are several alternative modifications open to us. To identify the modification most likely to lead to an increase in population size, we might simulate the evolution of the population under each of the alternatives. That is, we would build a model, implement it computationally, and compare the population sizes projected in all the different cases.

Simulation is one of the most widely used means of implementing mathematical models, and simulation models are now commonly used in the life, social, and management sciences. There are several special-purpose computer languages that facilitate the use of such models, and recent versions of popular spreadsheets include simulation capability. In this book we focus on the steps necessary to create, implement, and evaluate a simulation model, and we do so by considering several special situations. In Section 2.6 we illustrate the concepts discussed here with examples of actual situations in which simulation models can be useful. Then, in Chapter 4, we discuss the general problem of using simulations to study mathematical models.

1.6   Practical Aspects of Model Building

A mathematical model may be very useful to a mathematician and quite useless to a scientist. The value of a model to a mathematician lies in the structure and mathematical beauty of the model and in its relationships to other mathematical theories. The value of a model to a scientist lies in the utility of the model in explaining and predicting the events and phenomena being studied. In this text we will formulate and develop a number of different models; however, in most cases we will stop short of a full discussion of the important problem of evaluating the model. In this section, and again in Section 2.8, we shall try to fill this gap a little by considering the art and/or science of testing, evaluating, and using the mathematical models developed in the text. The treatment given here is not intended as a general introduction to the subject but rather as a natural part of the model-building activity introduced in this chapter. Our point of view is that of scientists who are considering particular mathematical models that they are interested in using. What should they do in order to decide whether the models are good ones—models worthy of further study and use? This question is difficult to answer completely; however, we do hope to introduce some of the tools and techniques that have been useful in the past and are likely to be used in the future.

This section is organized in two parts. We begin with a discussion of some of the heuristics of model evaluation. Then we consider the use of the statistical analysis of data in building and assessing models.

■ Intuitive Evaluations

A scientist who builds a model to aid in her studies usually has a good idea what she wants this model to do. She may be interested only in a convenient method of consolidating her data, in which case it is likely that a number of different models would be acceptable to her. However, in most cases the scientist wants her model to do more; it should also explain and predict, and perhaps contribute to the understanding of previously unanswered questions. In such circumstances, different models may give dramatically different results. The scientist must then decide which models she wishes to use and which ones she chooses to reject. There are many ways for her to make this decision. Some of the methods are mathematical and others are not. One of the most important nonmathematical techniques is based on the intuition of the model builder herself. This so-called *eyeball* technique consists of the intuitive feelings of the scientist about the assumptions and consequences of the model. Our scientist must decide for herself whether they seem to be correct—or at least have a reasonable chance of being correct.

■ Statistics for the Model-Building Process

The use of statistical techniques plays an important role in several aspects of model building. However, a thorough study of techniques for the treatment of data would be a major undertaking, and it would take us too far afield. Thus we shall discuss very briefly only a few of the applications of statistics in our work.

It is sufficient for us to think of mathematical statistics as a discipline concerned with extrapolating from data that describe a *subset* of a given population to obtain information about the *entire* population. Normally, one concentrates on a single characteristic of the population. For example, the population might be 20-year-old college students attending large state universities, and the subset might be 300 such students selected in some way from six such universities. The characteristic might be the probability that a subject uses a certain strategy in a certain game (for example, betting with a certain hand in a poker game). In a typical situation, an experiment would be conducted, and a value p of the probability would be obtained for each student. These values could be used to obtain a mean value $\hat{p}$ for the subset of 300 students. One reasonable statistical question: What is the relationship between the statistic $\hat{p}$ for the subset and the corresponding mean value $\bar{p}$ for the entire population? Problems of this sort are those of *parameter estimation*. They arise naturally because in order to make numerical predictions based on a model of a situation, one must have numerical values for the parameters occurring in the model. We discuss examples of parameter estimation in Section 2.8.

Another important role of statistical methods in mathematical modeling is in testing the model for correctness and accuracy: Does the model really do what it purports to? One approach to answering such questions is to accumulate data either from the results of controlled experiments or through selected observation of the real world and then evaluate the model by comparing the results predicted by the model with those resulting from a statistical analysis of the data. Several standard statistical tests, along with their many variants, are useful in this connection. Some of these will also be discussed in Section 2.8.

Obviously, the two problems just mentioned are not independent. One would hardly expect a model to predict accurate results if the parameters were carelessly chosen. Thus, when evaluating the predictions based on a particular model, it is essential to keep in mind the processes used to determine the parameters and the accuracy that can be expected. Also, one set of data is frequently used both to determine the parameters and to test the model, a part of the data being used for each purpose. In these situations, systematic bias in the data will affect both the parameters and the testing, and even though the model may predict results consistent with the data, defective data may obscure deficiencies in the model.

In discussing the testing of models, it is important to note that different models are, by virtue of their nature, tested in different ways. For some models it is natural that the axioms of the model be directly tested. For other models it is very difficult to test the axioms, but the theorems or conclusions of the model can be easily tested. In such cases, of course, one is indirectly testing the axioms of the model. The basic assumption here is that the theorems are true whenever the axioms are true (that is, when the proofs of the theorems are valid), and hence, if the theorems (predictions) are not verified in the real world, then the axioms cannot be correct statements about the real world.

It is also important to keep in mind that most testing of models is not an *either–or* situation that results in the acceptance or rejection of a model on the basis of a single test. Instead, much of the testing is for purpose of comparison to determine how well the model explains and predicts the real world. Frequently, there are a number of competing models, and experiments are conducted in an effort to determine which is the "best" at explaining and predicting the phenomena under study.

A statistical evaluation of the accuracy of a model is usually carried out using a standard measure of the discrepancy between the predicted and observed data. In this way, one obtains a numerical measure of the **goodness of fit** for each model. Naturally, if one model consistently gives a better fit than other models, then this model will be accepted and the others rejected. However, it often happens that one model will be the best to explain and fit certain sets of data, whereas another model will be the best at explaining and predicting other sets. Neither model can be rejected, because each is better under certain circumstances. Likewise, neither model should be completely accepted, because in certain cases each model is not the best available. Each can be conditionally accepted, studied, and used in those circumstances in which it is the appropriate choice. Naturally, a scientist would like to have a single model that is best at explaining all the known experimental results. However, such a model is not always available, and the scientist must work with the models at hand until better ones are developed.

■ **Exercises**

In these exercises, Σ denotes the axiom system involving teams and players introduced in Section 1.4.

1. Show that model 2 of axiom system Σ is a model of this system. That is, show that each of the axioms is satisfied.

2. Show that in axiom system Σ, if there are three distinct teams, then there are four distinct teams.

3. Let $P = \{A, B, C, D, E\}$ be a set of players, and define a set of teams T by $T = \{\{A, B\}, \{A, C\}, \{A, D\}, \{A, E\}, \{B, C\}, \{B, D\}, \{B, E\}, \{C, D, E\}\}$. Is this a model of axiom system Σ? If so, show that the axioms are satisfied. If not, show that at least one axiom is violated.

4. Let $P = \{A, B, C, D, E\}$ be a set of players, and define a set of teams T by $T = \{\{A, B\}, \{A, C\}, \{A, D\}, \{A, E\}, \{B, C, D, E\}\}$. Is this a model of the axiom system Σ? If so, show that the axioms are satisfied. If not, show that at least one axiom is violated.

5. For axiom system Σ, decide whether the following statement is true or false: If there are exactly five players, then there is exactly one team. If the statement is true, give a proof; if it is false, find a counterexample.

6. Show that in axiom system Σ, if there is a team consisting of exactly three players, then no team has fewer than three players.

7. Show that in axiom system Σ, if there are four distinct teams, then there are five distinct teams.

8. Add to axiom system Σ the axiom that there are exactly five players. How many different models are there for this new system? First, discuss what you mean by *different*.

9. Consider an axiom system Σ' consisting of axioms A_1, A_2, and A_3 of axiom system Σ and the following new axiom A_4': Given any team T and any player P not in T, then there is *at least one* team T' containing P and disjoint from T.

   (a) Show that model 2 is a model of axiom system Σ'.
   (b) How many different models of axiom system Σ' have exactly five players? Define what you mean by *different*.

10. Show that if there are no more than eight players in a model for axiom system Σ, then this model contains either one team or six teams.

11. Find a model for axiom system Σ that contains nine players and more than one team.

12. Show that in a model for axiom system Σ, if there is a player who belongs to exactly three teams, then no player belongs to more than three teams.

13. An axiom A in system Γ is said to be *independent* in Γ if there is a logical model of Γ and a logical model of the axiom system consisting of $\{\Gamma \setminus A, \sim A\}$. Let Σ' be axiom

system Σ together with the following axiom A_5: There are exactly five players. Show that axiom A_5 is independent in Σ'.

Remark $\Gamma \setminus A$ means Γ without axiom A, and $\sim A$ means the negation of axiom A.

14. Show that in axiom system Σ, axiom A_2 is independent.
15. Show that in axiom system Σ, axiom A_3 is independent.
16. Show that in axiom system Σ, axiom A_4 is independent.

CHAPTER 2

Model Building: Selected Case Studies

2.0   Introduction

Confidence in the use of mathematics to help us understand situations arising in the life, social, and management sciences rests, to a large extent, on a few quite successful models. In this chapter we use selected case studies to illustrate the ideas introduced in Chapter 1. These case studies, or expanded examples, were chosen for specific reasons. The first is a model of some aspects of elementary genetics. It illustrates the process described in Chapter 1: creating an axiom system and drawing conclusions from a study of that system. In many situations the predictions correspond well with observations of actual populations, and we may view these populations as logical models for the axiom system. The second case study, a collection of models for growth processes, illustrates a situation that can be studied using several different mathematical models. In particular, it illustrates the cycling of the model-building process: when a model leads to conclusions that are inconsistent with observations, it is necessary to refine or alter the model. Third, a discussion of models for social choice illustrates some of the problems associated with the construction of models that are not sets of equations. It also illustrates that sometimes the model-building process results in the conclusion that a desired goal cannot be attained. Following the material on genetics, growth models, and social choice mechanisms, we consider a small transportation problem, several simple examples of simulations, a problem involving waiting lines (queues), and some refined models for growth processes. We conclude the chapter with a discussion of estimating parameters. This chapter contains special cases of models that have been widely applied in situations arising in a variety of settings. Some of these topics will be discussed in greater detail in later chapters.

2.1   Mendelian Genetics

One of the significant early uses of a mathematical model to help researchers understand observations in the biological sciences was in elementary genetics, and the effort resulted in an excellent example of the modeling process. The original work postulated the existence of "units of heredity" (discrete units of information which govern inheritance) and the way these units behave during reproduction. In many situations, the predictions based on a study of the resulting mathematical system agreed well with observations. As biologists gained greater understanding of genomics and molecular biology, the view of the topic changed a great deal, but certain aspects of the original models continue to be quite useful. We describe here a version of the original experiments and the development of a model, emphasizing the aspects most closely related to our primary concern: using mathematical models to enhance our understanding of nature.

■ Some Observations

We begin by describing a simplified version of some basic experiments in plant biology, specifically in plant hybridization. Results of experiments similar to those described here, were published in 1866 by Gregor Mendel, an Augustinian monk who had attempted, without success, to pass the biology portion of the examination for a teaching license. The fundamental nature of his work went unnoticed for more than thirty years, and it was not until his results had been rediscovered several times that their significance was acknowledged.

We are interested in two types of characteristics of plants: physical characteristics and inheritance characteristics. By the latter, we mean the passage of characteristics or attributes from one generation to the next. We use the term *phenotype* to refer to physical characteristics and the term *genotype* to refer to inheritance characteristics. We consider a plant, such as the common garden pea plant, that can reproduce either by self-fertilization (pollen and flower on the same plant) or by cross-fertilization (pollen and flower on separate plants).

Suppose the plant has a characteristic, such as seed texture, that can occur in two forms: smooth and wrinkled. We denote these phenotypes by S and W for convenience. Our first assumption is that we can always distinguish between smooth and wrinkled peas. Although this may seem like a simple assumption, in many instances the classification task is not at all simple, and verification of the predictions of a model may be complicated by such problems. One of the experimental observations is that some smooth peas have the property that if they are allowed to reproduce by self-fertilization, or *selfing,* then each generation consists of only smooth peas. Likewise, for wrinkled peas. Such peas are called *pure-line* smooth and pure-line wrinkled, respectively.

Next, suppose that a pure-line smooth and a pure-line wrinkled pea are crossed; that is, pollen from one fertilizes the flower on the other. The result of this cross-fertilization is a smooth pea, but one that does not reproduce as a pure line. We refer to the result of such a cross as an F_1 pea, a pea in the *first filial generation.* All F_1 peas produced in this way are smooth; that is, they have the same phenotype as a pure-line smooth pea. However, they do not have the same genotype, and when such a pea reproduces by selfing, not all the peas in subsequent generations are smooth. In fact, if an F_1 pea reproduces by selfing, then $\frac{3}{4}$ of the offspring are smooth and $\frac{1}{4}$ are wrinkled. Furthermore, all wrinkled peas reproduce as pure-line wrinkled peas, and $\frac{1}{3}$ of the smooth peas reproduce as pure-line smooth peas. However, $\frac{2}{3}$ of the smooth peas reproduce as the F_1 peas.

Continued experimentation—for instance, crossing an F_1 pea with a pure-line smooth or a pure-line wrinkled pea—shows that any pea produced in an inheritance experiment is of one of these three types; nothing new appears in subsequent generations. We conclude that with respect to the texture characteristic, there are two distinct phenotypes (smooth appearance and wrinkled appearance) and three distinct genotypes (pure-line smooth, pure-line wrinkled, and F_1—the result of a cross between the two different pure lines).

Table 2.1

Phenotype	Fraction
SG	$\frac{9}{16}$
SY	$\frac{3}{16}$
WG	$\frac{3}{16}$
WY	$\frac{1}{16}$

Before beginning our construction of a mathematical model for the situation, we expand the example to include another characteristic. We use color, and we suppose that color occurs in two forms: green and yellow. We denote the phenotypes with respect to color by G (green) and Y (yellow). The experimental results are similar to the situation with the texture characteristic. There are two types of pure-line peas with respect to color: pure-line green peas and pure-line yellow peas. Again, the results of a cross between peas that are pure-line with respect to color are F_1 peas with respect to color: in this case, peas that have the phenotype G but do not reproduce as pure-line green peas. If an F_1 pea reproduces by selfing, then $\frac{3}{4}$ of the offspring are G and $\frac{1}{4}$ are Y. Furthermore, all Y peas reproduce as pure-line Y peas, $\frac{1}{3}$ of the G peas reproduce as pure-line G peas, and the remaining G peas reproduce as F_1 peas.

If we introduce notation to identify both texture and color for each pea, then the two-characteristic phenotypes are smooth–green, smooth–yellow, wrinkled–green, and wrinkled–yellow, or SG, SY, WG, and WY, respectively. We suppose, as was the case in the original experiments, that if a pea that is pure-line SG with respect to texture and color is crossed with a pure-line WY, and the resulting offspring (the F_1 peas in this experiment) reproduce by selfing, then the next generation (the F_2 peas) has the phenotypic distribution shown in Table 2.1.

We will return later to this two-characteristic situation and discuss the genotypic properties of the F_1 and F_2 peas.

■ A Real Model

The general observation based on data was that phenotype does not determine reproductive behavior but that, rather, all peas that reproduce similarly have the same phenotype. This observation led to the assumption that there are reproductive units that determine both inheritance and physical characteristics. To simplify the discussion we focus on the texture characteristic. Thinking first in biological terms, and following the ideas of Mendel, we imagine that each cell of each pea, other than those actually involved in reproduction, carries with it a fundamental unit that determines the texture of its descendants. This fundamental unit consists of a pair of *genes* (here we adopt standard terminology). Cells actually involved in reproduction, which we will consider later, carry only one gene. Because we assume there are two alternative forms of texture, we assume there are two alternative forms for the associated gene. We refer to the alternative forms of a gene as *alleles*. During reproduction the pair of genes splits, and one of them is contributed to the offspring. That is, the cells involved in reproduction, known as *gametes,* each carry only one gene (we are concerned only with one characteristic here). It is a fundamental assumption that the gene contributed by a parent during reproduction is selected arbitrarily and at random from the pair of the respective parent. The cell created during reproduction receives two gametes, one contributed by each parent. Each cell can be classified according to its genetic composition, and because every cell of a plant contains the same genes, the entire plant can be so classified.

To connect the reproductive behavior of a pea with physical characteristics, we suppose that a pea is smooth if it acquires a gene associated with that form of texture from either parent and that it is wrinkled if it acquires genes associated with that form from both parents. The phenomenon of a plant showing one form of a physical characteristic (here, smooth texture) even though it has genes associated with both forms is known as *dominance*. In this case, the smooth form is the *dominant* type of texture. The remaining one of each pair of alternative forms of a characteristic is known as *recessive*. Note that knowledge of the genotype of an individual and of the dominance relation enables one to determine the phenotype. The phenotype (here, the appearance of being smooth or wrinkled) does not, as we have seen, determine the genotype.

■ A Mathematical Model

To construct a mathematical model, we need to identify the essential features of the situation and create a mathematical structure that captures these features. Clearly, the crucial biology in the situation is reproduction, and we seek a way to represent this mathematically. Reproduction begins with two individuals (in selfing the two can be the same), and the result is an offspring. Again we begin by studying a single characteristic, seed texture. We use the symbols A and a to denote the alleles associated with the texture gene. In the real model proposed above, genes occur in pairs, so we consider the set of unordered pairs $V = \{AA, Aa, aa\}$. If we consider only reproduction and a single characteristic, then we can identify any pea with one of the elements of the set V. If we agree that the alleles denoted by A and those denoted by a correspond to smooth and wrinkled peas, respectively, then as a consequence of the dominance relation we have noted, peas with genotype AA and Aa will be smooth, and peas with genotype aa will be wrinkled.

Our goal is to construct a mathematical model that accounts for the observations described above on inheritance. To this end, we introduce some notation and our first definition. We begin the process with an assumption:

Assumption 2.1   Each gene occurs in two forms (alleles) denoted by A and a.

The set of pairs $\{AA, Aa, aa\}$ will be denoted by V. The symbol Aa denotes an unordered pair. We know, from the experiment of crossing F_1 peas described above, that the result of selecting a pea from among the offspring of specific parents can be a pea of any of the three distinct genotypes. Thus the result of reproduction is not a specific genotype but, rather, a probability distribution over the set of genotypes. In this situation there are three genotypes, and consequently a probability distribution can be viewed as a vector in R^3.

Assumption 2.2   Reproduction is a function from $V \times V$ to R^3.

Let r denote the reproduction function, and let r_1, r_2, and r_3 denote the coordinate functions. That is,

$$r : (u, v) \rightarrow [r_1(u, v) \quad r_2(u, v) \quad r_3(u, v)]$$

Table 2.2

$p(\alpha \mid u)$		u	
	AA	Aa	aa
α A	1	$\frac{1}{2}$	0
a	0	$\frac{1}{2}$	1

for every $(u, v) \in V \times V$. Our goal is to define r in such a way that predictions based on the resulting model correspond with observations. As we will show later, the following definition accomplishes this goal.

Definition 2.1 For each $u \in V$ and $\alpha \in \{A, a\}$, let $p(\alpha \mid u)$ be the conditional probability that α is selected when a random choice is made between the two letters that make up the symbol u, each choice assumed equally likely.

It follows that $p(\alpha \mid u)$ is given by Table 2.2.

Assumption 2.3 The reproduction function r satisfies

$$r_1(u, v) = p(A \mid u)p(A \mid v)$$
$$r_2(u, v) = p(A \mid u)p(a \mid v) + p(a \mid u)p(A \mid v)$$
$$r_3(u, v) = p(a \mid u)p(a \mid v)$$

A *probability vector* is a vector whose coordinates are nonnegative numbers that sum to 1.

THEOREM 2.1 The range of the function r is a set of probability vectors in R^3.

Proof. Because $p(\alpha \mid u) \geq 0$ for all $\alpha \in \{A, a\}$ and $u \in V$, it follows that the coordinates of $r(u, v)$ are nonnegative for all $u, v \in V$. Also, because $p(\alpha \mid u)$ is a probability and the only choices for symbols in u are A and a, $p(A \mid u) + p(a \mid u) = 1$. Consequently,

$$r_1(u, v) + r_2(u, v) + r_3(u, v)$$
$$= p(A \mid u)p(A \mid v) + p(A \mid u)p(a \mid v) + p(a \mid u)p(A \mid v) + p(a \mid u)p(a \mid v)$$
$$= p(A \mid u)[p(A \mid v) + p(a \mid v)] + p(a \mid u)[p(A \mid v) + p(a \mid v)]$$
$$= p(A \mid u) + p(a \mid u) = 1$$
■

Following the motivation for our definition of reproduction, we connect this mathematical definition with the real model by interpreting the coordinates of $r(u, v)$ as giving the genotypic distribution of the offspring resulting from a "mating" between u and v. That is, $r_1(u, v)$, $r_2(u, v)$, and $r_3(u, v)$ give the proportions of offspring with genotypes AA, Aa, and aa, respectively.

EXAMPLE 2.1 If $u = AA$ and $v = Aa$, then $r(u, v) = \left[\frac{1}{2} \quad \frac{1}{2} \quad 0\right]$.
Using the above interpretation, this means that if there is a mating between peas of genotype AA and Aa, then $\frac{1}{2}$ of the offspring are of genotype AA and $\frac{1}{2}$ are of genotype Aa. ■

EXAMPLE 2.2 If $u = AA$ and $v = aa$, then $r(u, v) = [0 \quad 1 \quad 0]$, and if $u = Aa$ and $v = Aa$, then $r(u, v) = [\frac{1}{4} \quad \frac{1}{2} \quad \frac{1}{4}]$.

Using our interpretation, this means that if there is a mating between two different pure lines, then all offspring are of genotype Aa. Furthermore, if a pea of genotype Aa, usually called a *hybrid,* mates with a pea with the same genotype, then $\frac{1}{4}$ of the resulting offspring are genotype AA, $\frac{1}{2}$ are Aa, and $\frac{1}{4}$ are aa. Using the assumptions about dominance, we know that the peas with genotype AA and Aa have the phenotype AA. Thus $\frac{3}{4}$ of the offspring from the selfing $Aa \times Aa$ have the phenotype AA, and $\frac{1}{4}$ have the phenotype aa—predictions that are consistent with the observations described earlier. ■

When we are considering a single characteristic, the change through time of the genotypic distribution that results from reproduction under selfing can be described by a simple recurrence relation. One way to specify this relation is with the use of matrices.

THEOREM 2.2 If the genotypic distribution at one generation is given by $\mathbf{x} = [x_1 \quad x_2 \quad x_3]$ (that is, the fractions of the population that are AA, Aa, and aa are x_1, x_2, and x_3, respectively), and if reproduction is by selfing, then the genotypic distribution at the next generation is given by the matrix product $\mathbf{xM}$, where $\mathbf{M}$ is the matrix

$$\mathbf{M} = \begin{bmatrix} 1 & 0 & 0 \\ \frac{1}{4} & \frac{1}{2} & \frac{1}{4} \\ 0 & 0 & 1 \end{bmatrix}$$

Proof. Under selfing, each pea with genotype AA produces only peas with genotype AA, and likewise, each pea with genotype aa produces only peas with genotype aa. On the other hand, under selfing, peas with genotype Aa produce peas with all three genotypes, and the proportions are $\frac{1}{4}AA$, $\frac{1}{2}Aa$, and $\frac{1}{4}aa$, as described above. Therefore, under selfing, a population with genotypic distribution $[x_1 \quad x_2 \quad x_3]$ at one generation has, at the next generation, the genotypic distribution

$$\left[x_1 + \tfrac{1}{4}x_2 \quad \tfrac{1}{2}x_2 \quad x_3 + \tfrac{1}{4}x_2\right] = x_1[1 \quad 0 \quad 0] + x_2[\tfrac{1}{4} \quad \tfrac{1}{2} \quad \tfrac{1}{4}] + x_3[0 \quad 0 \quad 1]$$

which is $\mathbf{xM}$.

Note that the fraction of hybrids is reduced by $\frac{1}{2}$ in each reproduction cycle. For example, if $\mathbf{x} = [\frac{1}{3} \quad \frac{1}{3} \quad \frac{1}{3}]$, then $\mathbf{xM} = [\frac{5}{12} \quad \frac{1}{6} \quad \frac{5}{12}]$. That is, if a population reproduces by selfing, then, over the long term, the percentage of hybrid plants declines. Indeed, that percentage tends to zero as the number of generations increases (Exercise 2). ■

EXAMPLE 2.3 A pure-line dominant pea and a pure-line recessive pea are crossed. Find the genotypic distribution of the first three filial generations, assuming reproduction by selfing after the initial cross.

The genotypic distribution of the first filial generation is $[0 \quad 1 \quad 0]$, as shown in Example 2.2. The genotypic distribution of the second filial generation is $[\frac{1}{4} \quad \frac{1}{2} \quad \frac{1}{4}]$, also shown in Example 2.2. It follows from Theorem 2.2 that the genotypic distribution of the third filial generation is $[\frac{3}{8} \quad \frac{1}{4} \quad \frac{3}{8}]$.

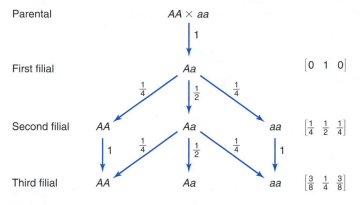

Figure 2.1

There is a useful alternative approach to determining genotypic distributions—namely, the use of a diagram usually called a **reproduction diagram.** The reproduction diagram for the self-reproduction in a population arising from an initial cross of AA and aa individuals is shown in Figure 2.1. The diagram is constructed as follows: The generations are listed vertically at the left of the diagram, and the parental cross is shown at the top. The first filial generation consists only of individuals with genotype Aa, and that is indicated by the 1 on the arrow connecting the parent generation with the first filial generation and by the vector [0 1 0] shown at the right. Genotypic distributions for the second and third filial generations (each under selfing) are shown at the right of the diagram. At each level the arrows show the possible genetic outcomes of selfing, and the labels on the arrows show the fraction of each genotype in the resulting filial generation. ∎

In addition to predicting the results of prescribed matings, including selfing, it is useful to be able to predict the results of matings that are random in a population whose genotypic distribution is known. We illustrate the technique with an example and continue the discussion in the exercises.

EXAMPLE 2.4 Suppose we have a population that has the genotypic distribution [.2 .8 0], and suppose this population reproduces by random mating. We find the genotypic distribution in the next generation.

Because the initial population contains individuals with genotypes AA and Aa, there are three possible matings: $AA \times AA$, $AA \times Aa$, and $Aa \times Aa$. The probability of two randomly selected individuals both having genotype AA is $(.2)(.2) = .04$. Likewise, the probability of both having genotype Aa is $(.8)(.8) = .64$. The probability of one having genotype AA and the other having genotype Aa is $2(.2)(.8) = .32$. The factor 2 results from the fact that there are two ways in which the genotypes can be assigned to the two individuals involved. Next, using the reproduction function, we have

$$r(AA, AA) = [1 \quad 0 \quad 0],$$
$$r(AA, Aa) = \begin{bmatrix} \frac{1}{2} & \frac{1}{2} & 0 \end{bmatrix}, \text{ and}$$
$$r(Aa, Aa) = \begin{bmatrix} \frac{1}{4} & \frac{1}{2} & \frac{1}{4} \end{bmatrix}$$

Finally, combining all this information, we conclude that the genotypic distribution in the next generation is

$$.04[1 \quad 0 \quad 0] + .32[\tfrac{1}{2} \quad \tfrac{1}{2} \quad 0] + .64[\tfrac{1}{4} \quad \tfrac{1}{2} \quad \tfrac{1}{4}]$$
$$= [.04 \quad 0 \quad 0] + [.16 \quad .16 \quad 0] + [.16 \quad .32 \quad .16]$$
$$= [.36 \quad .48 \quad .16]$$

Note that the genotypic distribution is a probability vector, as it must be. ∎

If, as a continuation of Example 2.4, the population reproduces again by random mating, then the same technique can be used to determine the genotypic distribution in the second filial generation. In this case all possible pairs can occur as parents. The unexpected result is that this distribution is again [.36 .48 .16]. Indeed, after the first reproductive cycle, the genotypic distribution does not change. This result, known as the **Hardy–Weinberg law,** is valid for any initial distribution of genotypes (see Exercises 3 and 4).

We conclude this section by returning to the situation with two characteristics discussed earlier. To apply the ideas developed here to this situation, we need to consider the possible relations between the genes corresponding to the two characteristics during reproduction. There are many possible relations. For example, it could be that the allele for smoothness always goes with the allele for the color green. Of course, this relation is inconsistent with the experimental results reported above, but it is a possible relation, and in fact there are situations in which the genes are linked in this way. Another possible relation is that the allele for texture is selected independently of the allele for color. This is the assumption we make here. Another possible relation is that the probabilities of the various pairs of alleles, one for texture and one for color, are not those arising from either complete linking or independent selection. This case of partial linking is explored in the exercises. Here we make the following assumption:

Assumption 2.4 During reproduction, the selection of an allele for texture is independent of the selection of an allele for color, and each is governed by a reproduction function as determined by Definition 2.1.

The set of possible genotypes in a situation with two characteristics is much more complicated than the single-characteristic case. Suppose that we continue to use AA, Aa, and aa to denote the three genotypes associated with texture and introduce BB, Bb, and bb to denote the three genotypes associated with color. Then a genotypic distribution in the two-characteristic situation is a 9-tuple whose coordinates give the proportions of individuals with genotypes $AABB$, $AABb$, $AAbb$, $AaBB$, $AaBb$, $Aabb$, $aaBB$, $aaBb$, and $aabb$, respectively. Assuming that smooth texture and green color are dominant, we have the relations between genotypes and phenotypes shown in Table 2.3.

Using the definition of reproduction and Assumption 2.4, we can determine the genotypic distribution of the offspring resulting from crossing any two individuals. In particular, if we begin with a cross of $AABB$ and $aabb$, then every individual in the first filial generation has genotype $AaBb$. Now, suppose an individual in the first filial generation reproduces by selfing. The result (Exercise 5) is that the genotypic distribution of the next

Table 2.3

Genotypes	Phenotype
$AABB, AABb, AaBB, AaBb$	smooth and green
$AAbb, Aabb$	smooth and yellow
$aaBB, aaBb$	wrinkled and green
$aabb$	wrinkled and yellow

Table 2.4

Phenotype	Proportion of Population
smooth and green	$\frac{1}{16} + \frac{1}{8} + \frac{1}{8} + \frac{1}{4} = \frac{9}{16}$
smooth and yellow	$\frac{1}{16} + \frac{1}{8} = \frac{3}{16}$
wrinkled and green	$\frac{1}{8} + \frac{1}{16} = \frac{3}{16}$
wrinkled and yellow	$\frac{1}{16}$

generation is

$$\begin{bmatrix} \frac{1}{16} & \frac{1}{8} & \frac{1}{16} & \frac{1}{8} & \frac{1}{4} & \frac{1}{8} & \frac{1}{16} & \frac{1}{8} & \frac{1}{16} \end{bmatrix}$$
$$AABB \quad AABb \quad AAbb \quad AaBB \quad AaBb \quad Aabb \quad aaBB \quad aaBb \quad aabb$$

where the genotypes are written below the respective coordinates of the distribution vector. Finally, using Table 2.3, we see that the predictions of the model correspond with the results of experiments reported in Table 2.1. This is confirmed in Table 2.4.

There are situations in which Assumption 2.4 does not hold and alternatives are more appropriate. An example of such a situation is the topic of Exercise 14.

■ Exercises 2.1

1. Genotypic frequencies associated with a single characteristic are tracked in a population that reproduces by selfing. At the nth observation, these frequencies are [.3 .2 .5].

   (a) What will the frequencies be at the $(n + 1)$st observation?
   (b) What were the frequencies at the $(n - 1)$st observation?
   (c) What will the frequencies be at the $(n + 2)$nd observation?
   (d) What were the frequencies at the $(n - 2)$nd observation?

2. Genotypic frequencies associated with a single characteristic are tracked in a population that reproduces by selfing. Use mathematical induction to show that if the initial genotypic distribution is [x y z], then the genotypic distribution in the nth generation is

$$\begin{bmatrix} x + y\frac{2^n - 1}{2^{n+1}} & y\frac{1}{2^n} & y\frac{2^n - 1}{2^{n+1}} + z \end{bmatrix}$$

3. Genotypic frequencies associated with a single characteristic are tracked in a population that reproduces by random mating. If the initial genotypic distribution is [.5 .5 0],

find the genotypic distribution in the second filial generation and show that this distribution does not change in future generations.

4. Genotypic frequencies associated with a single characteristic are tracked in a population that reproduces by random mating. Show that the genotypic frequencies are constant after the first filial generation. The result is the Hardy–Weinberg law.

5. Consider a situation in which the transmission of two characteristics is tracked from one generation to the next in the case where Assumption 2.4 holds. If an individual with genotype $AaBb$ reproduces by selfing, show that the genotypic distribution of the next generation is

$$\begin{bmatrix} \frac{1}{16} & \frac{1}{8} & \frac{1}{16} & \frac{1}{8} & \frac{1}{4} & \frac{1}{8} & \frac{1}{16} & \frac{1}{8} & \frac{1}{16} \end{bmatrix}$$

$$AABB \quad AABb \quad AAbb \quad AaBB \quad AaBb \quad Aabb \quad aaBB \quad aaBb \quad aabb$$

where the genotypes are written below the respective coordinates of the distribution vector.

6. Assume that an initial population has a genotypic distribution [.5 .2 .3] with respect to a single characteristic.

   (a) Find the genotypic distribution in the third filial generation under selfing.
   (b) Find the genotypic distribution in the third filial generation under random mating.

7. Genotypic frequencies associated with a single characteristic are tracked in a population that reproduces by random mating and in which the initial genotypic distribution is $\begin{bmatrix} \frac{1}{2} & \frac{1}{2} & 0 \end{bmatrix}$. Assume that selection is adverse with respect to the recessive form of the characteristic in that all individuals with genotype aa die prior to reproduction. Find the genotypic distribution in the first two filial generations immediately after reproduction.

8. Genotypic frequencies associated with a single characteristic are tracked in a population whose initial genotypic distribution is [.4 .6 0]. Suppose that in each generation, 10% of the population reproduces by selfing and 90% reproduces by random mating (which includes, of course, crossing of individuals with the same genotype). Find the genotypic distribution in the first filial generation.

9. Genotypic frequencies associated with a single characteristic are tracked in a population that reproduces by random mating and in which the initial genotypic distribution immediately prior to reproduction is [.4 .4 .2]. Assume that selection is adverse with respect to the hybrid form of the characteristic in that 25% of the individuals with genotype Aa die in the interval between successive reproductions. Find the genotypic distribution in the first two filial generations immediately after reproduction.

10. Genotypic frequencies associated with a single characteristic are tracked in a population that reproduces by random mating and in which the initial genotypic distribution immediately prior to reproduction is [.5 .3 .2]. Assume that selection is adverse with respect to the pure recessive form of the characteristic.

    (a) In particular, assume that only 60% of the individuals with genotype aa are capable of reproduction and that 100% of the individuals with genotypes AA and Aa are capable of reproduction. Find the genotypic distribution in the first two filial generations immediately after reproduction.
    (b) Suppose that a fraction f of the individuals with genotype aa are capable of reproduction and that 100% of the individuals with genotypes AA and Aa are

Table 2.5

$p(\alpha\|u)$	u		
	AA	Aa	aa
α A	1	$\frac{2}{3}$	0
a	0	$\frac{1}{3}$	1

capable of reproduction. If the genotypic distribution of the reproducing members of the first filial generation has about 5% of genotype aa, estimate the fraction f.

11. Genotypic frequencies associated with a single characteristic are tracked in a population that reproduces by random mating and in which the initial genotypic distribution immediately prior to reproduction is [.6 .4 0]. Assume that in the period between successive reproductions, any recessive allele mutates into the dominant form of the allele with probability .01, and the mutations of any two recessive alleles are independent.

   (a) Find the genotypic distribution in the first filial generation immediately after the first reproduction.
   (b) Find the genotypic distribution in the first filial generation immediately prior to the second reproduction.

12. The genotypic frequencies associated with a single characteristic are tracked in a population where reproduction is defined as in Assumption 2.3 except that the table in Definition 2.1 is replaced by Table 2.5.

   (a) Find the reproduction function. That is, find the genotypic distribution resulting from each of the six possible matings.
   (b) An initial population contains only individuals with genotype Aa. If reproduction is by random mating, find the genotypic distribution in the first two filial generations.
   (c) An initial population has genotypic distribution [x y z]. If reproduction is by selfing, find the genotypic distribution in the first two filial generations.

13. Consider a population in which two characteristics are monitored, and suppose they are texture and color. A plant that is pure dominant with respect to both characteristics is crossed with a plant that is pure recessive, and the resulting plant (hybrid with respect to both characteristics, a dihybrid) reproduces by selfing. Suppose that each form of the color gene is equally likely to be contributed from the parent to an offspring. Also suppose that the genes for texture and color are linked in the following way: If the dominant form of the color gene is contributed, then the dominant form of the texture gene is three times as likely to be contributed as the recessive form. If the recessive form of the color gene is contributed, then the dominant and recessive forms of the texture gene are equally likely to be contributed.

   (a) Find the genotypic distribution of the second filial generation—the result of the dihybrid cross.
   (b) Find the phenotypic distribution of the second filial generation—the result of the dihybrid cross.
   (c) How does this phenotypic distribution compare qualitatively with the 9:3:3:1 ratio predicted when the genes are not linked?

14. Consider a two-characteristic genetics model where the characteristics are texture (alleles S and W) and color (alleles G and Y); S and G are dominant. Suppose that during reproduction each form of the texture gene is equally likely to be selected if both are available. If the texture allele S is selected, then the color allele G is twice as likely to be selected as Y, assuming both are available, and if the texture allele W is selected, then each color allele is equally likely to be selected.

    A pure-line SG is crossed with a dihybrid.

    (a) Find the phenotypic distribution of the resulting population.
    (b) Find the genotypic distribution of the resulting population.

2.2 Models for Growth Processes

There are many situations where we are interested in the growth (increase) or decay (decrease) in the quantity of something of concern to us. Probably the most obvious are situations where we seek to predict the growth of human populations to assess the need for schools, housing, health care, and similar social services. Also commonly studied are the growth of animal populations for food production or wildlife preservation, the growth of colonies of bacteria or parasites to help us understand the development of diseases, the accumulation of trash in a city or university, the production of nuclear waste at an electric generating plant, and the growth of funds in a savings account. Clearly, there is a very broad range of questions where the study of a growth process may be helpful.

Just as there are many situations that involve growth, there are also a number of questions we can pose. We might be interested not only in the size but also in the composition of a population: how many individuals there are with certain characteristics. For instance, if we study the growth of a forest, we may be interested in the number of trees of a certain species, at or above a designated size. As another example, we might be interested in how many people have a certain strain of influenza and how the disease spreads geographically.

We have noted earlier the many decisions that must be made when a mathematical model for a situation is built. In particular, we must decide whether to build a deterministic or stochastic model, a discrete or continuous model, an analytic or computational model. We will use the study of growth processes to illustrate some of the consequences of making choices of this sort. We repeat a comment from Chapter 1 for emphasis: model building involves making assumptions. Assumptions that are appropriate for one growth process may be less appropriate, or even inappropriate, for another. Consequently, there is a variety of useful models, and the validity of the predictions of each model varies with the application. If the assumptions on which the model is based correspond to reality, then we expect the predictions to agree with observations. If our assumptions do not match well with reality because they either neglect or misrepresent phenomena that are present in nature, then a match between predictions and observations would not be expected.

For the purpose of this section, we consider a fairly simple situation. First, we assume we have a homogeneous population—that is, a population in which all individuals are the same. In some situations this is a very good assumption: All dollars in a savings account are the same, one barrel of waste from a nuclear generating plant is the same as another barrel, and one parasite of a specific type is much the same as another. In other situations the assumption may be less legitimate: There may be individual differences in a population of humans or animals that result in deviations between predictions based on a model assuming homogeneity and observations of an actual population. Indeed, one of the most common extensions of the models discussed here is to heterogeneous populations that can be subdivided into a number of homogeneous subpopulations—for instance, human populations that are subdivided into age cohorts. We consider simple versions of such models in Section 2.5.

Our next assumption is that we can measure the size of the population (number of individuals, biomass, or another appropriate measure) and that we are interested in the change of this size through time. In particular, we denote the size of the population at time t by $x(t)$, and we study the behavior of the function x in a suitable domain of times T.

■ Choices for Growth Models

In order to form a mathematical model to use in predicting future values of $x(t)$, we must make choices about the nature of the variable t and the function $x(t)$. These choices will determine the nature of the model to be used—that is, the nature of the mathematical assumptions in the model. The first decision is about time and, in particular, the possible values of the variable t. If the values of t are from a sequence, say $\{t_0, t_1, \ldots, t_k, \ldots\}$, then the model is said to be a **discrete-time model.** On the other hand, if t can be any value in an interval of real numbers, say $[a, b] = \{t : a \leq t \leq b\}$, where a and b are real numbers, then the model is said to be a **continuous-time model.**

The second choice that must be made for any growth model concerns the nature of the quantity x. If, for any given value of t, we assume that x is always completely determined, say by the initial value of the population size and the way the size of the population changes, then we say that the model is a **deterministic model.** For example, if a specific amount of money is deposited in a savings account, and if a specific interest rate is paid, with no other deposits or withdrawals, then the amount of money in the account at some time in the future can be determined exactly. A deterministic model is then appropriate. On the other hand, if we assume that x is a random variable, and that for each value of t it takes a value from a set, according to some probability distribution, then we say that the model is a *stochastic model.* For example, suppose that x is the number of cars at a tollbooth at time t. Then, for a fixed time (say 9:00 a.m.) the value of x will be different on different days, and a stochastic model is appropriate. In this section we begin by considering a *discrete-time, deterministic model* of an especially simple sort.

EXAMPLE 2.5   Suppose the set T is a set of equally spaced times. For instance, the population size is to be predicted on an hourly, monthly, or yearly basis. If we are measuring the size of a colony of bacteria, then hourly data may be appropriate; if the population is barrels of nuclear waste from a utility plant, then yearly measurements may be appropriate. Let $T = \{t_k\}$, and let x_k denote the size of the population at time t_k; that is, $x(t_k) = x_k$, $k = 0, 1, 2, \ldots$.

Model 1   Assume that the size of the population increases by a fixed amount A in each time interval $(t_k, t_{k+1}], k = 0, 1, 2, \ldots$. That is, we have the relation

$$(2.1) \qquad\qquad x_{k+1} - x_k = A, \quad k = 0, 1, 2, \ldots$$

or

$$(2.2) \qquad\qquad x_{k+1} = x_k + A, \quad k = 0, 1, 2, \ldots$$

Relations of this sort—that is, equations similar to Equation (2.2)—are known as *recurrence relations*.

Now, the Equations (2.2) for $k = 0, 1, 2, \ldots$, give expressions for x_1 in terms of x_0, for x_2 in terms of x_1, and so on. Successively substituting, we obtain a formula for x_2 in terms of x_0, and, in general, for x_k in terms of A, k, and x_0:

$$(2.3) \qquad x_k = x_0 + kA, \quad k = 1, 2, \ldots$$

The formal verification of Equation (2.3) involves a mathematical induction argument. The details are left for Exercise 1.

Model 2   For this model, assume that the size of the population changes by a fixed multiple of the beginning population size in each time interval $(t_k, t_{k+1}]$, $k = 0, 1, 2, \ldots$. That is, for some number r we have the relation

$$x_{k+1} - x_k = r x_k, \quad k = 0, 1, 2, \ldots$$

or, equivalently,

$$(2.4) \qquad x_{k+1} = x_k + r x_k = (1 + r) x_k, \quad k = 0, 1, 2, \ldots$$

Again, the Equations (2.4) for $k = 0, 1, 2, \ldots$, give expressions for x_1 in terms of x_0, for x_2 in terms of x_1, and so on. Successively substituting, we obtain a formula for x_2 in terms of x_0, $x_2 = (1 + r)^2 x_0$, and, in general, for x_k in terms of r and x_0:

$$(2.5) \qquad x_k = (1 + r)^k x_0, \quad k = 1, 2, \ldots$$

Remark   Equation (2.5) means that for $r > 0$, the population size increases as n increases, and that for $-1 < r < 0$, the population size decreases as n increases. Indeed, in the latter case, Equation (2.4) means that between t_k and t_{k+1} the population loses a fraction of its size at time t_k.

The graphs of the predictions of Models 1 and 2 are shown in Figure 2.2. Models 1 and 2 are usually referred to as models for arithmetic growth and geometric growth, respectively. Model 1 (arithmetic growth, the triangles in Figure 2.2) is very appropriate for certain situations where material is accumulating as a consequence of a process that operates at a constant rate. The material that is produced in one time interval does not influence the amount that is produced in subsequent time intervals. Model 2 (geometric growth) is appropriate where the material that is added (or subtracted) in a time interval plays the same role as the material present at the beginning of that time interval in subsequent time intervals. For instance, funds accumulating in a savings account are described by Model 2. The graph of the function $x(t) = x_0(1 + r)^t$ is shown as the continuous curve, $1 \le t \le 15$, in Figure 2.2. The diamonds on this curve are shown for integer values of t.   ■

EXAMPLE 2.6   (A continuation of Example 2.5)   Suppose that in an environment with sufficient resources, a population of bacteria grows as described by Model 2 with $r = .1$. That is, the population grows 10% in each unit of time. When will the population double in size?

To answer the question, we need to find a value of k for which $x_k = (1.1)^k x_0 = 2 x_0$, or $(1.1)^k = 2$. We have $(1.1)^7 = 1.949$ and $(1.1)^8 = 2.144$. Consequently, the population will double between the seventh and eighth observations.

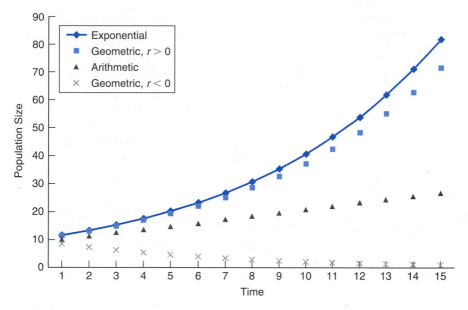

Figure 2.2   Deterministic Growth Models

It can be shown (see Exercise 2) that for a range of growth rates r (given as a percent), the number of time units necessary for a population to double in size is approximately $72/r$.   ∎

Next we turn to a continuous-time, deterministic model of the situation modeled in Example 2.5 in discrete terms.

EXAMPLE 2.7   Suppose that the actual size of the population can be approximated by a function, again denoted x, that is differentiable over the time interval of interest. Suppose that this interval is $[0, T]$. Also suppose that the rate of change of x is a known (continuous) function a and that the population size at time 0 is x_0.

Model 3   This model is a continuous version of the first discrete model. We suppose that the rate of change of the population size is given by a continuous function a:

$$\frac{dx}{dt} = a(t), \quad x(0) = x_0, \quad t \in [0, T]$$

With the rate of change as in this expression, if we integrate $\frac{dx}{dt}$, we obtain the population size as

$$x(t) = x_0 + \int_0^t a(s)\, ds, \quad t \in [0, T]$$

In the special case where a is a constant function, $a(t) = r$, for t in $[0, T]$, we have

(2.6) $$x(t) = x_0 + rt, \quad t \in [0, T]$$

The graph of this function is, of course, a straight line.

Another special case that occurs often in applications, especially biological applications, is where a is a periodic function. For example, suppose $a(t) = r[2 + \sin(2t)]$; then

$$x(t) = x_0 + r\left[2t - \frac{1}{2}\cos(2t)\right], \quad t \in [0, T]$$

That is, the population size is also a periodic function with the same period.

Model 4 If, as in the second discrete model, we assume that the rate of change is proportional to the population size, then the time evolution of the population x is described by

(2.7)
$$\frac{dx}{dt} = xa(t), \quad x(0) = x_0, \quad t \in [0, T]$$

and by integrating $\frac{1}{x}\frac{dx}{dt}$, we find that this problem has the solution

$$x(t) = x_0 \exp\left[\int_0^t a(s)\,ds\right], \quad t \in [0, T]$$

In the special case where a is a constant function, $a(t) = r$, for t in $[0, T]$, we have

(2.8)
$$x(t) = x_0 e^{rt}, \quad t \in [0, T]$$

The graph of this function, an exponential, is also shown in Figure 2.2. ∎

It is clear from Figure 2.2 that predictions based on the discrete model for geometric growth and the continuous model for the same situation are very similar. Indeed, if we were to construct discrete models with smaller and smaller time intervals between observations, then predictions based on the discrete model would become closer and closer to predictions based on the continuous model.

It is also clear, from Equations (2.5) and (2.8), that over long time intervals the predictions based on these models are unrealistic (assuming $r > 0$). When we use either Model 2 or Model 4, the predicted population size grows beyond all bounds if the time interval is long enough. Because such growth does not actually occur, we need to investigate modifications in our basic assumptions to bring predictions into closer agreement with reality. We continue the discussion in the context of biological populations because in this setting, modifications to the simple models discussed above are frequently necessary.

In most biological systems the growth rate of the population, the factor r in Model 2 of Example 2.5 and the function a of Model 4 in Example 2.6, is not a constant but depends on several factors, including the size of the population. In order to isolate the factor r for study, it is convenient to introduce the concept of a *relative growth rate,* defined by

$$\frac{x_{n+1} - x_n}{x_n} \quad \text{and} \quad \frac{1}{x}\frac{dx}{dt}$$

for the discrete case and the continuous case, respectively. Because of resource constraints (for instance, food, shelter, and space), the relative growth rate decreases as the population size increases, at least for large population sizes. We consider next a continuous model that incorporates an assumption of this sort.

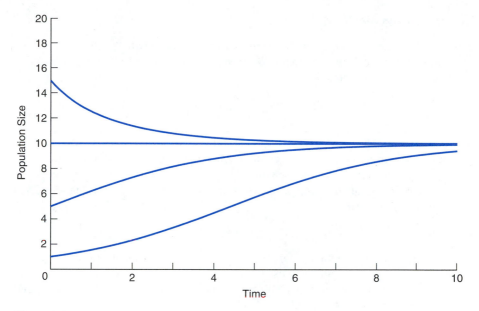

Figure 2.3

EXAMPLE 2.8   Suppose that M is the maximum size of a population that can be sustained with specific resources and that the relative growth rate is $k\left(1 - \frac{x}{M}\right)$. Then, the predicted population size x is the solution of the problem

$$(2.9) \qquad \frac{1}{x}\frac{dx}{dt} = k\left(1 - \frac{x}{M}\right), \quad t \geq 0, \ x(0) = x_0$$

This problem can be solved using standard techniques (partial fractions and direct integration), and the solution is

$$(2.10) \qquad x(t) = \frac{x_0 M}{x_0 + (M - x_0)e^{(-kt)}}, \quad t \geq 0$$

Graphs of this function are shown in Figure 2.3 for $M = 10$, $k = .5$, and three different initial population sizes ($x_0 = 1$, $x_0 = 5$, and $x_0 = 15$).   ■

   The model described in Example 2.8 is called the *continuous logistic model,* and it has often been used to describe the growth of an animal or fish population that is restricted in size by limited resources. Equation (2.9) is called the *logistic equation,* and the graph of the function given by (2.10) is called the *logistic curve.* It can be shown that if $x_0 \neq M$, then the solution x of Equation (2.9) never takes the value M. Phrased in geometrical terms, the graph of the function x never crosses the horizontal line $x = M$: If $x_0 < M$, then $x(t) < M$ for all $t > 0$, and if $x_0 > M$, then $x(t) > M$ for all $t > 0$.
   As our last example in this section, we consider a discrete model for the situation described by the logistic equation.

■ Discrete Logistic Model

Suppose that M is the size of the population that can be sustained with specific resources—the carrying capacity of the environment. We form a discrete model in which the relative growth rate is given by the equation

$$(2.11) \qquad \frac{x_{n+1} - x_n}{x_n} = k\left(1 - \frac{x_n}{M}\right)$$

where k is a constant. The constant k is sometimes referred to as the *intrinsic growth rate* of the population, the growth rate of the population when the resource constraint is not significant. It describes the rate of growth when the population size is very small compared to M.

We can make a formal change of variables (Exercise 4) and transform Equation (2.11) into an equation of the form

$$(2.12) \qquad p_{n+1} = bp_n(1 - p_n)$$

where $b = 1 + k$ and p_n is a multiple of x_n. We now study the sequence of values generated by Equation (2.12). Our goal is to use these values as (adjusted) population sizes, and consequently they must be nonnegative. We can achieve this by requiring $0 < p_0 < 1$ and $0 < b \leq 4$, and we impose these conditions.

Equation (2.12) has been widely studied as a population model, and it is simpler to study than Equation (2.11). However, the transformation of Equation (2.11) into Equation (2.12) has a cost. The information conveyed in a very straightforward way in Equation (2.11), especially the role of the growth rate k and carrying capacity of M, has been obscured in Equation (2.12) and some detail has been lost. A study of Equation (2.12) will be adequate for our purposes.

For notational simplicity, we revert to using lower case letters x for the population size. The equation studied in the remainder of this section is

$$(2.13) \qquad x_{n+1} = bx_n(1 - x_n), \quad n = 0, 1, 2, \ldots$$

Note that in Equation (2.13) the variable x has a different meaning than in Equation (2.11). In Equation (2.11) x_n is a population size; in (2.13) it is an adjusted relative population size. For simplicity, we continue to refer to x_n as a population size.

The solutions of Equation (2.13) depend on x_0 and on b, and they display quite different behavior for different values of b. Some examples are shown in Figure 2.4.

In Figure 2.4a the solutions associated with $b = 1.5$ and $x_0 = .1$ and with $b = 1.5$ and $x_0 = .5$ show behavior very similar to that of the solutions of the continuous model—namely, a monotone approach to a limiting value. Also in Figure 2.4a, the solution for $b = 0.7$ and $x_0 = .4$ corresponds to a solution of the continuous model with a negative value of k. However, the solution for $b = 2.9$ and $x_0 = .1$, shown by the diamonds in Figure 2.4b, illustrates the oscillation of successive values of x about a (normalized) population size of .655, and such oscillation is not predicted in the continuous logistic model for any choice of parameter values. In the discrete model with $b = 2.9$ and $x_0 = .1$, the sequence $\{x_n\}$ appears to be converging, and this convergence would be even clearer if we were to graph values of x_n for larger values of n. Finally, the solution for $b = 3.15$ and $x_0 = .1$, shown by the boxes in Figure 2.4b, also displays behavior very different from that of the continuous model: The population size appears to oscillate without converging to a limit. This is indeed the case. The odd iterates $\{x_{2n+1}\}$ approach one limit, and the even iterates

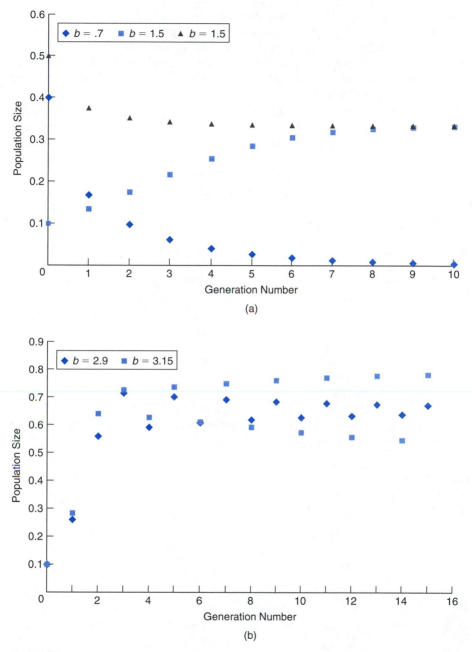

Figure 2.4

$\{x_{2n}\}$ approach another limit. For larger values of the parameter b, the situation becomes even more complex.

Oscillations have been observed in the size of populations over extended time intervals in both laboratory and field environments and for a variety of insects and animals.

Well-known examples include Australian blowflies, crabs in California, voles in central and northern Europe, wild sheep, and Canadian lynx populations. Consequently, there is interest in mathematical models that predict oscillations, at least for certain parameter values. Because of their simplicity, the models discussed here are not directly applicable to these situations, but more complex models that exhibit some of the same characteristics as the discrete logistic model have been used.

The predictions based on the continuous model follow from the solution of a differential equation. The predictions based on the discrete model, which differ significantly from those based on a continuous model, follow from a very different analysis. We will now provide some detail that gives an indication of how the solutions of Equation (2.13) behave for various values of b and x_0 and that supports the predictions reported above.

We describe the situation in geometrical terms, and we begin by introducing notation that simplifies the discussion. Define a function f by $f(x) = bx(1-x)$, for $0 \le x \le 1$. With this notation, Equation (2.13) becomes

$$(2.14) \qquad x_{n+1} = f(x_n), \quad x_0 \text{ given}, \; 0 \le x_0 \le 1, \; n = 0, 1, 2, \ldots$$

If b satisfies $0 \le b \le 4$, then the function f satisfies $0 \le f(x) \le 1$, so f maps the interval $[0, 1]$ into itself, and the sequence $\{x_n\}$ defined by Equation (2.14) is a sequence of nonnegative numbers. Such a sequence can reasonably be viewed as describing the time evolution of a population. It is clear from Equation (2.14) that, beginning with x_0, we define x_1 by $x_1 = f(x_0)$, x_2 by $x_2 = f(x_1)$, and so on. That is, the output at the nth step is the input (argument of the function f) at the $(n+1)$st step. Graphically, this process has the following meaning: The value y resulting from evaluating f at x_n, as shown in Figure 2.5, is x_{n+1}, and this value is to be used as the value of the argument for the $(n+1)$st step. This value can be obtained by projecting the point $(0, y)$ onto the ray from the origin with slope 1, obtaining the point (y, y), and using the first coordinate of this point as the new population value. This technique is illustrated in Figure 2.5.

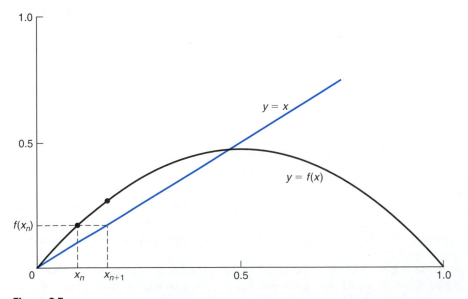

Figure 2.5

We now use the technique to study the behavior of the sequence $\{x_n\}$ for various values of b and various initial values x_0. We distinguish several cases:

Case 1: $0 < b \leq 1$ If b satisfies $0 < b < 1$, then from Equation (2.14) we have, because all population sizes are nonnegative,

$$x_{n+1} = f(x_n) = bx_n(1 - x_n) \leq bx_n$$

from which it follows that $x_{n+1} \leq b^{n+1}x_0$. Consequently, if $b < 1$, the sequence $\{x_n\}$ converges to 0. The same conclusion holds when $b = 1$, although the proof requires more care. This result is precisely what we should expect, given the meaning of the parameter b. Recall that $b = 1 + k$, where k is the relative growth rate. If $0 < b < 1$, then k is negative—the relative growth rate is negative, and we expect the population to decline.

Case 2: $1 < b \leq 2$ In this case there is a fixed point for the function f in the interval $(0, 1)$; that is, there is a number x^*, $0 < x^* < 1$, that satisfies $x^* = f(x^*)$. Solving the equation $x = f(x)$ for the fixed point, we find that $x^* = \frac{b-1}{b}$. The fixed point for $b = 1.5$ is $x^* = \frac{1}{3}$ and is illustrated in Figure 2.6.

The projection technique described in the discussion of Figure 2.5 is illustrated in Figure 2.6 for $b = 1.5$. The first few terms in the sequence $\{x_n\}$ are determined for two initial values: $x_0 = .1$ and $x_0 = .5$. These are the values graphed in Figure 2.4a. It can be shown that for values of b satisfying $1 < b \leq 2$, the sequence $\{x_n\}$ converges to $x^* = \frac{b-1}{b}$, and the convergence is monotonic with the possible exception of the first step.

For values of b above 2, the situation is more complex. We describe some results, but details of the arguments needed to verify them are omitted.

Case 3: $2 < b \leq 3$ For these values of b, the number $x^* = \frac{b-1}{b}$ is again a fixed point for the function f, and the sequence $\{x_n\}$ converges to x^*. However, in this case the convergence

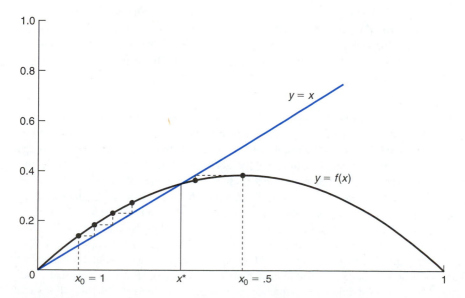

Figure 2.6

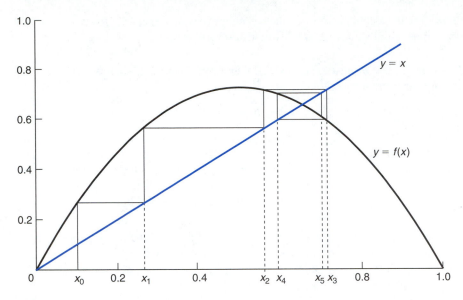

Figure 2.7

is not monotonic, but instead the values of x_n oscillate above and below x^*. The projection technique for this situation is illustrated in Figure 2.7 for $b = 2.9$. For $b = 2.9$ the graph of $\{x_n\}$ as a function of n is shown in Figure 2.4b.

For all values of b less than or equal to 3, the fixed point is said to be *stable*. This terminology results from the fact that if the initial value x_0 is $(b-1)/b$, then $x_n = x_0$ for all n, and if x_0 is different from $(b-1)/b$, then x_n approaches $(b-1)/b$ as n becomes large.

The value $b = 3$ is significant because for that value of b, the slope of the graph of the function f at $x = x^*$ is -1. For $b < 3$ the slope of the graph of the function f at x^* satisfies $|f'(x^*)| < 1$, and for $b > 3$ the slope of the graph of the function f at $x = x^*$ is less than -1. This provides information on the convergence of $\{x_n\}$ because it can be shown that for values of b for which $|f'(x^*)| < 1$, the sequence $\{x_n\}$ converges to x^* for any x_0, $0 < x_0 < 1$, whereas for values of b for which $|f'(x)| > 1$, the sequence $\{x_n\}$ does not converge (except for $x_0 = x^*$).

Case 4: 3 < b As the parameter b increases through the value 3, the asymptotic behavior of the sequence $\{x_n\}$ changes significantly. For values of b just slightly larger than 3 and initial values other than the fixed point, the sequence $\{x_n\}$ diverges, but each of the subsequences $\{x_{2n}\}$ and $\{x_{2n+1}\}$ converges, each to a different limit. This is illustrated in Figure 2.4b for the value $b = 3.15$. For $b = 1 + \sqrt{6}$ there is a further change. For values of b just slightly larger than $1 + \sqrt{6}$, the sequence $\{x_n\}$ does not converge, but in this case each of the subsequences $\{x_{4n}\}$, $\{x_{4n+1}\}$, $\{x_{4n+2}\}$, and $\{x_{4n+3}\}$ converges, each to a different limit. This is illustrated in Figure 2.8, where the sequence $\{x_n\}$ for $b = 3.5$ (shown as diamonds in Figure 2.8) has the property that with $x_0 = .1$, the four subsequences $\{x_{4n}\}$, $\{x_{4n+1}\}$, $\{x_{4n+2}\}$, and $\{x_{4n+3}\}$ converge to .81733, .52256, .88569, and .35941, respectively. As the parameter b increases, the behavior of the sequence $\{x_n\}$ becomes increasingly complex. There is a critical value of b, about 3.828, beyond which the sequence $\{x_n\}$ displays no regular behavior. For such values of b, the behavior of the sequence $\{x_n\}$ is said to be *chaotic*. Chaotic behavior of the

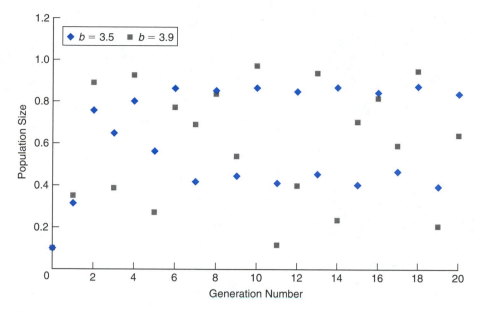

Figure 2.8 Population Changes for Values of *b* near 4

model is illustrated in Figure 2.8 with $b = 3.9$. The asymptotic behavior of $\{x_n\}$ would be much clearer in Figure 2.8 if the graph were extended for many more values of n. However in that case, because of the number of points plotted, it becomes difficult to sort out the various subsequences.

Of course, if the initial value is the fixed point, then $x_n = x^* = (b - 1)/b$ for all n. However, any initial deviation from the fixed point, no matter how small, leads to the behavior described. That is, the fixed point is *unstable*.

We can gain insight into the behavior of the population in the long run—that is, for large values of n—with a diagram. In Figure 2.9, values of the parameter b are shown on the horizontal axis, and information on the long-range behavior of $\{x_n\}$ is shown on the vertical axis. For a specific value of b there is a single value for the long-run population size if $\{x_n\}$ converges. For $0 \le b \le 1$ the sequence $\{x_n\}$ converges to 0, and that is shown in the figure. For $1 \le b \le 3$ the curve in Figure 2.9 is the graph of $(b - 1)/b$, which is the limit of $\{x_n\}$ for b in this interval. For values of b larger than 3, the sequence $\{x_n\}$ does not converge (except in the special case when $x_0 = (b - 1)/b$), but for values of b less than the critical values when chaos begins, subsequences do converge. For values of b slightly larger than 3, the subsequences $\{x_{2n}\}$ and $\{x_{2n+1}\}$ both converge, and the two limits are the intersection of a vertical line based at b with the curve. For example, if $b = 3.15$, then the subsequences of odd and even iterates both converge, and the limits are .534 and .784 (compare Figure 2.4b). Which subsequence converges to which limit depends on the value of x_0. For b slightly larger than $1 + \sqrt{6}$, there are four convergent subsequences, and the graph in the figure intersects the vertical line based at b in four points, the limits of the four subsequences. The figure becomes increasingly complex for larger values of b, and in Figure 2.9 it is shown only for values of b less than 3.5.

We see from these examples that the predictions of discrete and continuous models may be very similar, or they may be very different. Continuous models of the form studied here

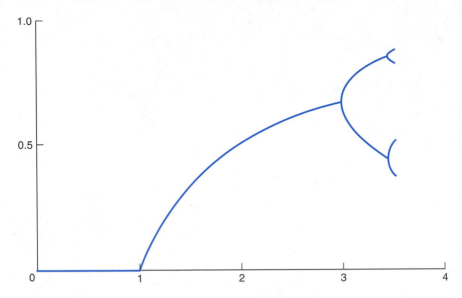

Figure 2.9

have the property that the rate of change of the population at time t depends on the size of the population at that time. In contrast, the discrete models are such that the population size at the nth observation is determined by the population size at the previous observation—that is, at a different time. Some of the predictions of these discrete models are also predictions of continuous models with delays. These are models in which the rate of change of the population at time t depends on the population size at time $t - \tau$ (that is, at a time τ units in the past). There are situations in which continuous delay models are much more appropriate than the continuous models studied here.

In many situations the effects of random events on the population growth must be included in the model, and we return to that topic in Section 2.6 and Chapter 4.

Exercises 2.2

1. Assume that Equation (2.1) holds, and give a mathematical induction proof that Equation (2.3) holds for $k = 1, 2, 3, \ldots$.

2. Consider the discrete-time model for growth in which the amount added in each time interval is proportional to the population size at the beginning of the interval: Model 2 and Equation (2.5). For growth rates r between 2.5% and 25%, graph the number $n = n(r)$ of time periods necessary for the initial population to double in size. Also graph the function $72/r$ for the same range of values of r. For what values of r is the relative error between $n(r)$ and $72/r$ less than 2%?

3. Verify that the function x defined by Equation (2.10) is a solution of the initial value problem given as expression (2.9).

4. Show that with an appropriate change of variables involving x_n, k, and M, Equation (2.11) can be transformed into Equation (2.12).

5. Let g be a continuous function defined on an interval $[0, T]$. Show that no positive solution of a problem of the form $\frac{dx}{dt} = g(x)$, $x(0) = x_0$ can oscillate on that interval.

6. Consider the discrete model described by

$$x_{n+1} = bx_n(1 - x_n^2), \quad x_0 \text{ given}$$

   Discuss the behavior of $\{x_n\}$, including results similar to those developed in this section. That is, find a range of parameter values b for which the sequence $\{x_n\}$ generated by this recurrence relation can be used as a population model. Discuss the convergence of the sequence $\{x_n\}$ for various values of b. In particular, identify the ranges of values of b for which the sequence converges to 0, converges monotonically to a nonzero limit, and converges with oscillations.

7. For the model of Exercise 6, find a value of the parameter b for which each of the subsequences of odd and even iterates, $\{x_{2n+1}\}$ and $\{x_{2n}\}$, converges but to different limits. Find the limits.

8. Consider the discrete model described by

$$x_{n+1} = bx_n \exp(-5x_n), \quad x_0 \text{ given}$$

   Discuss the behavior of $\{x_n\}$ as in Exercise 6 for this model.

9. Consider the discrete model described by

$$x_{n+1} = \frac{b(x_n)}{(1 + x_n)^2}, \quad x_0 \text{ given}$$

   Discuss the behavior of $\{x_n\}$ as in Exercise 6 for this situation.

10. In the discrete model, $x_{n+1} = bx_n(1 - x_n)$, and in the models of Exercises 6 and 8, the parameter b can be viewed as a reproduction factor for small population sizes. In each of these cases the population dies out (that is, x_n approaches 0 as n becomes large) if the reproduction factor is small but does not die out if the reproduction factor is large enough. This conclusion holds for any initial population size.

    Some populations have the characteristic that for all values of the reproduction factor, or its equivalent, the population dies out when the initial population size is small enough. Examples include animal populations distributed over a wide geographic area. The model described by the following equation has this characteristic.

$$x_{n+1} = bx_n^2 \exp(-5x_n), \quad x_0 > 0 \text{ given}$$

    (a) Find a value of b for which there is some initial value x_0 for which the population does not die out.
    (b) For this value of b, show that if x_0 is small enough, then the population does die out.

11. For the model of Exercise 10, find the smallest value of b, call it b^*, for which the population does not die out for some choice of the initial population size.

    (a) For $b = b^*$ find an initial population size for which $\{x_n\}$ does not converge to 0.
    (b) For $b = 1.2b^*$ there is a range of initial population sizes—that is, a set of values of x_0 for which the population dies out. Find this range.

12. Consider the continuous model described by

$$\frac{dx}{dt} = kx \left[1 - \left(\frac{x}{M} \right)^2 \right], \quad \text{for } t > 0, \ x(0) = x_0$$

Find an explicit solution of this equation and describe the geometry of the solutions. Compare these solutions with those of the standard logistic model for selected parameter values.

2.3 Social Choice

The study of the ways in which groups of people, or societies, choose to make decisions that affect the entire group has traditionally been viewed as the purview of social scientists: sociologists, political scientists, and economists. However, when the problems are described in precise terms, they can be studied using mathematical methods, and the results are interesting to mathematicians and philosophers as well as to social scientists. The goal of this discussion is to illustrate the problems, largely through examples, and to give one of the major mathematical results in the area.

The problems discussed in this section arise in a variety of familiar settings, such as

- A group of voters and a slate of candidates
- A city council and a set of planning options
- A legislative committee and a bill with a set of amendments
- A board of directors and a set of options for expansion of the company
- A group of friends planning to go out for dinner and a set of restaurants
- An office staff and a set of options for a new copy machine
- A state board of education and a set of curriculum alternatives

The general situation is one in which we have a group of people and a set of alternatives on which the people have opinions.

We suppose that each individual has a preference ranking on the set of alternatives. That is, using these preferences, each individual can order the alternatives in a list such that if alternative X precedes alternative Y in the list, then the individual prefers X to Y. A set of individual preference rankings, one for each member of the group, is called a **group preference schedule.** One goal of our work is to consider the problem of constructing a **group preference ranking** from the individual preferences (that is, from the group preference schedule). If such a group preference ranking can be created, then one could call the alternative at the top of the group list the alternative preferred by the group. However, it is important to note that such a group ranking may not be possible, and moreover, even if it is possible, the alternative at the top of the list may not be one that would win a majority vote in an election among all options. Thus a second goal of our work is to consider other possible ways of picking a winner, especially when none of the alternatives would receive a majority vote in an election among all alternatives. As we proceed, we will identify properties of the decision process that correspond to our intuitive ideas about the characteristics such decision processes should have.

We begin with some assumptions about preferences. First, we assume that individual preferences are *transitive;* that is, if an individual prefers X to Y and prefers Y to Z, then the individual prefers X to Z. Situations where such transitivity does not hold are unusual, and though they are sometimes interesting from a social or psychological perspective, will not be considered here. Also, it is customary to assume that group preferences should be transitive; we require all group decision processes to yield preference rankings that have this property. Thus, if the group prefers X to Y and prefers Y to Z, then it should prefer X to Z.

In general, we make no explicit assumptions about the strength of preferences. That is, preference is simply a relation and not a relation with a strength. An exception to this is the approach discussed in Example 2.12.

We require that between any two alternatives the individual preference relation be defined and that exactly one of the two possible relations holds. That is, if the alternatives are X and Y, then either X is preferred to Y or Y is preferred to X. The case of indifference— the case where neither X nor Y is preferred to the other—will not be considered here. With attention to additional details, much of the discussion that follows could be modified to include indifference. However, our goals can be achieved in the simpler setting where there is always a preference between two alternatives, and we shall restrict our discussion to that setting. Moreover, in practice, situations where indifference is possible sometimes exhibit phenomena that differ in essential ways from the simpler settings. For example, an individual might be indifferent between alternatives X and Y and between alternatives Y and Z but prefer X to Z.

Before proceeding, it is useful to comment again on the use of simple situations to illustrate the important aspects of matters such as this. Actual situations in the social sciences are normally very complex, and efforts to study real models reflecting this complexity lead to equally complex mathematical ideas and discussions. Consequently, when we are interested in a concept, it is customary to begin by considering a relatively simple model, an idealized situation that illustrates the concept even though this idealized situation may not include all attributes of the actual situation. The real models we consider are simple and describe idealized situations, although there are real-world events that are much like these examples. We comment more on what one learns from such real models, these simple examples, later in the discussion.

An essential first step in a study of a specific situation is to determine the goal (or goals) of the study. As we have noted, one goal is to develop a method for taking a group preference schedule and determining a group preference ranking. In our first example we approach this problem by making pairwise comparisons, and we determine the preferred alternative of a pair by simple-majority vote. If this method results in a transitive group preference ranking, then we can call the most preferred alternative (the one at the top of the group list) the "winner." However, we will show that there are situations in which it is not possible to obtain a group preference list, and yet there is an alternative that "should" be called the winner. For example, if a majority of the voters have the same alternative at the top of their preference lists, then that alternative would win a majority of votes in an election and thus is, in the usual sense of the word, the winner. Yet that does not imply that there is a transitive group preference list.

EXAMPLE 2.9   Suppose that we have a group of three people, labeled A, B, C, and a set of three alternatives, labeled X, Y, Z. For this example, suppose the individual preference rankings are as follows:

$$A \text{ prefers } X \text{ to } Y \text{ to } Z.$$
$$B \text{ prefers } Y \text{ to } Z \text{ to } X.$$
$$C \text{ prefers } Z \text{ to } X \text{ to } Y.$$

We represent these individual preferences in table form:

A	B	C
X	Y	Z
Y	Z	X
Z	X	Y

Using the approach described above (pairwise comparisons and simple-majority rule), we see that both A and C prefer X to Y, and therefore, because the vote is 2 to 1, the group should prefer X to Y. Also, both A and B prefer Y to Z, and therefore the group should prefer Y to Z. Therefore, on the basis of this information, we would propose that the group preference ranking should be X preferred to Y preferred to Z. However, both B and C prefer Z to X, and therefore the group should prefer Z to X. We conclude that the proposed group preference ranking is not transitive: The group prefers X to Y to Z to X. As we noted earlier, this intransitive, or cyclic, behavior is normally considered unacceptable for a preference ranking. We conclude that even in this simple situation, the majority rule decision process can lead to unacceptable preference rankings. This phenomenon is sometimes called a *voting paradox*. ∎

What are some ways to cope with the results of this example? One response is to hope that the example is so artificial and atypical that the phenomena it illustrates would never actually occur. Unfortunately, this optimistic view is false on two counts. First, the phenomena do occur. Indeed, empirical results based on experimental situations (college students ranking European cities) and actual events (university elections) illustrate the occurrence of intransitive group preferences. Second, as the number of individuals and alternatives increases, if the individual preferences are relatively widely distributed (in a sense that can be made precise), then such intransitivity can be shown to be typical rather than the exceptional behavior. That is, for many groups and sets of preferences, the group preferences determined by pairwise majority rule voting are intransitive.

To illustrate the latter point, consider again the simple situation of three individuals and three alternatives. Then each individual has 6 different preference rankings—that is, 6 ways in which the 3 alternatives can be listed: 3 choices for the alternative listed first, 2 choices for the alternative listed second, and 1 choice for the alternative listed last. Because there are 3 individuals, there are $6 \times 6 \times 6 = 6^3 = 216$ different preference schedules for the group. The likelihood of intransitive group preferences depends on how the voters select their individual preference rankings. For instance, if we know that two of the voters have the same preference ranking, then that preference ranking will be the preference ranking for the group, and intransitivity will not occur. As another example, if two voters have alternative X as their top choice, then intransitive group preferences will never occur.

On the other hand, if individuals select their individual preferences at random, then the situation is more complicated, and intransitive group preferences can occur. Indeed, if an individual makes a random selection among the 6 preference rankings, then the probability of a specific selection is $\frac{1}{6}$. If each individual selects a preference ranking independently, then the probability of a specific preference schedule being selected by the group is $\frac{1}{216}$. There are 12 different preference schedules for the group, which lead to intransitivity when simple majority voting is used to determine a group preference ranking. The preference schedule given in Example 2.9 is one example. If the individual preferences of individuals A and B are interchanged, which gives a different group preference schedule, then the result is again intransitivity. There are 6 different preference schedules that can be obtained from the schedule in Example 2.9 by reassigning individual preferences. Also, there are 6 different preference schedules in which one individual has the preference ranking

$$X \text{ preferred to } Z \text{ preferred to } Y$$

(a fact you are asked to verify in Exercise 9). Consequently, if individuals in the group select their individual preference rankings independently and at random, then with probability 12/216 the result is intransitivity—a voting paradox.

In actual voting situations, it is likely that the voters have opinions that fall somewhere between complete agreement and random choice. We pursue this idea by introducing the concept of a *culture* for our group. Intuitively, we can view a culture as a set of opinions common to all of the voters. More precisely, a culture for the group of three individuals voting among three alternatives is a probability vector $\mathbf{c} = [c_1 \quad c_2 \quad c_3 \quad c_4 \quad c_5 \quad c_6]$ where

$$c_1 = \text{ probability an individual prefers alternative } X \text{ to } Y \text{ to } Z$$
$$c_2 = \text{ probability an individual prefers alternative } X \text{ to } Z \text{ to } Y$$
$$c_3 = \text{ probability an individual prefers alternative } Y \text{ to } X \text{ to } Z$$
$$c_4 = \text{ probability an individual prefers alternative } Y \text{ to } Z \text{ to } X$$
$$c_5 = \text{ probability an individual prefers alternative } Z \text{ to } X \text{ to } Y$$
$$c_6 = \text{ probability an individual prefers alternative } Z \text{ to } Y \text{ to } X$$

Using this terminology, a culture in which each individual makes a random selection is $\mathbf{c} = [\frac{1}{6} \quad \frac{1}{6} \quad \frac{1}{6} \quad \frac{1}{6} \quad \frac{1}{6} \quad \frac{1}{6}]$. A culture in which an individual always prefers X to Y, and otherwise makes choices at random, is $\mathbf{c} = [\frac{1}{3} \quad \frac{1}{3} \quad 0 \quad 0 \quad \frac{1}{3} \quad 0]$. A culture in which there is bias for X over Y (say that preference rankings with X preferred to Y are twice as likely as preference rankings with Y preferred to X), and other choices are random, is $\mathbf{c} = [\frac{2}{9} \quad \frac{2}{9} \quad \frac{1}{9} \quad \frac{1}{9} \quad \frac{2}{9} \quad \frac{1}{9}]$. The likelihood of intransitive preference rankings in cultures defined in various ways is pursued in the exercises.

For 3 individuals and 3 alternatives, there are 3 possibilities: the process of pairwise comparisons yields a transitive group preference rating, with a majority winner, or there is a paradox, or neither of these outcomes occurs (Exercise 10). An example of the latter is

$$A \text{ prefers } X \text{ to } Y \text{ to } Z.$$
$$B \text{ prefers } Y \text{ to } Z \text{ to } X.$$
$$C \text{ prefers } Z \text{ to } Y \text{ to } X.$$

In this case there is a transitive pairwise ordering ($Y > Z > X$) but no majority winner. For numbers of alternatives greater than 3, there are other possibilities. For example, if there are 4 alternatives—W, X, Y, Z—then the individual preference rankings

A prefers W to X to Y to Z.

B prefers W to Y to Z to X.

C prefers W to Z to X to Y.

result in both a majority winner (W) and a paradox among the remaining alternatives. And the individual preference rankings

A prefers W to X to Y to Z.

B prefers W to Y to Z to X.

C prefers W to Z to Y to X.

result in a majority winner (W), but no alternative has a majority vote for second choice, and there is no paradox. Thus, when considering simple-majority decision processes with larger number of alternatives, it is important to specify the goal precisely (Exercises 11 and 12).

In light of this discussion about the difficulties encountered with simple-majority voting, we look for other ways to achieve our primary goal of finding ways for groups to make decisions. Another response to the paradox is to change the goal slightly—for example, to identify only a most preferred alternative rather than a ranking of all alternatives. We do so by introducing the concept of sequential voting: a sequence of votes where at each vote, a choice is to be made between two alternatives. In any situation with an odd number of individuals, this process always yields a result, and this result can be taken as a most preferred alternative. However, as we show below, this method also has problems.

EXAMPLE 2.10 Suppose that a most preferred alternative is to be identified by first selecting between X and Y and then selecting between the winner in the first step and Z (see Figure 2.10). We see that for the individual preference rankings given in Example 2.9 and the decision process defined here, the result of this process is Z. Indeed, the vote between X and Y yields X, and the subsequent vote between X and Z yields Z. These results are shown in Figure 2.10a.

There is, however, a difficulty with this sequential-voting approach: The resulting preferred alternative depends on the sequence in which the alternatives are considered. Indeed, if the first vote is between X and Z, and the subsequent vote is between the winner of that round and Y, then the group prefers Y (Figure 2.10b). Finally, if the first vote is between Y and Z, and the second vote is between the winner of that round and X, then the group prefers X (Figure 2.10c). Thus we are in the unhappy situation of having a group preference that depends on the sequence in which the votes were taken. ∎

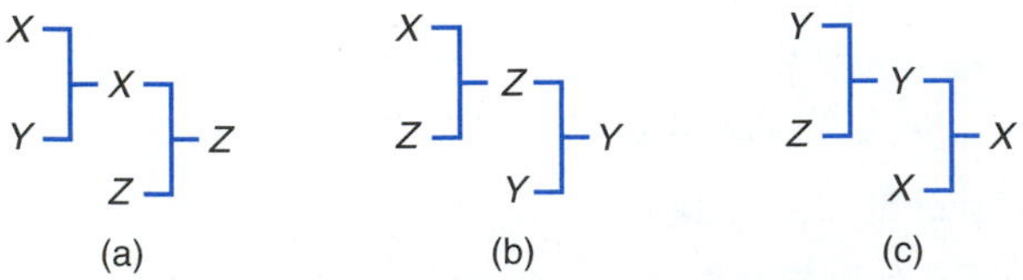

Figure 2.10 Three Different Sequential-Voting Schemes for the Same Preference Schedule

It is not uncommon to see versions of this approach in the political arena, where a great deal of time and effort may be devoted to determining the sequence in which issues are to be brought to a vote—that is, on the structure of the voting process. For example, the outcome of the debate and voting on a bill that has many amendments is often determined by the order in which the amendments are considered, and politicians devote a great deal of energy to orchestrating the sequence that is most favorable to their goals. Another feature of sequential voting, and of other decision processes as well, is that it may be desirable for an individual to vote in ways that are inconsistent with his or her preference rankings if the result of doing so leads to an outcome that is more preferable than that which would be obtained as a result of voting strictly according to his or her preferences. This technique, sometimes called *sophisticated* or *strategic voting,* can lead to very complex behavior, as each voter tries to anticipate the behaviors of the others. Unless we explicitly state otherwise, we always assume that individuals vote according to their preference rankings, a practice sometimes called *sincere voting*.

It will be useful to continue our discussion of sequential voting by expanding the ideas illustrated in Example 2.10. We will do so in a way that illustrates sophisticated voting as well as other features of sequential voting.

EXAMPLE 2.11   Suppose we have three individuals but, this time, four alternatives: W, X, Y, Z. Also, suppose the individual preference rankings are as follows:

A	B	C
X	Y	W
Y	Z	X
Z	W	Y
W	X	Z

The group makes a decision through sequential voting by selecting first between alternatives X and Y, then between the winner and W, and then between the winner and Z. The result of the first vote is X, the result of the second vote is W, and the result of the third vote is Z (see Figure 2.11a). Thus, using this method the group selects Z as its most preferred alternative.

Note that the group selected Z even though

- not one of the individuals has Z as a most preferred alternative, and
- every individual prefers Y to Z.

This decision-making process clearly has shortcomings!

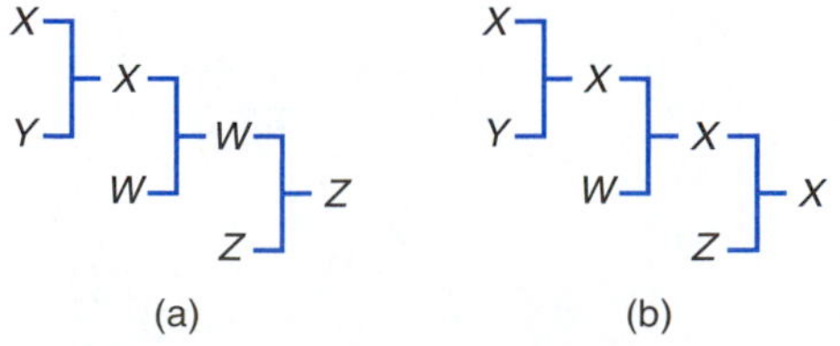

(a)   (b)

Figure 2.11

Next, we use this situation to illustrate sophisticated voting. Suppose that the actual preference ranking for B has the preferences for X and W reversed. That is, we assume that individual B has the preference ranking

(2.15) Y preferred to Z to X to W

We assume that the preference rankings for A and C are as shown in the table. In this case, using the same sequential voting scheme (see Figure 2.11b), we conclude that the group selects X as its most preferred alternative. That is, if individual B votes according to his true preferences, then the alternative preferred by the group, alternative X, is less preferable to B than alternative Z, which the group would have selected if B had voted according to the preference ranking shown in the table. Individual B benefitted by voting in ways that are inconsistent with his preferences, and sophisticated voting would achieve results preferred by B to those resulting from sincere voting. We emphasize that for B to reap the benefit of doing sophisticated voting, it was necessary for him to make accurate assumptions about the behavior of A and C. In actual situations, this is the point at which the analysis becomes quite complicated.

Finally, note that in both cases, the case where B has the preference ranking shown in the table and the case where he has that shown in (2.15), each individual has the same relative preferences for X and Z: In each case A prefers X to Z, B prefers Z to X, and C prefers X to Z. Yet the two decision processes yield a group preference of Z to X in one case and of X to Z in the other. Again, this is behavior that a legitimate decision process ought not display.   ■

We have illustrated some of the problems with simple-majority rule and sequential-voting decision processes. We turn next to another approach to the problem: assigning points to alternatives on the basis of their relative rankings and defining a group preference ranking by adding the points assigned to each alternative by all individuals. In particular, suppose each individual assigns 1 point to the least popular alternative, 2 points to the second least popular, and so on. For a specific alternative, add the points assigned by all individuals. The alternative with the most points is the most preferred, the alternative with the second largest number of points is the second most preferred, and so on. This method is known as the Borda count group decision process. We observe that this decision process has an implicit relative strength of preferences: The relative strengths of all preferences are the same.

EXAMPLE 2.12 We illustrate the technique by considering three individuals and the preference rankings of Example 2.11. Assigning points as described above, we have

A	B	C	Points
X	Y	W	4
Y	Z	X	3
Z	W	Y	2
W	X	Z	1

The group preference ranking is obtained by adding the points assigned to each alternative:

Alternative	Sum of Points
X	$4 + 1 + 3 = 8$
Y	$3 + 4 + 2 = 9$
Z	$2 + 3 + 1 = 6$
W	$1 + 2 + 4 = 7$

We conclude that the group preference ranking is

Y preferred to X to W to Z ∎

EXAMPLE 2.13   Consider a group of seven individuals and four alternatives with the following individual preference rankings.

			Individual			
A	B	C	D	E	F	G
X	Y	W	X	W	Y	W
Y	Z	X	Y	X	Z	X
Z	W	Y	Z	Y	W	Y
W	X	Z	W	Z	X	Z

Adding the points for each alternative yields this result:

Alternative	Points
X	19
Y	20
Z	13
W	18

We conclude that the group preference ranking is

Y preferred to X to W to Z

Now suppose that the least preferred alternative, Z, is no longer an option. Assuming that the elimination of alternative Z does not change the relative preferences of the remaining alternatives by any individual, the individual preferences are

			Individual			
A	B	C	D	E	F	G
X	Y	W	X	W	Y	W
Y	W	X	Y	X	W	X
W	X	Y	W	Y	X	Y

Adding the points for each alternative, we have

Alternative	Points
X	14
Y	13
W	15

We conclude that the group preference ranking among the three alternatives X, Y, and W is that W is preferred to X and X is preferred to Y.

This group preference ranking is exactly the reverse of that of the same alternatives when Z is an option. Thus, adding or removing an alternative completely changes the relative rankings of the remaining alternatives, a result that seems inconsistent with what we expect of a decision process. ∎

By considering a few examples, we have identified shortcomings of some common decision processes. We could expand our set of examples a great deal (others are described in the exercises), and indeed, many other decision processes have been proposed and investigated. Upon study, however, every proposed decision process was found to have shortcomings similar to those we have illustrated. There is, as we shall see, a very good reason for this.

To conclude our discussion, we look at the situation from the perspective of the assumptions that underlie the decision process, the basis of the mathematical modeling process described in Chapter 1. That is, for this discussion we focus on the assumptions (or axioms) that describe the characteristics we expect of an acceptable decision process. There are several sets of properties that could be used—that is, several other axiom systems—and the set given here is one of the simpler versions. Other versions may be formulated in different ways, may be more general, and may be better suited for the study of specific aspects of social choice.

First, we expect the decision process to produce a group preference ranking for every set of individual preferences. Recall that a set of individual preference rankings, one for each individual in the group, is a preference schedule for the group. With this terminology, our first property is

1. For each preference schedule, the decision rule must produce a transitive group preference ranking.

Next, we expect the decision process to preserve preferences that are common to all members of the group.

2. If alternative X is preferred to alternative Y by every member of the group, then the decision process selects for the group a preference ranking in which alternative X is preferred to alternative Y.

Also, we suppose that if we look only at the group ranking of a subset of the alternatives, then that ranking should depend only on the relative ranking of those alternatives by the individuals.

3. Let V be a subset of the alternatives. The decision process should determine a group preference ranking in which the relative ranking of the alternatives in V depends only on the relative ranking of those alternatives by the individuals.

We illustrate Property 3 with an example.

EXAMPLE 2.14   Suppose we have five voters, four alternatives, and the following two preference schedules.

	Individual				
	A	B	C	D	E
	X	X	Y	Z	W
Preference	Z	Y	X	W	X
Schedule 1	Y	Z	W	X	Y
	W	W	Z	Y	Z

	Individual				
	A	B	C	D	E
	Z	X	Y	W	W
Preference	X	W	X	Z	Y
Schedule 2	Y	Y	W	Y	Z
	W	Z	Z	X	X

Because the relative ranking of X and W is the same for each voter, a decision process should yield the same relative ranking of X and W for the two schedules.

For instance, if we use the Borda count decision process in this example, then for schedule 1 the Borda counts for X and W are 16 and 11, respectively. However, for schedule 2 the Borda counts for X and W are 12 and 14, respectively. Consequently, the Borda count decision process violates Property 3. ∎

Finally, we expect all individuals to have the same say in the decision of the group. In particular, we assume that there is no dominant individual.

4. There is no individual I such that for every preference schedule, the decision process results in a group preference ranking that is the same as that of individual I.

The difficulty in finding decision rules that have properties that seem important is a consequence of the following theorem, which was proved by the economist Kenneth Arrow in 1952.

THEOREM 2.3   (Arrow)   There is no decision process which satisfies all four of the conditions given above. ∎

That is, any method of assigning a group preference ranking to each of the possible preference schedules of a group of individuals must violate (at least) one of the four properties.

With the proof of Arrow's theorem, the work in social choice theory shifted from efforts to find acceptable decision processes to efforts to identify the strengths and weaknesses of various processes, discussions of the role of the various properties asked of a decision rule, and ways to weaken the properties so that it is possible to find decision rules.

▉ Exercises 2.3

1. There are five individuals and three alternatives with the preference rankings shown on next page.

	Individual				
	A	B	C	D	E
Preference Ranking	X	X	Y	Z	Y
	Y	Z	X	Y	Z
	Z	Y	Z	X	X

(a) Using pairwise comparisons and a simple-majority vote, determine (if possible) a group preference ranking and the most preferred alternative.

(b) Use a sequential-voting scheme as in Figure 2.10a to find the most preferred alternative.

2. There are five individuals with the preference rankings for four alternatives shown below.

	Individual				
	A	B	C	D	E
Preference Ranking	X	W	Y	Z	Y
	Y	Z	W	W	Z
	Z	Y	X	X	W
	W	X	Z	Y	X

(a) Use the sequential-voting method of Figure 2.11a to determine the preferred alternative for the group.

(b) Suppose alternative W is eliminated. Use the sequential-voting method of Figure 2.10b to determine the preferred alternative for the group.

3. Use the preference schedule of Exercise 1 and the Borda count decision process of Example 2.12 to determine the preference ranking for the group.

4. Use the preference schedule of Exercise 2 and the Borda count decision process of Example 2.12 to determine the preference ranking for the group.

5. Suppose there are five individuals with the preference rankings for four alternatives shown in Exercise 2. Find a method of changing the preferences for voter E such that sequential voting can be used to obtain different winners by starting the voting with different pairs.

6. There are 21 individuals and 3 alternatives. The numbers of individuals with various preference rankings are as shown below.

	Number of Individuals				
	5	3	3	6	4
Preference Ranking	X	X	Z	Z	Y
	Y	Z	X	Y	X
	Z	Y	Y	X	Z

(a) Using pairwise comparisons and a simple-majority vote, determine (if possible) a group preference ranking and the most preferred alternative.

(b) Use sequential voting as in Figure 2.10a to find the most preferred alternative.

7. The *plurality method* of making a group decision is to select, as the most preferable of the alternatives, the one that is the first choice of the largest number of individuals.

   (a) Using the data of Exercise 6, use the plurality method to determine the most preferred alternative for the group.

   (b) If alternative Y is eliminated, what happens to the group preference?

8. There are five individuals with the preference rankings shown below for four alternatives.

		Individual			
	A	B	C	D	E
Preference Ranking	X	W	Y	Z	Y
	Y	Z	W	W	Z
	Z	X	X	X	W
	W	Y	Z	Y	X

Does a pairwise comparison and simple-majority decision process result in a group preference ranking?

9. Suppose there are three individuals, there are three alternatives, and the decision process is to determine pairwise group preferences using a simple-majority vote.

   (a) Show that there are 168 different preference schedules for which this method yields a group preference ranking and a majority vote winner.

   (b) Show that there are 12 different preference schedules that result in intransitivity.

   (c) Show that there are 36 different preference schedules that result in neither a majority winner with a group preference ranking nor intransitivity.

10. In the situation of Exercise 9, use a computer to generate group preference schedules at random, and then use these data to estimate the probability that there is no majority vote winner when individual preference rankings are selected at random.

11. Suppose there are five individuals, there are three alternatives, and the decision process is to determine pairwise group preferences using simple-majority vote.

    (a) Find the number of different group preference schedules for which this method yields a group preference ranking and a majority vote winner.

    (b) If individual preference rankings are selected at random, find the probability that a group preference ranking cannot be determined using this method.

12. In the situation of Exercise 11, find the probability that if individuals select their preference rankings at random, then the group preferences will be intransitive.

13. Suppose there are three individuals, there are four alternatives, and the decision process is to determine pairwise group preferences using simple-majority vote. Find the number of different group preference schedules for which this method yields a group preference ranking.

14. Suppose there are three individuals, there are four alternatives, and the decision process is to determine pairwise group preferences using simple-majority vote. Find examples

of preference schedules for which

(a) There is an alternative that is a majority vote winner, and there is no intransitivity.

(b) There is an alternative that is a majority vote winner, and there is intransitivity among the remaining three alternatives.

(c) There is intransitivity among all four alternatives.

(d) There is intransitivity among three (but not all four) alternatives, and there is no alternative that is a majority vote winner.

15. Suppose there are seven individuals, there are three alternatives, and the decision process is to determine pairwise group preferences using simple majority vote. If individual preference rankings are selected at random, find the probability that this method yields a group preference ranking and a majority vote winner.

16. Suppose there are five individuals and three alternatives.

(a) For how many group preference schedules is there a majority vote winner?

(b) If the individual preference rankings are selected at random, find the probability that there is a majority vote winner.

17. Suppose there are three individuals and three alternatives, and suppose that the culture is such that no individual prefers Y to X, but otherwise individual preferences are selected at random.

(a) Find the culture c that describes this situation.

(b) Find the probability that in this situation, the group preferences are intransitive.

18. Suppose there are three individuals and three alternatives, and suppose that the culture is such that each individual is twice as likely to prefer X to Z as Z to X. Individual preferences with respect to Y are equally likely; that is, each individual is just as likely to have a preference in which X is preferred to Y as to have one in which Y is preferred to X, and similarly for Y and Z.

(a) Find the culture c that describes this situation.

(b) Find the probability that in this situation, the group preferences are intransitive.

19. Suppose there are three individuals and three alternatives, and suppose that the culture is such that one individual selects individual preferences at random and the other two individuals select individual preferences as follows: Each of them is twice as likely to prefer Z to Y as to prefer Y to Z. Individual preferences with respect to X are equally likely; that is, each individual is equally likely to have a preference in which X is preferred to Z as to have one in which Z is preferred to X, and similarly for X and Y.

(a) Find the culture c that describes this situation.

(b) Find the probability that in this situation, the group preferences are intransitive.

20. There are five individuals with the following preference rankings for four alternatives.

	Individual				
	A	B	C	D	E
	Y	X	W	Z	Y
Preference	X	Z	Y	W	X
Ranking	Z	W	X	X	W
	W	Y	Z	Y	Z

A group decision process is to determine the most preferred alternative using sequential voting by first voting between X and Y, then between the winner and Z, and finally between the winner and W.

(a) Show that with these individual preferences and the decision process just defined, the most preferred alternative is W.

(b) Alternative W is ranked fairly low by individual E. Suppose that E knows that all others will vote according to their preference rankings. Is there a sophisticated voting scheme for E such that alternative Y will emerge as the most preferred? That is, can E adopt a bogus preference ranking that will result in alternative Y being selected as the most preferred?

(c) Consider the same question for alternative X.

21. There are 51 voters and 5 alternatives. Each voter selects one of the following individual preference rankings.

- Fifteen voters select the preference ranking A preferred to D to E to B to C.
- Twelve voters select the preference ranking B preferred to E to C to D to A.
- Thirteen voters select the preference ranking C preferred to E to B to D to A.
- Eleven voters select the preference ranking D preferred to B to E to C to A.

(a) Which is the most preferred alternative if the decision process is simple-majority rule?

(b) Which is the most preferred alternative if the decision process is Borda count?

(c) Which is the most preferred alternative if the decision process is plurality (see Exercise 7)?

2.4 Moving Mobile Homes

In this section we shall consider a situation in which the basic concern is with scheduling or allocation. A general theory of such decision processes, which will include a consideration of problems similar to this one as special cases, will be developed in Chapter 5. Although it will be necessary to develop some new mathematical ideas for the general study, we shall be able to handle the necessary technical details of this problem using only basic algebra.

■ The Situation

Mr. Wheeler owns two factories at which he assembles mobile homes, and in addition he operates three regional distribution centers at which the homes are sold. Over a period of time he has kept careful records, and consequently he has a good idea how many homes will be sold each month at each of the centers. Also, he is able to estimate production rates, and therefore he knows (at least approximately) how many homes will be available each month at the factories. Finally, he is able to obtain commitments from mobile home movers to ship homes from his factories to his lots at specified rates. The rates, of course, vary with the origin and destination of the home. Mr. Wheeler's goal is to determine the most economical allocation of homes from his factories to the distribution centers. This is to be interpreted as minimizing the shipping costs while fulfilling demand.

■ A Real Model

As a first step in his study of the situation, Mr. Wheeler examines the data he has collected, and he concludes that he can estimate the supply and demand accurately enough to assume they are known quantities. Of course, he does not know *exactly* what supply and demand will be, and if the range of possibilities is large, then it might be an oversimplification to assume that they are known exactly. Similar comments apply to the shipping costs, which for this model we assume to be known constants. In Chapter 5 we consider some situations similar to this one in which quantities such as supply and demand are not deterministic parameters. Here, we take the following information as known:

$$d_j = \text{demand at distribution center } j, \quad j = 1, 2, 3$$
$$s_i = \text{supply at factory } i, \quad i = 1, 2$$
$$c_{ij} = \text{cost of moving one mobile home from factory } i \text{ to}$$
$$\text{distribution center } j, \quad i = 1, 2; \; j = 1, 2, 3$$

The next matter to be considered is the cost of moving several homes. Although there may be minor efficiencies in moving several homes over the same route, it seems reasonable to assume that the cost of moving M mobile homes over a route is M times the cost of moving one home over that route. The total cost of moving homes is taken to be the sum of the costs of moving homes over each route.

In this work Mr. Wheeler has created a real model. He has simplified his problem by ignoring certain features of the real world, such as the uncertainties of supply and demand, and by making assumptions regarding others, such as the way the total transportation cost was obtained. He hopes that with these simplifications, he can determine the least expensive shipping schedule. The merits of his conclusions ultimately rest on the validity of the assumptions.

Let us now consider the situation in more detail. The problem involves six unknown quantities: the number of homes to be shipped from each of the two factories to each of the three distribution centers. If we let f_{ij} be the number of homes to be moved from factory i to distribution center j, then $f_{11}, f_{12}, f_{13}, f_{21}, f_{22},$ and f_{23} are to be determined. The total number of homes moved to center j is $f_{1j} + f_{2j}$. Because it is required that the number of homes moved to each center satisfy the demand there, we have the inequalities

$$\begin{aligned} f_{11} + f_{21} &\geq d_1 \\ f_{12} + f_{22} &\geq d_2 \\ f_{13} + f_{23} &\geq d_3 \end{aligned}$$

(2.16)

Also, the total number sent from each factory must not exceed the number produced there, so we have

$$\begin{aligned} f_{11} + f_{12} + f_{13} &\leq s_1 \\ f_{21} + f_{22} + f_{23} &\leq s_2 \end{aligned}$$

(2.17)

At this point it is clear that the problem has a natural side condition. If the demand at each center is to be met, then there must be at least $d_1 + d_2 + d_3$ homes available at the factories. Thus the condition $d_1 + d_2 + d_3 \leq s_1 + s_2$ is a necessary one for the problem to be solvable. In this section we choose to assume that $d_1 + d_2 + d_3 = s_1 + s_2$. In Chapter 5 we shall drop this assumption, and we shall also consider larger numbers of factories and centers.

Finally, we consider the quantity in which Mr. Wheeler is most interested—namely, the cost. With each set of f_{ij} one can associate a total moving cost C. For notational convenience we let $\mathbf{y} = [f_{11} \quad f_{12} \quad f_{13} \quad f_{21} \quad f_{22} \quad f_{23}]$. Note that $\mathbf{y}$ is a vector, and therefore the order of the f_{ij} is crucial. In view of the assumptions regarding moving costs, we have

$$(2.18) \qquad C(\mathbf{y}) = c_{11}f_{11} + c_{12}f_{12} + c_{13}f_{13} + c_{21}f_{21} + c_{22}f_{22} + c_{23}f_{23}$$

The problem can now be phrased in precise terms in the following way:

Find numbers $f_{11}, f_{12}, f_{13}, f_{21}, f_{22}, f_{23}$ such that the inequalities (2.16) and (2.17) are satisfied and so that the value of the function C defined by Equation (2.18) is as small as possible. A problem of this form is known as a two-by-three transportation problem.

In general, the systems (2.16) and (2.17) are true inequalities. However, here we made a special assumption that supply is equal to demand, and using this, we can show that (2.16) and (2.17) are actually equalities. Indeed, adding the inequalities in system (2.16) and those in (2.17), we obtain

$$f_{11} + f_{12} + f_{13} + f_{21} + f_{22} + f_{23} \geq d_1 + d_2 + d_3$$

and

$$f_{11} + f_{12} + f_{13} + f_{21} + f_{22} + f_{23} \leq s_1 + s_2$$

Now, $s_1 + s_2 = d_1 + d_2 + d_3$, and thus each of these inequalities must be an equality, and consequently each of the inequalities in (2.16) and (2.17) must be an equality as well.

As a final side condition of our problem, note that each of the f_{ij} must be a nonnegative integer. That each must be an integer is clear; we do not want to break a mobile home into pieces. That each must be nonnegative reflects the fact that homes are to move from factories to distribution centers and not in the other direction.

■ A Mathematical Model

At this point, we could proceed to study the situation in its present formulation. However, we prefer first to formulate a mathematical model in quite precise terms. This model will be generalized and studied in some depth in Chapter 5.

A mathematical model for this situation should incorporate the concepts identified by Mr. Wheeler as important ones, and it should also incorporate the assumptions and simplifications introduced in the real model. The quantities and concepts appearing in the mathematical model will be abstractions of their counterparts in the real model. Thus we replace the commonplace notions of supply and demand by an undefined term called a *data vector*. Similarly, the shipping schedule is replaced by an *allocation vector*, and moving costs are replaced by a *cost vector*. These terms and the relationships among them are governed by certain axioms, and the theory is obtained from them by using standard mathematical arguments. The object of the study is to obtain information on a quantity that represents the total cost of moving mobile homes. One such mathematical model is the following:

Undefined Terms   data vector, allocation vector, cost vector

Axioms

A_1: A data vector is an ordered set of five nonnegative integers

$$\mathbf{d} = [s_1 \quad s_2 \quad d_1 \quad d_2 \quad d_3]$$

A_2: $s_1 + s_2 = d_1 + d_2 + d_3$

A_3: An allocation vector is an ordered set of six nonnegative integers

$$\mathbf{y} = [f_{11} \quad f_{12} \quad f_{13} \quad f_{21} \quad f_{22} \quad f_{23}]$$

A_4: A cost vector is an ordered set of six nonnegative real numbers

$$\mathbf{c} = [c_{11} \quad c_{12} \quad c_{13} \quad c_{21} \quad c_{22} \quad c_{23}]$$

A_5: A data vector and an allocation vector satisfy the following:

$$f_{11} + f_{21} = d_1 \qquad f_{12} + f_{22} = d_2 \qquad f_{13} + f_{23} = d_3$$
$$f_{11} + f_{12} + f_{13} = s_1 \qquad f_{21} + f_{22} + f_{23} = s_2$$

Definition 2.2 The total transportation cost associated with the cost vector $\mathbf{c}$ and the allocation vector $\mathbf{y}$ is the number $\mathbf{y} \cdot \mathbf{c}$ defined by

$$\mathbf{y} \cdot \mathbf{c} = c_{11}f_{11} + c_{12}f_{12} + c_{13}f_{13} + c_{21}f_{21} + c_{22}f_{22} + c_{23}f_{23}$$

Using this mathematical model of the situation, we can state the question facing Mr. Wheeler as follows: Given a data vector $\mathbf{d}$ and a cost vector $\mathbf{c}$, find an allocation vector $\mathbf{y}$ that makes the total transportation cost as small as possible.

We now solve Mr. Wheeler's problem in the context of the model. To begin our search for the desired allocation vector $\mathbf{y}$, we examine the equations that the coordinates of $\mathbf{y}$ must satisfy. These equations can be written in the form

$$\begin{aligned}
f_{11} \quad\quad\quad + f_{21} \quad\quad\quad &= d_1 \\
f_{12} \quad\quad\quad + f_{22} \quad\quad &= d_2 \\
f_{13} \quad\quad + f_{23} &= d_3 \\
f_{11} + f_{12} + f_{13} \quad\quad\quad\quad &= s_1 \\
f_{21} + f_{22} + f_{23} &= s_2
\end{aligned}$$

Axioms A_1 to A_5 can be used to show that these equations are not independent. In fact, it is clear that the last equation is the sum of the first three minus the fourth, and it is sufficient to solve only the first four equations as a system. Thus the system of equations we must solve has four equations and six variables. From this point of view, we now have at least two unknowns whose values will not be determined by solving the system. We consider f_{22} and f_{23} as parameters, and we show that we can solve for the other unknowns in terms of f_{22} and f_{23}. If we carry out the necessary algebra, we obtain

(2.19)
$$\begin{aligned}
f_{11} &= d_1 - s_2 + f_{22} + f_{23} \\
f_{12} &= d_2 - f_{22} \\
f_{13} &= d_3 - f_{23} \\
f_{21} &= s_2 - f_{22} - f_{23}
\end{aligned}$$

Using the information that each of the f_{ij} must be nonnegative, we obtain the following constraints on f_{22} and f_{23}:

(2.20)
$$0 \leq f_{22} \leq d_2$$
$$0 \leq f_{23} \leq d_3$$
$$s_2 - d_1 \leq f_{22} + f_{23} \leq s_2$$

Next we express the cost function C in terms of the parameters f_{22} and f_{23}. We obtain

$$C = A + Bf_{22} + Df_{23}$$

where A, B, and D are constants given by

$$A = [c_{11}(d_1 - s_2) + c_{12}d_2 + c_{13}d_3 + c_{21}s_2]$$
$$B = [c_{11} + c_{22} - c_{12} - c_{21}]$$
$$D = [c_{11} + c_{23} - c_{13} - c_{21}]$$

Thus this "two-by-three" transportation problem can be phrased as the problem of finding the proper choice of the parameters f_{22} and f_{23} to satisfy conditions (2.20) and to make C as small as possible. The specific choices for f_{22} and f_{23} that should be made in a given problem are determined by the relative magnitudes of the constants B and D. For example, if B is negative and D is positive, then one should choose f_{22} large and f_{23} small, because these choices make C small. There are many such cases to consider. We work out one numerical example in detail, and others are the topic of Exercise 1.

Let the demands, supplies, and moving costs be as follows:

$d_1 = 6$	$d_2 = 9$	$d_3 = 10$	$s_1 = 15$	$s_2 = 10$	
$c_{11} = 70$	$c_{12} = 70$	$c_{13} = 60$	$c_{21} = 80$	$c_{22} = 60$	$c_{23} = 80$

The constants A, B, and D have the values $A = 1750$, $B = -20$, and $D = 10$, so the cost function C is

$$C = A + Bf_{22} + Df_{23} = 1750 - 20f_{22} + 10f_{23}$$

and the constraints (2.20) for these data are

$$0 \leq f_{22} \leq 9$$
$$0 \leq f_{23} \leq 10$$
$$4 \leq f_{22} + f_{23} \leq 10$$

The set of points (f_{22}, f_{23}) that satisfy these constraints is known as the *feasible set* for the problem. The feasible set for the data given here is graphed in Figure 2.12, and the cost function is graphed in Figure 2.13 for values of (f_{22}, f_{23}) in the feasible set.

Our task is to find values of the parameters f_{22} and f_{23} for which the value of the cost function is as small as possible. To do so we choose f_{22} as large as possible, $f_{22} = 9$, and f_{23} as small as possible, $f_{23} = 0$. This gives $C = 1750 - 20(9) = 1570$.

Hence the minimum shipping cost, \$1570, is achieved by the allocation vector whose coordinates are

$$f_{11} = 5 \qquad f_{21} = 1$$
$$f_{12} = 0 \qquad f_{22} = 9$$
$$f_{13} = 10 \qquad f_{23} = 0$$

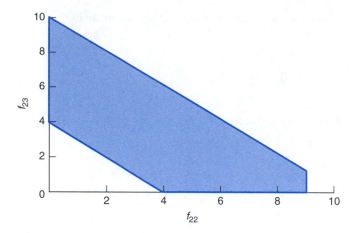

Figure 2.12

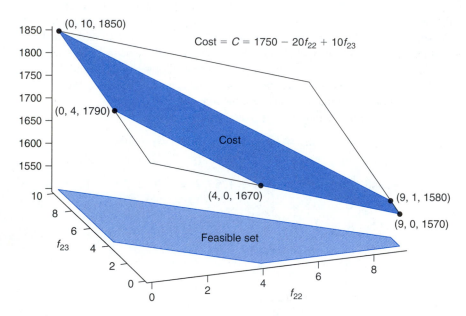

Figure 2.13

We note that in general a problem of this sort need not have a unique solution. However, in this case the solution is unique (Exercise 4).

It is interesting to examine the special case that corresponds to the condition $d_3 = 0$. Thus there are actually only two distribution centers, and the problem reduces to a two-by-two transportation problem. It is easily shown that conditions (2.19) and (2.20) become, respectively,

(2.21)
$$f_{11} = s_1 - d_2 + f_{22}$$
$$f_{12} = d_2 - f_{22}$$
$$f_{13} = s_2 - f_{22}$$

and

(2.22) $$\max(0, d_2 - s_1) \leq f_{22} \leq \min(d2, s2)$$

The cost function C now has the form

$$C = A + Bf_{22}$$

where A and B are constants (Exercise 2). In this case the solution to the problem of finding a least-cost shipping schedule is given by the following result.

THEOREM 2.4   The allocation vector that minimizes the cost function C is the vector determined by (2.21) and the choices of f_{22} specified by the following conditions:

1. If $B > 0$, then $f_{22} = \max(0, d_2 - s_1)$.
2. If $B < 0$, then $f_{22} = \min(d_2, s_2)$.
3. If $B = 0$, then f_{22} is any value such that

$$\max(0, d_2 - s_2) \leq f_{22} \leq \min(d_2, s_2)$$ ■

The proof of this result is the topic of Exercise 9.
As an application of Theorem 2.4, consider the two-by-two problem with the data

$$d_1 = 8 \qquad d_2 = 2 \qquad s_1 = 4 \qquad s_2 = 6$$
$$c_{11} = 8 \qquad c_{12} = 4 \qquad c_{21} = 6 \qquad c_{22} = 3$$

Then $B = 1$, and hence $f_{22} = \max(0, -2) = 0$. Therefore, the shipping schedule is given by $f_{11} = 2$, $f_{12} = 2$, $f_{21} = 6$.
It is interesting to note that even though the least expensive shipping rate is between factory 2 and distribution center 2, there are no shipments along this route—that is, $f_{22} = 0$.

Exercises 2.4

1. Solve the two-by-three transportation problems with the following data:
   (a) $d_1 = 8$, $d_2 = 12$, $d_3 = 10$, $s_1 = 15 = s_2$, $c_{11} = 50$, $c_{12} = 80$, $c_{13} = 40$, $c_{21} = 50$, $c_{22} = 40$, $c_{23} = 40$.
   (b) Same as (a) with $c_{23} = 30$.
   (c) Same as (a) with $c_{22} = 90$ and $c_{23} = 50$.
   (d) $s_1 = 4$, $s_2 = 5$, $d_1 = 2$, $d_2 = 4$, $d_3 = 3$, $c_{11} = 3$, $c_{12} = 5$, $c_{13} = 1$, $c_{21} = 6$, $c_{22} = 7$, $c_{23} = 6$.
   (e) Same as (d) with $c_{13} = 4$.

2. Formulate the two-by-two transportation problem and show that it is the same problem as the two-by-three problem with $d_3 = 0$.

3. Formulate a transportation problem with three factories and three distribution centers; denote the supply at the third factory by s_3. Show that if $s_3 = 0$, then the problem is the same as the two-by-three problem formulated in this section.

4. Show that the solution of the sample two-by-three problem worked in the text is unique. *Hint:* Show that for all least-cost schedules f_{22} and f_{23} are as given.

5. Formulate an *m*-by-*n* transportation problem.

6. Discuss the usefulness of Theorem 2.4 for problems modeled as two-by-two transportation problems. Under what conditions will this theorem give useful information, and when might it give misleading results? More precisely, give three aspects of the real-world situation that seem to be adequately incorporated in the model and three that are not.

7. Discuss the two-by-three transportation model with axiom A_2 replaced by axiom A_2': $d_1 + d_2 + d_3 \leq s_1 + s_2$. How might the introduction of a fictitious distribution center be used to convert this into the previous model?

8. How should the transportation model of this section be modified to account for a partial shipping strike that limits the total number of mobile homes that can be sent over certain routes?

9. Prove Theorem 2.4.

10. In the case of a two-by-three transportation problem, show that when the data vector has integer coordinates, then the allocation vector that solves the problem always has integer coordinates. (There are similar optimization problems when integer values of the data do not guarantee integer values of the solution, an issue we consider in Chapter 5.)

2.5   A Stratified Population Model

The model-building process involves making assumptions, creating a model based on these assumptions, and checking predictions based on the model with observations of the actual situation being studied. In Section 2.2 we developed models for growth processes under a variety of assumptions. In Section 2.6 we will consider examples of stochastic models where the assumption of deterministic reproduction rates will be replaced with information on the probabilities of various reproduction rates. However, in all these situations, our models are based on the assumption that the population being studied can be viewed as homogeneous; that is, every individual plays the same role as every other individual. There are many situations where this assumption leads to predictions that differ from observations. For instance, in a population where reproduction takes place only for individuals in a few subgroups, or in a population where survival differs significantly from one subgroup to another (as in human and many animal populations), the assumption of homogeneity may be inappropriate.

We consider a situation in which the homogeneity assumption is replaced by the assumption that the population can be stratified. In a stratified population the entire population can be divided into subgroups, and individuals move from one subgroup to another as time passes. For example, it is common to study human populations by grouping individuals into age groups; it is common to study the student body at a college by grouping students according to class standing; it is common to study forests by grouping trees into size groups; and it is common to study certain insect populations by grouping individuals into various life stages. The fundamental characteristic of such models is that each individual either moves from one stratum to another in a regular way or leaves the system—departs or dies.

A variant of this model that we consider later includes the possibility of an individual being in the same stratum on successive observations.

Suppose that there are m subgroups or strata, that observations are made at discrete times, and that the number of individuals in the ith subgroup at observation k is denoted by $x_i(k)$, for $i = 1, 2, \ldots, m$. The population at the kth observation can be described by a vector $\mathbf{X}(k)$, where

$$(2.23) \qquad \mathbf{X}(k) = \begin{bmatrix} x_1(k) \\ x_2(k) \\ \vdots \\ x_m(k) \end{bmatrix}$$

Our goal is to study the distribution of individuals among the subgroups as time changes—that is, to study the sequence $\{\mathbf{X}(k)\}$ as k increases.

A particularly simple situation is that of age stratification in a population where there is no in-migration or out-migration. We assume that the range of ages in each stratum is the same and is equal to the interval between observations. In this case, an individual in age group j at observation k either dies or is in age group $j+1$ at observation $k+1$. Let s_j denote the survival probability that an individual who is in the jth age group at observation k is in the $(j+1)$st age group at observation $k+1$. Note that the notation appears to mean that the survival probability is independent of the observation, and this is our assumption. More refined models might include time-dependent survival rates, but we do not. This survival information can be summarized as follows:

$$x_2(k+1) = s_1 x_1(k), \quad x_3(k+1) = s_2 x_2(k), \ldots, \quad x_m(k+1) = s_{m-1} x_{m-1}(k)$$

Also, suppose that individuals in the first age group arise from reproduction of individuals in other age groups. In particular, suppose that between observations k and $k+1$, each individual in the jth age group will contribute f_j individuals to the first age group, for $j = 1, 2, 3, \ldots, m$. Because we are constructing a deterministic model, the reproduction assumption must be formulated in these terms, although in practice, each f_j is interpreted as an average value. This assumption leads to the relation

$$x_1(k+1) = f_1 x_1(k) + f_2 x_2(k) + \cdots + f_m x_m(k)$$

Again we suppose that the reproduction rates are independent of the observation.

Information on both survival and reproduction can be included in a matrix representation of the dynamics of the population. We have

$$(2.24) \qquad \mathbf{X}(k+1) = \mathbf{A}\mathbf{X}(k)$$

where $\mathbf{X}(k)$ is as given by Equation (2.23) and

$$(2.25) \qquad \mathbf{A} = \begin{bmatrix} f_1 & f_2 & f_3 & \cdots & f_{m-1} & f_m \\ s_1 & 0 & 0 & \cdots & 0 & 0 \\ 0 & s_2 & 0 & \cdots & 0 & 0 \\ & & & \ddots & & \\ 0 & 0 & 0 & \cdots & s_{m-1} & 0 \end{bmatrix}$$

Because the matrix $\mathbf{A}$ contains reproduction and survival rates, it is usually called the matrix of vital rates for the population.

The study of the time evolution of a population whose initial age distribution is $\mathbf{X}(0)$ thus becomes the study of the sequence $\{\mathbf{X}(k)\}$ defined by (2.23) and (2.24) where the initial vector is $\mathbf{X}(0)$ and the matrix $\mathbf{A}$ is given by (2.25).

EXAMPLE 2.15   Suppose we have four age groups, and the initial population distribution $\mathbf{X}(0)$ and the matrix $\mathbf{A}$ of vital rates are given by

$$\mathbf{X}(0) = \begin{bmatrix} 40 \\ 20 \\ 10 \\ 5 \end{bmatrix}, \quad \mathbf{A} = \begin{bmatrix} 0 & 2 & 5 & 10 \\ .8 & 0 & 0 & 0 \\ 0 & .5 & 0 & 0 \\ 0 & 0 & .2 & 0 \end{bmatrix}$$

A straightforward matrix multiplication shows that

$$\mathbf{X}(1) = \begin{bmatrix} 140 \\ 32 \\ 10 \\ 2 \end{bmatrix}, \quad \mathbf{X}(2) = \begin{bmatrix} 134 \\ 112 \\ 16 \\ 2 \end{bmatrix}, \quad \mathbf{X}(3) = \begin{bmatrix} 324.0 \\ 107.2 \\ 56.0 \\ 3.2 \end{bmatrix} \quad ■$$

In situations where a model based on (2.24) and (2.25) is applicable, it can be shown that in many cases of interest, the long-range behavior of the sequence $\{\mathbf{X}(k)\}$ depends on the largest positive eigenvalue of the matrix $\mathbf{A}$ and the associated eigenvector. Recall that the *eigenvalues* of a matrix $\mathbf{A}$ are the solutions of the algebraic equation (called the characteristic equation of the matrix $\mathbf{A}$)

$$det\,(\mathbf{A} - \lambda\mathbf{I}) = 0$$

If $\mathbf{A}$ is an $m \times m$ matrix, then the characteristic equation is an mth-degree polynomial equation for λ. The following fact can be proved using techniques from linear algebra:

If there are two consecutive positive reproduction rates in the set $f_1,\ f_2,\ f_3,\ \ldots,\ f_m$, then there is a unique positive eigenvalue λ_0, and for each initial vector $\mathbf{X}(0)$ we have

$$(2.26) \qquad\qquad \mathbf{X}(k)/\lambda_0^k \to c\mathbf{W}, \quad \text{as } k \text{ increases}$$

where c is a constant depending on $\mathbf{X}(0)$, and $\mathbf{W}$ is the eigenvector corresponding to λ_0. That is, $\mathbf{AW} = \lambda_0\mathbf{W}$. It will be convenient to assume that $\mathbf{W}$ is normalized so that the sum of the coordinates is equal to 1. In the context of population dynamics, $\mathbf{W}$ gives the fraction of the population in each age group.

EXAMPLE 2.16   For the matrix $\mathbf{A}$ of Example 2.15, the eigenvalue λ_0 and the eigenvector $\mathbf{W}$ are

$$\lambda_0 = 1.73686, \quad \mathbf{W} = \begin{bmatrix} .6217 \\ .2864 \\ .0824 \\ .0095 \end{bmatrix} \quad ■$$

It follows from (2.26) that if $\lambda_0 < 1$, then in the long run the population dies out; if $\lambda_0 > 1$, then in the long run the population increases indefinitely; and if $\lambda_0 = 1$, then in the long run the population stabilizes with the relative distribution of individuals in each age group given by **W**. If $\lambda_0 > 1$, then the population does not approach a limit, but in the long run the relative sizes of the subgroups are given by **W**.

EXAMPLE 2.17   For the matrix **A** of Example 2.15, the vectors in the sequence $X(k)/\lambda_0^k$ for $k = 1, 2, 3, 5, 10$, and 50 are

$k = 1$	$k = 2$	$k = 3$	$k = 5$	$k = 10$	$k = 50$
80.61	44.42	61.84	56.84	57.70	57.73
18.42	37.13	20.46	26.64	26.62	26.59
5.76	5.30	10.69	8.20	7.64	7.65
1.15	0.66	0.61	0.68	0.88	0.88

As k increases, the vectors $X(k)/\lambda_0^k$ are approaching a limit; the limit is a multiple of the eigenvector **W**. ■

EXAMPLE 2.18   An age-stratified population modeled as in Equation (2.24) has the following matrix of vital rates:

$$\begin{bmatrix} 0 & 1 & 6 \\ .4 & 0 & 0 \\ 0 & .2 & 0 \end{bmatrix}$$

With these vital rates the positive eigenvalue for the matrix **A** is .95, and consequently the model predicts a population decline of about 5% between successive observations. Suppose that the goal is to stabilize the population by altering the environment to improve survival rates, and suppose that changing survival rates between successive stages requires different environmental interventions depending on the stages involved. That is, a specific intervention affects a single survival rate. Suppose it is determined that survival from stage 2 to stage 3 is to be enhanced. Estimate the smallest change in the survival rate s_2 that is required to yield a population that does not die out. Also, determine the third coordinate of the resulting stable population distribution.

Our task is to find the smallest value of s_2 for which the positive eigenvalue for the matrix

$$\mathbf{A} = \begin{bmatrix} 0 & 1 & 6 \\ .4 & 0 & 0 \\ 0 & s_2 & 0 \end{bmatrix}$$

is at least 1. The positive eigenvalue of the matrix **A** is the positive solution of the equation $det(\mathbf{A} - \lambda\mathbf{I}) = 0$, which is $2.4s_2 + .4\lambda - \lambda^3 = 0$. Setting $\lambda = 1$ in this equation, we have $s_2 = .25$. We conclude that the survival rate from stage 2 to stage 3 must be at least .25 for the population to survive. Thus the survival rate from stage 2 to stage 3 is to be increased from .2 to .25, a 25 percent increase. The third coordinate of the solution W of $\mathbf{AW} = \mathbf{W}$ is 0.0667. ■

The foregoing discussion applies to more general situations than the examples of age-stratified populations. Another situation where the general idea is helpful is that of stage-stratified populations. Here, it is possible for an individual to remain in the same stage from one observation to the next. An important example of this is a stage-stratified model for a forest where the stages are specified by the sizes of trees. In this situation, a tree can be in the same stage for several observations. In the case of stage stratification, we modify the matrix $\mathbf{A}$ by adding nonzero entries on the main diagonal. We let r_i denote the fraction of individuals in the ith stage at observation k that will still be in the ith stage at observation $k + 1$, for $i = 1, 2, 3, \ldots, m$. In this case we have

$$(2.27) \qquad \mathbf{A} = \begin{bmatrix} r_1 + f_1 & f_2 & f_3 & \cdots & f_{m-1} & f_m \\ s_1 & r_2 & 0 & \cdots & 0 & 0 \\ 0 & s_2 & r_3 & \cdots & 0 & 0 \\ & & & \vdots & & \\ 0 & 0 & 0 & \cdots & s_{m-1} & r_m \end{bmatrix}$$

EXAMPLE 2.19 Suppose that the matrix $\mathbf{A}$ of Example 2.18 is modified to include retention as follows: in the first (10%), third (20%), and fourth (50%) stages. The resulting matrix is

$$\mathbf{A} = \begin{bmatrix} .1 & 2 & 5 & 10 \\ .8 & 0 & 0 & 0 \\ 0 & .5 & .2 & 0 \\ 0 & 0 & .2 & .5 \end{bmatrix}$$

For this situation the eigenvalue λ_0 and the eigenvector $\mathbf{W}$ are

$$\lambda_0 = 1.836, \quad \mathbf{W} = \begin{bmatrix} .629 \\ .274 \\ .084 \\ .013 \end{bmatrix}$$

With $\mathbf{X}(0) = \begin{bmatrix} 40 \\ 20 \\ 10 \\ 5 \end{bmatrix}$, the first three iterations of Equation (2.24) for this matrix yields

$$\mathbf{X}(1) = \begin{bmatrix} 144 \\ 32 \\ 12 \\ 4.5 \end{bmatrix}, \quad \mathbf{X}(2) = \begin{bmatrix} 183.40 \\ 115.20 \\ 18.40 \\ 4.65 \end{bmatrix}, \quad \mathbf{X}(3) = \begin{bmatrix} 387.24 \\ 146.72 \\ 61.28 \\ 6.00 \end{bmatrix} \qquad \blacksquare$$

EXAMPLE 2.20 A population of sea turtles is stratified into four strata: eggs and hatchlings, juveniles, young adults, and adults. If observations are made yearly, then the matrix of vital rates is

$$\mathbf{A} = \begin{bmatrix} 0 & 0 & 10 & 25 \\ .1 & 0 & 0 & 0 \\ 0 & .2 & .2 & 0 \\ 0 & 0 & .4 & .2 \end{bmatrix}$$

The largest eigenvalue for the matrix **A** is .866, and consequently the population dies out at a rate of about 13% per year. It is possible to provide greater protection for the eggs and hatchlings by monitoring and safeguarding the nesting area, and it is believed that the survival rate of eggs and hatchlings might be doubled through these efforts. Is it possible to stabilize the turtle population through this mechanism alone? If so, how much must the survival rate from eggs and hatchlings to juveniles be increased to stabilize the population?

To check whether the population can be stabilized by improving the survival from the first stratum to the second, we find the eigenvalue of the matrix of vital rates with the rate s_1 replaced by .2. This eigenvalue is 1.03, and consequently, the population eventually increases by about 3% per year. We can find the value of s_1 for which the largest positive eigenvalue is 1 by finding the characteristic equation with s_1 as a parameter, setting λ equal to 1, and solving for s_1. The characteristic equation with s_1 as a parameter is

$$\lambda^4 - 0.4\lambda^3 + 0.04\lambda^2 - 2s_1\lambda - 1.6s_1 = 0$$

and setting $\lambda = 1$ in this equation gives $.64 - 3.6s_1 = 0$. Solving the latter equation for s_1 yields $s_1 = 1.6/9 = 0.1778$ (with four-digit accuracy). As an alternative, we can proceed by finding the largest eigenvalue for systematically selected values of s_1. For example, we know that the largest positive eigenvalue of the matrix with $s_1 = .1$ is less than 1 and the largest eigenvalue of the matrix with $s_1 = .2$ is larger than 1. We check for the eigenvalue for $s_1 = .15$ and find that it is less than 1 (= .958). Continuing, we find the largest eigenvalue for $s_1 = .17$ is less than 1, whereas the largest eigenvalue for $s_1 = .18$ is larger than 1. After several additional trials, we find that the largest eigenvalue will be equal to 1 for $s_1 = .1778$ (with four-digit accuracy).   ■

Exercises 2.5

1. The matrix of vital rates for a three-stage stratified population model follows. Find the positive eigenvalue and the associated (normalized) eigenvector for this model.

$$\begin{bmatrix} 0 & 2 & 7 \\ .5 & 0 & 0 \\ 0 & .3 & 0 \end{bmatrix}$$

2. The matrix of vital rates for a four-stage stratified population model follows. Find the positive eigenvalue and the associated (normalized) eigenvector for this model.

$$\begin{bmatrix} 0 & 2 & 7 & 10 \\ .5 & .2 & 0 & 0 \\ 0 & .3 & 0 & 0 \\ 0 & 0 & .2 & 0 \end{bmatrix}$$

3. A general model for a three-stage age-stratified population has the following matrix of vital rates. Show that if the vital rates are all positive, this matrix always has exactly one positive eigenvalue and an associated eigenvector with positive entries.

$$\begin{bmatrix} 0 & f_2 & f_3 \\ s_1 & 0 & 0 \\ 0 & s_2 & 0 \end{bmatrix}$$

Table 2.6

Stage	f_i	s_i
1	0	.997
2	.001	.998
3	.085	.998
4	.306	.997
5	.400	.996
6	.281	.995
7	.153	.992
8	.064	.989
9	.015	.983
10	.001	

4. A model of a three-stage age-stratified population has the following matrix of vital rates.

$$\begin{bmatrix} 0 & 1 & 6 \\ .4 & 0 & 0 \\ 0 & .2 & 0 \end{bmatrix}$$

This matrix leads to predictions in which the population declines by about 5% between successive observations (see Example 2.18). Suppose that the goal is to stabilize the population by altering the environment to improve reproduction rates, and suppose that an environmental intervention that changes the reproduction rates affects them both by the same percentage. Estimate the smallest percentage change in the reproduction rates that is required to yield a population that does not die out.

5. A ten-stage model for human population growth has the reproduction and survival rates as shown in Table 2.6. Using these data, estimate the long-term rate of increase and the long-term distribution of the population in the ten age groups.

6. A model of a three-stage age-stratified population has the following matrix of vital rates.

$$\begin{bmatrix} 0 & 2 & 5 \\ .4 & 0 & 0 \\ 0 & .3 & 0 \end{bmatrix}$$

You can change exactly one of the survival rates by 10%. Which rate should you change to increase the long-term growth rate of the population as much as possible?

7. A population has individuals in three strata; individuals in the first, second, and third strata are labeled small, medium, and large, respectively. The population has the matrix of vital rates shown below. Suppose observations are made each year, and the population in the nth year immediately after reproduction is denoted $X(n)$. Suppose that a fraction h of the large individuals are harvested each year *before* the reproduction reported in $X(n + 1)$. That is, if we denote by $X(n)'$ the result of subtracting a fraction h of the large individuals from $X(n)$, then we have $X(n + 1) = AX(n)'$. If the goal is to select

h such that the population does not die out, how large can h be? If there are initially 100 individuals in each stratum, how many can be harvested each year?

$$\begin{bmatrix} 0 & 2 & 5 \\ .4 & .1 & 0 \\ 0 & .3 & 0 \end{bmatrix}$$

8. In the situation discussed in Example 2.20, the largest eigenvalue, λ_0, of the matrix of vital rates depends on the survival rate s_1. Graph λ_0 as a function of s_1 for $.05 \le s_1 \le .2$. Also, determine the long-run fraction of the population in the adult stratum for the same range of values of s_1.

9. In many situations where a stratified population model is appropriate, the vital rates are not constants but instead depend on time, population size, or environmental factors. Consider a situation that can be modeled as a stratified population model with three strata. Suppose that the fertility rates are assumed to be constants and that the survival rates depend on the population size as in the matrix of vital rates

$$A(\mathbf{X}_n) = \begin{bmatrix} 0 & 2 & 5 \\ s_1(\mathbf{X}_n) & 0 & 0 \\ 0 & s_2(\mathbf{X}_n) & 0 \end{bmatrix}$$

where $s_1(\mathbf{X}_n) = s_2(\mathbf{X}_n) = 2/(2 +$ size of the population at nth generation). If $\mathbf{X}(0) = \begin{bmatrix} 1 \\ 2 \\ 1 \end{bmatrix}$, determine the percentage of the population in each stratum in the long run.

Does the result change for $\mathbf{X}(0) = \begin{bmatrix} 1 \\ 0 \\ 1 \end{bmatrix}$? (*Hint:* Compute $\mathbf{X}(n)$ for successive values of n until the population distribution changes little from one value of n to the next.)

10. As in Exercise 9, suppose we have a stratified population model in which the survival rates depend on population size as shown in the matrix of vital rates

$$A(\mathbf{X}_n) = \begin{bmatrix} 0 & 2 & 3 & 5 \\ s_1(\mathbf{X}_n) & 0 & 0 & 0 \\ 0 & s_2(\mathbf{X}_n) & 0 & 0 \\ 0 & 0 & s_3(\mathbf{X}_n) & 0 \end{bmatrix}$$

where

$$s_1(\mathbf{X}_n) = .8 \exp[-.2(\text{size of the population at } n\text{th generation})]$$
$$s_2(\mathbf{X}_n) = .5 \exp[-.5(\text{size of the population at } n\text{th generation})]$$
$$s_3(\mathbf{X}_n) = .3 \exp[-.7(\text{size of the population at } n\text{th generation})]$$

Suppose the initial distribution of the population is half in each of the first two strata. Find the long-term distribution of the population among the four strata.

2.6 Simulation Models in Athletics, Marketing, and Population Studies

In Section 1.5 we briefly discussed the idea of using computer simulation to study mathematical models that involve uncertainty. We now consider three examples that show how simulations can be used. In Chapter 4 we will consider the simulation process in greater detail.

EXAMPLE 2.21 (Shooting Free Throws) Hack Schameal is a star basketball player who is an outstanding player except when shooting free throws. He makes an average of 40% of his free-throw attempts. Coaches of other teams have adopted the imaginative strategy of fouling him as much as possible, and now he shoots about 25 free throws each game. His own coach would like to know whether it helps or hurts the team to have Hack in the game. His coach asks two important questions:

1. On average, how many points will Hack score on free throws in each game, and what is the variation in this number?
2. Does the average number of points he scores on free throws depend on his success early in the game? That is, if he begins shooting badly, should the coach take him out of the game?

There is an easy method to answer questions 1 and 2, but unfortunately, it involves assumptions about Hack's free throw shooting that may not be valid. For example, if Hack's free throws are independent events, then, in a game in which he shoots 25 free throws and has a probability .4 of making each one, the average number he makes per game is $25(.4) = 10$. Also, the standard deviation—one measure of the variation from the average—is $\sqrt{25(.4)(.6)} = \sqrt{6}$. If the free throws are independent events, then it does not matter how he begins a game; his average number of points remains 10. This discussion and the answers to the questions depend on the assumption that the free throws are independent, and this may not be a valid assumption.

In contrast to the assumption of independence, suppose that after two consecutive misses, Hack gets nervous and his probability of making the next shot drops to .2. Alternatively, perhaps when he makes two consecutive shots, his confidence increases and his probability of making the next shot increases to .7. In situations where we make assumptions like these, the simple approach described above does not apply, and different methods are needed to answer the questions. One such method is to use simulation to "play" many games in which Hack shoots 25 free throws and different assumptions are made about his success in making free throws. In Chapter 4 we will discuss the details of using simulation in the situation described here. For the remainder of this example, we restrict our discussion to one simple version of such assumptions to illustrate the nature of simulation models.

Suppose our goal is to use a simulation in the case where the free throws are independent events, and the probability of success on each shot is .4. This is the case where the theory tells us that the average number of free throws made is $25(.4) = 10$. In a simulation of this simple situation, a trial consists of 25 computer-generated free throws. A computer-generated free throw can be thought of as a random number in the interval [0, 1]—that is,

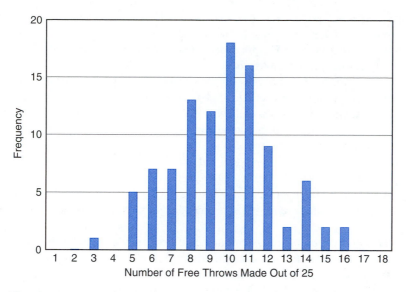

Figure 2.14

a number selected by a computer program in such a way that any number in the interval [0, 1] is equally likely to be selected. If the number is between 0 and .4, then we say the attempt is successful, and if the number is in the interval [.4, 1], then we say the attempt is unsuccessful. In each trial of 25 free throws we count the number of successful attempts, and we record that number. A specific implementation of a simulation of this sort consisting of 100 trials resulted in the data recorded on the histogram in Figure 2.14. For these data the average number of successful free-throw attempts is 9.68 and the standard deviation is 2.68.

On the basis of this simulation, we expect that the average number of points Hack will score on free throws in a game is between 9 and 10. However, it is important to note that sometimes he only makes 3 and other times he makes 16. Of course, these results are a consequence of many random events (actually the selection of $100 \times 25 = 2500$ random numbers), and if we repeated the simulation we would almost certainly get a different average. Note that the simulation gives us an "answer," a prediction of 9.68 points per game, that differs from the answer based on probability theory: $25(.4) = 10$. One possible source of the discrepancy is that our simulation consists of only 100 trials. If we were to do a simulation with 1000 trials, then we would expect the results of the simulation to be closer to the theoretical value 10. In fact, a specific implementation of the simulation with 1000 trials resulted in an average number of points equal to 10.09 and a standard deviation of 2.46. In Chapter 4 we discuss the question of how many trials are necessary to achieve a specified level of accuracy. Also in Chapter 4, we discuss in more detail a slightly more complex version of this situation.   ■

EXAMPLE 2.22   (Collecting Tokens)   It is a common marketing strategy to include "tokens" in packages of a product; if a customer collects one of each token in a set, then the customer receives a prize. The usual situation is that the prize is substantial, and to constrain the cost to the producer, one or more of the tokens are rare. To make the situation precise, suppose that a cat food producer includes a plastic letter in each bag of cat food. The letters are C, A, T, and S, and suppose that every customer who collects all four letters receives a free bag of cat food. From the perspective of the manufacturer, it is important to know how the distribution of tokens is related to the expected number of bags of cat food that a customer needs to buy before receiving a free bag.

The construction of a mathematical model for the situation is straightforward. For the purpose of this discussion, the only characteristic of a bag of cat food that interests us is what token it contains. Consequently, we can associate with each bag exactly one of the symbols in the set $\{C, A, T, S\}$. A collection of bags can be viewed as a collection of symbols. We assume that if a customer buys a bag of cat food, then the likelihood of getting a particular symbol is the frequency with which that symbol is added to food bags. That is, we assume that the tokens appear in bags of food purchased by consumers with the same frequency with which they are added by the manufacturer. This means, for instance, that the distribution of bags of cat food to various retail stores is independent of the tokens packaged in the bags.

Suppose that three of the four tokens occur with equal frequency and the fourth is rare in comparison with the other three. For this example, suppose that each of the tokens C, T, and S occurs with frequency .3 and that the token A occurs with frequency .1. With respect to the collection of tokens, we can imagine consumer behavior as follows: A consumer buys a bag of cat food and checks which token it contains. If the consumer has a complete set, the prize is awarded and the scenario is finished. If not, then the consumer buys another bag of cat food, and the process continues. If we are to "simulate" these actions, then we need a process that selects symbols at random from the set $\{C, A, T, S\}$ with the frequencies given in Table 2.7.

A record of the token received is maintained, and the record is updated after each "purchase"—that is, after each selection of a symbol. Once a symbol of each of the four types has been selected, the simulation ends and the total number of symbols selected is recorded. This set of activities represents one "run" of the simulation. For instance, the sequence selected might be

$$T,\ T,\ C,\ T,\ A,\ C,\ T,\ C,\ S$$

and the total number of symbols selected is 9.

Clearly, the number of symbols selected before one of each has been collected varies from one run to another, and in order to have confidence in the results, we determine the average number of symbols selected over a large number of runs. We illustrate the reduction in variation that (usually) accompanies an increase in the number of runs in a simulation by giving data on simulations consisting of 10, 100, and 1000 runs. In particular, in Table 2.8 we give data—based on the frequencies in Table 2.7—for 10 repetitions of 10 runs each, for 10 repetitions of 100 runs each, and for 10 repetitions of 1000 runs each. In each case we

Table 2.7

Symbol	Frequency
C	.3
A	.1
T	.3
S	.3

Table 2.8

	Number of Runs per Simulation		
	10	100	1000
	10.000	11.500	12.136
	10.400	12.040	11.839
	10.200	10.900	11.794
Average Number	14.500	11.810	12.053
of Selections	12.600	11.790	12.168
per Simulation	13.200	11.120	12.118
	14.100	11.150	12.234
	12.200	10.050	11.932
	17.400	12.740	12.156
	13.000	12.810	11.997
Average for 10 Simulations	12.760	11.591	12.043

determine for each run the number of symbols selected in order to have one of each type, and then we compute the average over the specified number of runs.   ■

The reduction in variation in the results as the number of runs in a simulation increases is clearly illustrated in Table 2.8. How many runs are necessary to produce acceptable results is an important question. We return to this and several related questions, and we provide examples of the application of simulation models in other fields, in Chapter 4.

The situation modeled in Example 2.22 is a very simple one. Indeed, it is possible to find the desired results by analytical means (Exercises 8 and 9). However, it illustrates the main ideas, and it is easy to modify the situation slightly to one where analytical methods cannot be used.

EXAMPLE 2.23   (A Stochastic Population Model)   The population models discussed in Sections 2.2 and 2.5 are deterministic models. However, each of them can be modified to incorporate probabilistic behavior by assuming that one or more of the parameters are random variables with known distributions. In this discussion, we consider a relatively simple example. Other examples are considered in Chapter 4.

Consider a model of a homogeneous population—similar to Model 2 of Section 2.2— in which the rate of increase r is not constant but, rather, varies randomly with a known distribution. Such a model might be appropriate with an animal population in which r is a net rate (birth rate minus death rate) of increase and r varies with the amount of food available. Specifically, suppose that the initial population size is 100 and that the rate of increase r takes the values $-.01, 0, .01,$ and $.02$ with the probabilities given in Table 2.9.

The reproduction rate r is a random variable with expected value .005. Consequently, if we were to construct a deterministic model of this situation using the expected value of r as the deterministic rate of increase, then using Equation (2.5), we would expect the population at the kth observation to be about $100(1.005)^k$, for $k = 1, 2, \ldots$. In particular, we might estimate the population after 2 observations (that is, 2 observations following the

Table 2.9

Value	Probability
−.01	.3
0	.1
.01	.4
.02	.2

initial observation at which the population size is 100) to be $100(1.005)^2$, which is about 101. This deterministic approach, however, makes no use of the more detailed information we have about the rate of increase, and it provides no information on the distribution of values of the population size on the second observation. A stochastic model would give us this additional, more detailed information, and in the case of 2 observations, we can construct such a model using a tree diagram. If we do so, we find that the possible values of the population size and their respective probabilities are as shown in Table 2.10.

We conclude from Table 2.10 that the expected population size is about 101 and the standard deviation is about 1.589. Thus we have obtained more refined information using a stochastic model. However, the process we used (a tree diagram) is feasible only for small numbers of observations. Indeed, the number of branches on the tree is 4^k, where k is the number of observations, and even considering that more than one branch may yield the same value of the population size, the situation becomes complex quickly. For 2 observations, the tree has 16 branches and there are 10 distinct values of the population size. For 3 observations, the tree has 64 branches and there are 20 distinct values of the population size. With 10 observations, we have a tree with over a million branches—scarcely a feasible way to approach the problem.

Because the rate of increase is a random variable, the population size at any observation will also be a random variable. Pick a fixed observation number, say $k = 20$ for definiteness, and let P be the random variable whose values are the possible population sizes at the 20th observation. We develop estimates of the expected value of P and its distribution by

Table 2.10

Population Value	Probability
98.01	0.09
99.00	0.06
99.99	0.24
100.00	0.01
100.98	0.12
101.00	0.08
102.00	0.04
102.01	0.16
103.02	0.16
104.04	0.04

Mean =	101.0025
Standard deviation =	1.589094

simulation. We note that if we were to construct a deterministic model of this situation using the expected value of r as the deterministic rate of increase, then using Equation (2.5), we would estimate the population at the 20th observation to be $100(1.005)^{20}$, which is about 110.49.

The simulation process is much like that described in Examples 2.21 and 2.22. Each trial begins with our selecting one of the values of r at random, with the probabilities given in Table 2.9. Call this value r_1. The value of the population at the first observation is

$$x_1 = 100(1 + r_1)$$

Then we select a second value of r at random, again using the probabilities of Table 2.9. We denote the second value by r_2. The value of the population at the second observation is x_1 multiplied by $1 + r_2$, where r_2 is the value just selected. That is,

$$x_2 = x_1(1 + r_2)$$

The process continues for 20 observations—that is, until x_{20} has been computed. This completes one trial of the simulation.

Suppose the simulation continues for 100,000 trials; in other words, one run of the simulation consists of 100,000 trials. Each trial results in a number, the population size resulting from the specific sequence of rates of increase selected in the trial. To help determine the distribution of P, we cluster the values for the trials by dividing the range into convenient intervals. In this situation, all values of x_{20} fall between 90 and 135. We subdivide this interval into 9 subintervals, each subinterval 5 units long, and we count the number of values of x_{20} in each subinterval. The results are shown in Table 2.11.

We use the fraction of the total number of values of x_{20}—the total is 100,000—that lie in each subinterval as an estimate of the probability that P takes a value in that subinterval. For instance, the probability that P takes a value in the subinterval $95 \leq P < 100$ is .02949. The histogram for Table 2.11 is shown in Figure 2.15, and this is an approximation of the graph of the density function of the random variable P.   ∎

It is important to recognize that if the process is simulated again—that is, if another run of 100,000 trials is performed—then the results will be slightly different. The data in a new version of Table 2.11 will be different, and the new graph of the density function

Table 2.11

Interval	Number of Values of x_{20}	Probability
[90, 95)	81	.00081
[95, 100)	2949	.02949
[100, 105)	14234	.14234
[105, 110)	28622	.28622
[110, 115)	35621	.35621
[115, 120)	14218	.14218
[120, 125)	3775	.03775
[125, 130)	477	.00477
[130, 135)	23	.00023

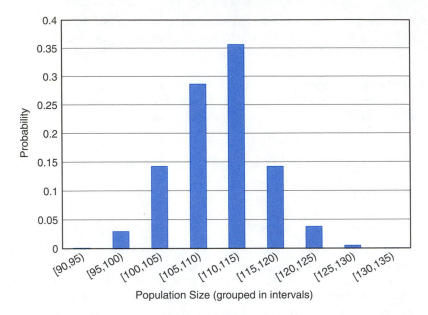

Figure 2.15   **Histogram for Population Simulation Model.**

will be slightly different from Figure 2.15. To illustrate, Table 2.12 provides the average population size for ten simulation runs of 10 trials each, ten runs of 100 trials each, and ten runs of 1000 trials each. Note that the variation among the averages of the runs of 10 trials each is substantially larger than the variation in the averages of the runs of 1000 trials each. This illustrates the value of including a sufficiently large number of trials in the simulation run.

Also, if the data in Table 2.9 are altered, the results will be different. Even if a new Table 2.9 is used with the same expected value for r, the estimate for the density function will differ. However, if we have already set up a simulation, it is a simple matter to make these changes and to run it again with the new data.

Table 2.12   Average population size

Runs of 10 Trials Each	Runs of 100 Trials Each	Runs of 1000 Trials Each
109.481	110.595	110.328
112.745	110.495	110.305
110.229	110.444	110.773
108.984	110.628	110.548
111.776	110.934	110.614
110.572	110.329	110.746
111.578	110.070	110.692
110.228	110.342	110.600
111.234	110.749	110.705
113.294	110.880	110.732

▮ Exercises 2.6

Several of the following exercises involve the use of random numbers. This topic is explored in Chapter 4, but for the present, it will be adequate to use one of the standard spreadsheet packages (such as Excel) to generate the numbers needed. With Excel, we can generate random numbers in the interval from 0 to 1 by using the function command "=RAND()".

1. A simulation of the experiment of flipping a fair coin until there are two consecutive heads can be conducted as follows. Pick a specific entry in a list of random numbers as a starting point, and interpret each number between 0 and .5 as a flip that results in a head and each number between .5 and 1 as a flip that results in a tail. Using the numbers in the list consecutively, continue until you have two consecutive numbers between 0 and .5—two consecutive heads—and then stop.

    The average number of flips before there are two consecutive heads can be estimated using simulation by repeating the process described above. On each repetition, begin with the random number in your list that immediately follows the last one used in the preceding simulation. If the process is repeated 10 times, and the number of flips before two consecutive heads appear is recorded on each repetition, then the average of the 10 numbers you recorded will be your estimate of the average number of flips before two consecutive heads appear.

    Use this method to estimate the average number of flips before two consecutive heads appear.

2. Use the technique described in Exercise 1 to simulate the experiment of flipping a fair coin until there are a total of three heads. Use 10 repetitions of the simulation to estimate the average number of flips before there are a total of three heads.

3. (a) Repeat Exercise 1 if the coin is assumed to be unfair with $\Pr[H] = \frac{1}{3}$.
   (b) Repeat Exercise 2 if the coin is assumed to be unfair with $\Pr[H] = \frac{1}{3}$.

4. An experiment consists of flipping a fair coin repeatedly, and each time it comes up heads you receive $1, and each time it comes up tails you receive $2. The experiment stops as soon as you have at least $10.

   (a) Use the technique described in Exercise 1 to simulate the experiment.
   (b) Use 10 repetitions of the experiment to estimate the average number of flips before you have at least $10.

5. Use the technique described in Exercise 1 to simulate the experiment of rolling a fair die until there are three consecutive odd numbers.

6. Use the technique described in Exercise 1 to simulate the experiment of rolling a fair die until a 1 and a 2 occur in that order.

7. Use a list of random numbers as in Exercise 1 to conduct a simulation of the token-collecting situation described in Example 2.22. For instance, assume that each random number between 0 and .3 corresponds to a C token, each random number betweem .3 and .4 corresponds to an A token, each random number between .4 and .7 corresponds to a T token, and each random number between .7 and 1 corresponds to an S token. Simulate the collection of all four tokens 10 different times, and find the average number of tokens collected.

8. Suppose that the token-collecting situation of Example 2 is modified such that there are only two different tokens, ⋆ and ♦, and such that ⋆ occurs with probability .8 and ♦ with probability .2. One run of a simulation consists of selecting tokens successively until you have both the ⋆ and the ♦ and then recording the total number of tokens selected. Use the process described in Exercise 1 to conduct a simulation consisting of 10 runs, and determine the average number of tokens selected before you have both the ⋆ and the ♦.

9. In the situation of Exercise 8, find the expected number of tokens selected using probability theory.

10. In the situation of Example 1, suppose that if Hack makes his first free throw, then the probability of his making any subsequent free throw is .5, and if he misses his first free throw, then the probability of his making any subsequent free throw is .3. In each game, the probability of his making the first free throw is .4. If he shoots 25 free throws, what is the average number of points he scores from free throws?

    (a) Use the ideas of Exercise 1 to conduct a simulation of this situation.
    (b) Use 10 runs of the simulation to estimate the average number of points scored.
    (c) Use a tree diagram and probability theory to answer the question.

11. Suppose you have an unfair coin with $\Pr[H] = .3$. Use the technique described in Exercise 1 to simulate the experiment of flipping the coin four times. For each run of the simulation, record "Yes" if there are both heads and tails, and "No" otherwise. Repeat the simulation 10 times, and use $(\frac{1}{10})$(number of times "Yes" was recorded) as an estimate of the probability that both heads and tails occur when the coin is flipped four times.

12. In the situation of Exercise 11, find the probability that both heads and tails occur in four flips of the unfair coin by using probability theory.

2.7 Waiting in Line Again!

We all spend more time waiting in lines than we think is necessary or desirable. For example, commuters wait in line at tollbooths, grocery shoppers wait in line to check out, and students may wait in line to use a special computer or to get tickets to a rock concert. The study of *queues*—our official name for a system that includes units being served and (possibly) units waiting to be served—is both an interesting mathematical topic and a very important practical concern for business and industry. In this section we describe a specific problem that can be investigated using the concepts and methods of queuing theory, and we examine enough of the theory to solve a special case of the problem.

■ The Setting

A software company, let's call it Wordfast, plans to market a new word processing program, and one of its main selling points is to be that registered users of the program will have access to toll-free help 24/7 (24 hours a day, 7 days a week). To provide this help, the company must hire and train operators, who will then staff a central phone location. The problem for the company is to decide how many operators are needed so that "most" callers will receive help in a "reasonable" amount of time (if all operators are busy, then the caller is put on hold, and all callers on hold form a queue). Unfortunately, demand (the number of incoming calls) is not deterministic. Customers call when they have a problem, not according to a set schedule. Also, the nature of each call and the amount of time it will take to answer it are not predictable. However, the company does have access to the data from similar operations that it has carried out in the past (from some records kept internally and others provided by the phone company), so it has a basis for making assumptions about the distribution of calls and the length of calls. Of course, in making such assumptions, the company is forming a mathematical model. This mathematical model will then be used to help answer the questions posed about the real-world situation.

■ Assumptions Used in the Model

In order to study any queue, one needs to know or make assumptions about three of its features:

1. **The arrival scheme.** How do units (calls, people, cars, or whatever) arrive for service? Is there a pattern to the arrivals (for example, everyone has an appointment and they arrive on time) or are arrivals random? If they are random, what is the average time between arrivals and what is the distribution of arrival times?
2. **The queue discipline.** How do units line up for service? Is it first come, first served (FCFS), last come, first served (LCFS), or perhaps some sort of priority system (largest unit in line is served next)?
3. **The service scheme.** How do units receive service? Is service provided by a single server or are there multiple servers? Do all arrivals go into a single queue, or is there a separate queue for each server? Does it always take the same amount of time to service a unit or is the service time a random variable? If it is a random variable, what is the probability distribution of this random variable?

In addition to these three features, it is also important to know the capacity of the system being studied. In particular, one needs to know the maximum number of servers who can work at one time, the maximum number of units that can be in the queue at the same time, and the size of the pool from which the units are drawn.

We now discuss these assumptions in the context of the problem faced by Wordfast.

Wordfast knows that when the new word processing program is first introduced, demand for the product will be small, and it is likely that only a single operator will be needed to provide help. Company planners also believe (on the basis of experience) that the times between successive arrivals have an exponential distribution and that the times it takes to provide service (the length of the interval between when service begins and when it is finished—not including the time spent in line) also have an exponential distribution. This is a major assumption by the company, and in general, it would need to be checked by collecting data about arrival times and service times. Finally, the phone system is designed so that all calls are handled first come, first served. Thus the planners know the general characteristics of the queuing system to be studied. Also, on the basis of data collected in similar situations, they estimate that the average time between arrivals (calls) is five minutes and the average time to complete a call is three minutes. They would like to know the average time that a caller will wait before talking to the operator. In other words, how long should a caller expect to be on hold?

Before addressing the problem we have just posed, it is important to note why it really is a problem. Indeed, if calls arrive, on the average, every 5 minutes, and if they last, on the average, only three minutes, it would seem that no queue will ever exist. However, the key phrase here is "on the average." Sometimes the time between consecutive calls is only one minute, and sometimes a single call may last 10 minutes. Thus it is possible that a registered user who calls will be third or fourth in line and may have to wait several minutes before talking to the operator. The company wants to know how often this sort of thing will happen.

To answer the key question about average waiting time, it is necessary for us to work with the exponential distribution and to recall certain facts about this distribution. First, however, in order to relate the assumption that the arrival distribution is exponential to the actual arrivals, we need to introduce some notation about arrival times. For a given queue, we are interested in both the arrival times, denoted by $t_1, t_2, \ldots, t_n$, and the interarrival intervals, defined by $T_i = t_{i+1} - t_i$, $i = 1, 2, 3, \ldots, n$. We let A be a random variable whose values are the lengths of the interarrival intervals. Thus A records the number of units of time between arrivals. In most applications, A is a continuous random variable, a random variable that takes values in the interval $(0, \infty)$, and it is important to know as much as possible about the probability density function of A. We let $a(t)$ denote this density function. Thus the relationship between A and $a(t)$ is given by

$$\Pr[A \leq x] = \int_0^x a(t)\, dt$$

In general, one needs to collect data to determine the nature of the function $a(t)$. However, in special cases, such as the one for Wordfast, one may know enough about the arrival scheme to make good assumptions about $a(t)$ without needing to collect data. Wordfast assumes that $a(t)$ is exponential. Thus there is a constant λ such that $a(t) = \lambda e^{-\lambda t}$.

Service times are also random for most queues, and we let S be the random variable that records the number of units of time needed to complete service. The distribution function for S is denoted by $s(t)$, and, just as with the arrival distribution, one must either determine the nature of $s(t)$ by collecting data about service times or by making assumptions about $s(t)$. In the case of Wordfast, we again make the assumption that $s(t)$ is an exponential distribution. Thus there is a constant μ such that $s(t) = \mu e^{-\mu t}$. We are now ready for our review of the exponential distribution.

■ Comments on the Exponential Distribution

The most common choice of the distribution function for the time between arrivals (the random variable A) is the exponential distribution. The probability density function for the exponential distribution with parameter λ is given by $f(t) = \lambda e^{-\lambda t}, t \geq 0$. One of the reasons why the exponential distribution is such a common choice for the arrival distribution is that this distribution obeys the no-memory property. This property states that the probability that the next interarrival time will be in a certain interval does not depend on earlier interarrival times. To be precise, if A obeys the exponential distribution, then for any nonnegative numbers t and h,

$$\Pr[A > t + h \mid A \geq t] = \Pr[A > h]$$

Thus the fact that a certain amount of time has elapsed since that last arrival does not change the probability that the next arrival will occur in a certain interval of time. This property greatly simplifies some of the computations needed to describe queues, and, remarkably enough, it has been shown to fit many real-world queuing situations. In fact, the first real successes of models for queues were related to phone systems, and they used the exponential distribution.

A direct computation, using the definition of *mean* and *variance,* shows that the mean of the exponential distribution is $\frac{1}{\lambda}$ and the standard deviation is also $\frac{1}{\lambda}$. Also, if the distribution for a random variable A is the exponential distribution with parameter λ, then the probability that A takes a value less than or equal to a given number x is

$$\Pr[A \leq x] = \int_0^x \lambda e^{-\lambda t}\, dt = 1 - e^{-\lambda x}$$

Thus, for any positive number x, $\Pr[A > x] = 1 - \Pr[A \leq x] = e^{-\lambda x}$.

■ Predictions

In order to answer the key question posed by Wordfast management (on average, how long will customers spend on hold?), we need to study the behavior of the queue that consists of the calls that arrive in the phone system. For general arrival distributions, $a(t)$, and/or general service distributions, $s(t)$, this is a very difficult problem, and few results are known. In fact, many such queues are studied by simulating the queue with a computer and examining the results of the simulations. The fitness center example of Chapter 4 illustrates this approach. However, in the case of queues for which $a(t)$ and $s(t)$ are exponential, the behavior of the queue has been described in great detail, and we simply state and apply the results.

■ Comments on Queues with Exponential Arrivals and Service

For a single-server queue with arrival distribution $a(t) = \lambda e^{-\lambda t}$ and service distribution $s(t) = \mu e^{-\mu t}$, $\lambda < \mu$, the behavior of the queue will reach an equilibrium state. In other words, after an initial period following the first arrival, the probabilities of the queue containing a specific number of people will not change with time. In this setting the following results hold:

1. The probability that there are a total of n units in the queuing system (being served or in line) is $p_n = (\lambda/\mu)^n (1 - \lambda/\mu)$.
2. The average total time spent in the queuing system is $w = 1/(\mu - \lambda)$.
3. The average time spent waiting in line to be served is $(\lambda/\mu)w$.

In order to use results 1–3, we must know the values of λ and μ for the Wordfast queue. If $a(t) = \lambda e^{-\lambda t}$, and if the average time between arrivals is 5 minutes, then $\frac{1}{\lambda} = 5$, because the mean time between successive arrivals given by $a(t)$ is $\frac{1}{\lambda}$. Thus $\lambda = \frac{1}{5}$; there is 1 arrival per 5 minutes or 12 arrivals per hour. In the same way, using the fact that the average time for service (getting your question answered) is 3 minutes, we see that $\mu = 20$ service completions per hour.

Using results 1–3 and the values of λ and μ, given above, we can now answer questions about the Wordfast telephone queue. In particular, using formula 1, we see that the probability that there are n customers in the phone system (getting a question answered or waiting on hold) is

$$p_n = \left(\frac{12}{20}\right)^n \left(1 - \frac{12}{20}\right) = \left(\frac{3}{5}\right)^n \left(\frac{2}{5}\right)$$

Thus the probability that no calls are being processed is $p_0 = \frac{2}{5}$. The person answering the phone will be idle 40% of the time. Using formulas 2 and 3, we see that the average time in the system is $w = \frac{1}{(20-12)} = \frac{1}{8}$ of an hour, or 7.5 minutes and that the average time on hold is $\left(\frac{12}{20}\right)w = 4.5$ minutes. Thus, even though the person answering the calls is idle about 40% of the time, customers spend an average of 4.5 minutes on hold. Of course, many customers get their question answered immediately, and a few may be on hold for a long time.

■ Remarks about Other Distributions

Wordfast assumed that the arrival and service distributions were exponential. If either of these distributions is not exponential, then formulas 1–3 do not hold, and other methods must be used to describe the behavior of the queue. We mention one other interesting case and discuss a method of describing the queue in that case.

Suppose that instead of a person answering the phone at Wordfast, there is a recorded message that takes exactly three minutes to play. Thus, suppose that arrivals are still exponential with a mean time between arrivals of 5 minutes, but service is deterministic and always takes 3 minutes. What happens to the queue in this case, and how long is the average caller on hold?

As we consider the behavior of the queue under our new assumptions, at first glance this seems to be a simpler setting, because service times do not vary from one caller to another. However, we also note immediately that the no-memory property is violated. Not only does the time when service is completed depend on when service starts, but this time always occurs exactly three minutes after the starting time. In such cases, it is often very difficult to derive formulas such as 1–3. One method that is often used is to have a computer generate arrivals, complete service, and record the amount of time spent in the queue for each arrival. If this is done many times and arrivals are generated to fit the known arrival distribution, then the average time spent in the queue, as calculated by the computer, will be very close to the average in real life. Exactly how many times the computer should generate an arrival and how it should generate arrivals are two of the topics considered in Chapter 4.

■ Exercises 2.7

1. Suppose the person answering the calls speeds up with experience, and the average time to answer a call is reduced to 2.5 minutes. How long, on average, are customers in the queuing system, and how long are they on hold waiting for service?

2. Suppose demand increases and calls come in with an average time between them of 4 minutes. If the service time is still 3 minutes on average, what is the average time spent in the system and the average time spent on hold?

3. A single-bay drive-through car wash has arrivals, with exponentially distributed inter-arrival times, at a rate of 10 per hour. The manager believes that the maximum time a

customer will spend in the car wash is 10 minutes, including the time in line and the time car is being washed, and she wants to find a service rate that keeps total customer average time in the system less than 10 minutes.

  (a) If the service times are exponentially distributed, what service rates μ will achieve the manager's goal?

  (b) Graph the average time at the car wash as a function of μ for $10 < \mu < 20$.

4. In the situation of Exercise 3, suppose that the goal is to be sure that the time in the line waiting for the car wash to begin is less than 5 minutes. What service rates will achieve this goal?

5. Derive a formula for the average length of the queue in terms of λ and μ.

6. Use the facts given in Section 2.7 to show that the exponential distribution exhibits the no-memory property.

7. Suppose that for a single-server queue with exponential arrivals and exponential service distributions, the arrival rate λ suddenly doubles to 2λ, while the service rate μ remains unchanged. Suppose also that the ratio $\frac{\lambda}{\mu}$, which was $\frac{1}{3}$, is now $\frac{2}{3}$. How does the average time spent in the queue change, and how does the average number of units in the queue change?

8. Suppose calls arrive at a call center at random at a rate of 15 per hour. The arrival times of calls from 6:00 to 7:00 a.m. are shown in Table 2.13. In that table, the times are shown in fractions of an hour measured from 6:00 a.m.

  (a) If each call is serviced with a recorded message 3 minutes long, find the maximum length of the queue waiting to be serviced.

  (b) With a 3-minute message as in part (a), find the average number of callers waiting for service between 6:00 and 7:00 a.m.

  (c) How does the maximum queue length change if the message is increased to 4 minutes? How does the average numbers of calls waiting change? What if the message is reduced to 2.5 minutes?

9. In the setting of Exercise 8, suppose that each call is serviced by an operator whose standard response requires 3.5 minutes. However, the operator adjusts the response according to the number of calls waiting. Accordingly, the response takes 3.5 minutes

Table 2.13

Arrival (Call) Number	Arrival Time (in hours after 6 a.m.)	Arrival (Call) Number	Arrival Time (in hours after 6 a.m.)
1	0.0464	10	0.5511
2	0.0498	11	0.5535
3	0.1335	12	0.6089
4	0.1510	13	0.6310
5	0.1909	14	0.6407
6	0.2040	15	0.6491
7	0.2782	16	0.7286
8	0.3861	17	0.9040
9	0.4163	18	0.9265

if, when a call is completed, there is 1 call waiting; the response takes 3 minutes when there are 2 calls waiting; and the response takes 2.5 minutes when there are 3 or more calls waiting.

(a) Find the maximum length of the queue waiting to be served.

(b) Estimate the average waiting time for an incoming call before the service begins.

10. In the setting of Exercise 8, suppose that each call is serviced by an operator and that the 18 calls with arrival times given in Table 2.13 have the lengths of service times shown (in hours) in Table 2.14.

(a) Find the maximum length of the queue waiting to be served.

(b) Estimate the average waiting time for an incoming call before the operator begins service.

11. Describe, in words, how you would organize a computer simulation of the situation where the arrivals are exponential but the service times are deterministic with a constant service time.

12. Suppose that the following set of times of arrivals and of service are observed. Use these times to estimate λ and μ.

Times of arrival: 9:02, 9:03, 9:07, 9:08, 9:09, 9:15, 9:17, 9:27, 9:38, 9:51,
                    9:57, 9:59, 10:10, 10:12, 10:13

Service times: 3, 5, 2, 1, 7, 10, 2, 2, 1, 1, 2, 3, 1, 4, 8

13. Using the setting of this section, suppose that λ and μ are both replaced by $k\lambda$ and $k\mu$, where $k > 1$. How do the following quantities change?

(a) p_n = the probability that there are n customers in the system (being served or in line)

(b) the average time spent in the system

(c) the average time spent waiting to be served

Table 2.14

Arrival (Call) Number	Length of Service (in hours)	Arrival (Call) Number	Length of Service (in hours)
1	0.0327	10	0.0197
2	0.0121	11	0.0279
3	0.2187	12	0.0298
4	0.1281	13	0.0265
5	0.0946	14	0.0421
6	0.0389	15	0.0224
7	0.0170	16	0.0191
8	0.0340	17	0.0436
9	0.0472	18	0.1004

2.8   Estimating Parameters and Testing Hypotheses

Many of the models constructed in this book cannot be tested until certain parameters are determined. We now consider the problem of estimating parameters, and we give two useful methods: the *method of maximum likelihood* and the *method of minimum discrepancy*. It should be emphasized that many other techniques have been successfully used, and some of these can be found in Bernard Flury's *A First Course in Multivariate Statistics* (Springer, 1997).

■ Maximum Likelihood

The philosophy behind the method of maximum likelihood is as follows. Appropriate data are collected, and it is hypothesized that the data depend on certain parameters in a specific way. Next, the probability (or density function in the nondiscrete case) that these particular data would occur in a study of this sort is determined, and then this probability is expressed as a function of the parameters. Finally, the possible parameter values are examined to determine whether there are values of the parameters that maximize this probability. That is, one searches for specific parameter values such that the probability of obtaining these data is greater than or equal to the probability of obtaining the same data from a system in which the parameters take on other values. Parameters selected in this way are said to be determined by the **maximum likelihood principle.**

The following examples illustrate the method.

EXAMPLE 2.24   Suppose that a model of learning uses a parameter p that is the probability that a student in psychology P007 at Big State University can identify a concept in a certain limited time. We shall assume that all the students have the same ability to identify this concept. Also, suppose that 100 students are randomly selected and that 20 of them are able to identify the concept in the allotted time. Theoretically, the parameter p may take any value in the interval $[0, 1]$. For each choice of p, the probability that exactly 20 of 100 students will *independently* make the identification is equal to

$$f(p) = \binom{100}{20} p^{20}(1 - p)^{80}$$

This formula defines a polynomial function f that depends on p, $0 \le p \le 1$. The maximum value of f on this interval can be found by examining the values of $f(\tilde{p})$ for $\tilde{p} = 0, 1$ and $\tilde{p} \in \{p : \frac{df}{dp}(p) = 0\}$. The equation $\frac{df}{dp} = 0$ is

$$\binom{100}{20} (20p^{19}(1 - p)^{80} - 80p^{20}(1 - p)^{79}) = 0$$

or

$$p^{19}(1 - p)^{79}[(1 - p) - 4p] = 0$$

Thus $\{p : \frac{df}{dp}(p) = 0\} = \{\frac{1}{5}, 0, 1\}$, and we need to compare $f(0)$, $f(1)$, and $f(\frac{1}{5})$. Because $f(0) = f(1) = 0$ and $f(\frac{1}{5}) > 0$, it is clear that the maximum value of f on

[0, 1] is assumed for $p = \frac{1}{5}$. Thus $\frac{1}{5}$ is the estimate for p determined by the method of maximum likelihood. Note that this is also the sample proportion because $\frac{20}{100} = \frac{1}{5}$. In this setting, if N students are tested and n succeed, then the maximum likelihood estimate of p is n/N. ∎

The next example is somewhat more complicated.

EXAMPLE 2.25 Recall our example in Section 2.6 of Hack Schameal, the star basketball player who has trouble shooting free throws. It is very likely that the probabilities needed to answer the coach's key question (Should I leave him in the game or take him out?) vary from game to game. Suppose the coach decides to estimate these parameters using the data obtained from Hack's first five free throws of each game. Also suppose that the coach is confident that the probability of Hack's making the first free throw is always .4, but two critical probabilities change from game to game. They are $p_1 = $ the probability that Hack makes the next free throw given that he made his last one, and $p_2 = $ the probability that Hack makes the next free throw given that he missed his last one.

The coach wants a method of estimating p_1 and p_2 on the basis of the results of the first five free throws. We show how the method of maximum likelihood would work in a special case.

Special Case Suppose that the results on Hack's first five free throws are 1–0–0–1–0, where 1 means he made the shot and 0 means he missed the shot. Using the definitions of p_1 and p_2, and recalling that the probability of making the first shot is .4, we have

$$f(p_1, p_2) = \Pr[1\text{--}0\text{--}0\text{--}1\text{--}0 \mid p_1, p_2] = .4(1 - p_1)(1 - p_2)(p_2)(1 - p_1)$$
$$= .4(1 - p_1)^2 p_2(1 - p_2)$$

Using the method of maximum likelihood, we need to find the maximum of $f(p_1, p_2)$ for $0 \leq p_1 \leq 1$ and $0 \leq p_2 \leq 1$. The graph of f is shown in Figure 2.16. From this graph and the basic tools of calculus, we see that the maximum exists for $p_1 = 0$ and $p_2 = \frac{1}{2}$. ∎

In the general case, the method of maximum likelihood is used in the following way: Consider a model that has n parameters $p_1, \ldots, p_n$ to be determined. Let $L(\mathbf{x}; p_1, \ldots, p_n)$ be the function that associates with each point $(p_1, p_2, \ldots, p_n) \in \mathcal{A} \subset R^n$ the probability of obtaining given data, $\mathbf{x}$. Here $\mathcal{A}$ is the set of admissible (possible) values for the parameters. Note also that we are using the context of a discrete problem (a countable set of possible data vectors $\mathbf{x}$). If the context is that of a continuous problem, then L is a probability density function. The object is to find a maximum of L for $(p_1, \ldots, p_n) \subset \mathcal{A}$. If L satisfies certain differentiability conditions, then the method of maximum likelihood leads to a consideration of the values of L for those points $\tilde{\mathbf{p}} = (\tilde{p}_1, \ldots, \tilde{p}_n)$ that satisfy the system of equations

$$\frac{\partial L}{\partial p_i} = 0, \quad i = 1, 2, \ldots, n$$

or that lie on the boundary of the set $\mathcal{A}$.

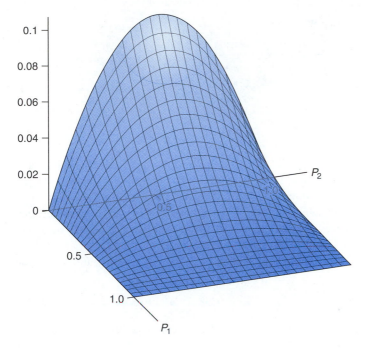

P_2

P_1

Figure 2.16

■ Minimum Discrepancy

We turn now to a second method of estimating parameters, the **method of minimum discrepancy.** It is similar in spirit to the maximum likelihood technique. Here too, data are collected by conducting experiments or by other suitable means, and these data are compared with the results predicted by the model. However, in the present method the parameters are to be chosen such that an appropriate measure of the difference between the data and the predictions of the model is as small as possible. The criteria used to measure this difference will vary from model to model, but for important theoretical reasons, the *least-squares* and *chi-square* measures are often used. We shall illustrate these measures and the method by examples.

EXAMPLE 2.26   Consider a model in which the following relation between age and learning ability arises. Subjects of age 4, 6, 8, and 10 years are asked to perform as many tasks of a certain type as possible in a unit time. For example, the task might be to identify as many objects as possible from a list that the student studied earlier. It is hypothesized that at these ages, the number of tasks performed is a linear function of age—that is,

$$\text{Number of tasks} = m \times (\text{age}) + b$$

where m and b are parameters that cannot be determined within the model. Let n be the number of tasks performed by a subject in a unit time, and let t be the age of the subject. Suppose that the observed data are as shown in Table 2.15.

   The problem is to determine values of m and b such that the function $n = n(t)$, defined by $n(t) = mt + b$, "best" fits the data obtained from observation. Of course, this is the

Table 2.15

Age	Observed n
4	15
6	17
8	29
10	34

Table 2.16

Age	Predicted n
4	$4m + b$
6	$6m + b$
8	$8m + b$
10	$10m + b$

well-known problem of linear regression—namely, that of finding the slope and intercept of a line that "best" fits a given data set in the plane. We give details here because we want to use the same method in more complicated settings. For the **method of least squares,** the best function is the one for which the sum of the squares of the differences between the observed and the predicted values is minimized. Thus we seek m and b such that Σ(observed value − predicted value)^2 is as small as possible.

The predicted values of n are as shown in Table 2.16. Thus we consider the sum of the squares of the differences,

$$S(m, b) = (15 - 4m - b)^2 + (17 - 6m - b)^2 + (29 - 8m - b)^2 + (34 - 10m - b)^2$$

and select m and b such that $S(m, b)$ is a minimum. Because m and b are unrestricted and because S is large for $|m|$ and $|b|$ large, the desired minimum is obtained by solving the system of equations $\partial S/\partial m = 0$, $\partial S/\partial b = 0$. This system has the form

$$28m + 4b = 95$$
$$108m + 14b = 367$$

and the solution is $m = 3.45$ and $b = -.40$. Thus the linear function that agrees best with the observed data in the sense of least squares is

$$n(t) = 3.45t - 0.40$$

Note that once these parameters have been estimated, the model can be used to make predictions about children of other ages. For example, the model would predict that a 9-year-old would be able to perform 30 tasks and a 12-year-old would be able to perform 41 tasks. These predictions could be tested. Note also that this linear model may not be a good model to explain the data obtained. It may well be that a nonlinear model fits the data better and provides much better predictions. (There may also be theoretical reasons for using a nonlinear model.) Later in this section we consider the question of comparing models. ■

This method can be used to fit functions other than linear ones to observed data. For example, one could use this technique to determine growth constants in models for population growth. Simple assumptions regarding growth lead to models predicting that the size of the population at time t is given by ae^{rt}, where a and r are parameters. Experimental data taken at several times can be used to determine the values of a and r that are best in the least-squares sense (Exercise 3).

The sum of the squares of the differences between the observed values and the predicted values is one measure of the discrepancy between actual and predicted results. Another useful measure is the **chi-square measure,** written χ^2. Briefly, this is a weighted measure of the squares of the differences between observed and predicted frequencies. If the predicted frequencies depend on parameters, then the values assigned to the parameters influence the size of the χ^2 measure. Parameters selected such as to make the χ^2 measure as small as possible are said to have been selected according to the method of minimum discrepancy with the χ^2 measure.

EXAMPLE 2.27   (See also Section 2.1)   Consider a situation in botany in which one is concerned with the frequency of occurrence of flowers of a certain color. We suppose that it is possible to classify all flowers of the sort being investigated as to color, and we take the classifications to be the desired color and all others. Let r be the (unknown) proportion of flowers of the desired color. To estimate r, we examine three populations of flowers of the given type, each population consisting of 50 flowers. Suppose that the observed ratio of desired colors to other colors is 33:17 for the first population, 30:20 for the second, and 27:23 for the third. Because $50r$ is the predicted frequency, the model predicts a ratio of $50r{:}50(1-r)$ in each case. The χ^2 measure is defined by

$$\sum_{\text{all observations}} \frac{(\text{Observed frequency} - \text{Predicted frequency})^2}{\text{Predicted frequency}}$$

Thus in the present case we have

$$\chi^2(r) = \frac{(33-50r)^2}{50r} + \frac{(30-50r)^2}{50r} + \frac{(27-50r)^2}{50r}$$

This expression can be minimized by using the standard techniques of the calculus. Thus, because a minimum clearly does not occur at $r = 0$, and for $r = 1$ the value is much bigger than the value for $r = \frac{1}{2}$, we compute $d\chi^2/dr$ and find those values of r for which $(d\chi^2/dr)(r) = 0$ (Exercise 4). We obtain

$$r^2 = \frac{(33)^2 + (30)^2 + (27)^2}{3(50)^2}$$

or

$$r = 0.602$$

where the last figure is correct to three decimal places. Note that this is close to, but not identical to, the average of the three sample values of r.   ∎

■ Testing Hypotheses and Comparing Models

The usual method of evaluating a model is to compare the results of experiments or observations with predictions based on the model. Frequently this evaluation is concerned with the parameters of an assumed probability distribution for a random variable. Occasionally it is concerned with the form of the distribution. In these cases, statistical tests are appropriate. We are interested both in the test to be used to obtain a comparison and in the criteria used to measure the results of the test. The following discussion is quite selective, but the selection was made so as to give a sample that is fairly representative of the techniques in common use.

To test some feature of a model, we test a specific statement regarding this feature. By a *test* of a statement we mean a determination of whether the statement should be accepted or rejected. The method of determination makes use of the results of experiments or observations, and it relates these results to the predictions of the model. The test also gives probability statements concerning the likelihood of errors in accepting or rejecting the statement. The specific statement being considered is known as a *hypothesis based on the model* or a *hypothesis of the model*. If the hypothesis specifies completely and exactly both the form of the probability distribution and all its parameters, then it is said to be a *simple hypothesis*; otherwise, it is said to be *composite*.

There are often several aspects of a model that can be tested. For example, as noted earlier, one can test the underlying assumptions (the axioms) directly, or one can test the predictions (theorems) directly and thereby test the underlying assumptions indirectly. Note the nature of this second method. The theorems are true statements in the mathematical model; there is no question about their validity, so there is no sense in testing them in the model. However, the theorems or predictions can be tested in the real world. There is no guarantee that the predictions of the model will exactly match reality. This is what we hope, but we have no right to assume it. There are a number of standard tests and criteria in common use to help us decide whether a model is a good one.

We now consider the question of testing a hypothesis in more detail. We begin by noting that different types of errors can occur in evaluating the results of a test of a hypothesis. First, it is possible to conclude that the hypothesis is false even though it is actually true, and second, it is possible to conclude that the hypothesis is true even though it is actually false. Errors of the former type are often called Type I errors, and those of the latter type are often called Type II errors. The characteristics of the experiment, the statistical test to be used, and the criteria for accepting or rejecting the hypothesis all contribute to the occurrence of the two types of errors. Ideally, we would like to use a test that simultaneously minimizes the chances for both types of errors. In general, however, it is not possible to find a test and reasonable criteria for acceptance that accomplish this. It is customary to proceed by agreeing in advance to an acceptable probability of a Type I error (0.05 and 0.01 are commonly accepted values) and then selecting a test that gives a tolerable Type II error.

It is useful to consider an example illustrating some of these concepts. First, however, we introduce one of the tests that is often used to check the accuracy of a statistical hypothesis having to do with a mathematical model: the chi-square test of predicted frequencies. This test is used to determine whether predictions of the model concerning the frequencies of certain events are supported by experimental evidence. The object of the present discussion is simply to indicate how such a test can be used. A discussion of its probabilistic and statistical basis is properly in the domain of statistics, and we refer the interested reader to the book by Bernard Flury cited at the beginning of this section. Suppose that the model predicts that a set of mutually exclusive and exhaustive events will occur with frequencies e_i, $i = 1, 2, \ldots, n$ (these are called the *expected frequencies*) and that in an appropriate experiment, the observed frequencies of these same events are o_i, $i = 1, 2, \ldots, n$. The chi-square (χ^2) measure of the difference between the observed and expected frequencies was defined above, and we reproduce the definition here for convenience:

$$\chi^2 = \sum_{i=1}^{n} \frac{(o_i - e_i)^2}{e_i}$$

EXAMPLE 2.28   Consider the genetics model of Section 2.1. Suppose that we are concerned with two distinct attributes, say color and texture, each of which occurs in two forms. If the two colors are yellow and green, green dominant, and the two textures are smooth and wrinkled, smooth dominant, then the phenotype distribution of the first filial generation resulting from a dihybrid cross is in the ratio 9:3:3:1, with 9/16 of the members of phenotype green–smooth, 3/16 of phenotype yellow–smooth, 3/16 of phenotype green–wrinkled, and 1/16 of phenotype yellow–wrinkled. A dihybrid cross is a mating of two individuals each of which is a hybrid with respect to each of the two genes. Thus, if the two alternative forms of each of the two genes are denoted by A, a and B, b, respectively, then the mating is a cross between individuals of genotype $Aa Bb$. The ratios 9:3:3:1 are predicted on the basis of a model similar to that constructed in Section 2.1 (see also Exercise 5 of Section 2.1). Suppose that an experiment is conducted in which two individuals of genotype $Aa Bb$ are crossed and the resulting descendants are classified according to color and texture. Assume that Table 2.17 summarizes the experiment by giving the observed and predicted phenotypes of 480 individuals.

According to the definition, we have

$$\chi^2 = \frac{(257 - 270)^2}{270} + \frac{(86 - 90)^2}{90} + \frac{(103 - 90)^2}{90} + \frac{(34 - 30)^2}{30}$$
$$= 0.626 + 0.178 + 1.87 + 0.533 = 3.22$$

What interpretation is to be given to the result $\chi^2 = 3.22$? Recall that the original question was to determine whether the model provided an accurate description of the real-world situation or, at least, whether the predictions of the model were consistent with observations. One expects that a small value of χ^2 corresponds to a good fit between experiment and prediction, but this is not precise enough for our purposes. One prefers probability statements concerning the likelihood of obtaining discrepancies at least as large as the value of χ^2 actually obtained. Such probability statements can be obtained from the derived value of χ^2 and standard statistical tables. In this example the probability associated with the value 3.22 of χ^2 is 0.36. That is, the statistical test implies that if the predicted values of the frequencies were in fact the actual ones and if the experiment described above were repeated many times, then 36% of the time, purely by chance, one would obtain deviations at least as large as those just observed. On this basis one must decide whether it is reasonable to conclude that the observed data are in agreement with the predicted 9:3:3:1 ratio. In this connection, it is useful to have in mind a clear statement of the hypothesis being tested and a criterion for evaluating the results of the test. In this example, it is natural for the hypothesis to be the statement that the four phenotypes occur with the ratios 9:3:3:1. A possible criterion is that the hypothesis will be rejected if $\chi^2 \geq \alpha$, where α is chosen such that $\chi^2 \geq \alpha$ will occur by chance with probability ≤ 0.05 when the hypothesis is correct.

Table 2.17

Phenotype	Observed Numbers	Predicted Numbers
Green–smooth	257	270
Yellow–smooth	86	90
Green–wrinkled	103	90
Yellow–wrinkled	34	30

If $\chi^2 < \alpha$, then there are two options to consider. We may decide to accept the hypothesis, or we may conclude that although the experimental results do not warrant rejection of the hypothesis, judgment on whether to accept the hypothesis is better reserved until further testing is complete. For this example, the critical value of α is 7.8, and hence, with this criterion, the above experiment would indicate that we should not reject the hypothesis. Instead, the hypothesis would be conditionally accepted pending the results of additional tests. ∎

The situation just discussed illustrates an evaluation of the predictions of the model. One can sometimes also use the chi-square method to test the validity of assumptions. In Exercise 5 we shall consider testing the assumption on the demand made in the linear programming problem of Section 2.4.

These examples are illustrations of the process of testing hypotheses where the intent is to accept or reject the hypothesis on the basis of the test. However, as we noted in the introduction to this section, most model testing is not of this form. Instead, the test is usually designed to obtain a quantitative measure of the ability of the model to explain the phenomenon under study. In this way the model can be compared with other models designed to study the same phenomena. As an illustration of the process of comparing models, we now turn to an example of the use of the least-squares measure to compare three different models for the growth of a population of fruit flies.

EXAMPLE 2.29 Let time be measured in days, and let $P(t)$ be the size of a population of fruit flies on day t. The three models that we shall consider are the linear model, the exponential model, and the logistic model. Each of these models depends on certain parameters, and in terms of these parameters the models can be expressed as follows:

Linear: $\quad P_1(t) = mt + b$ $\qquad\qquad\qquad m, b$ parameters

Exponential: $\quad P_2(t) = Ae^{(kt)}$ $\qquad\qquad\quad A, k$ parameters

Logistic: $\quad P_3(t) = \dfrac{MP_0}{P_0 + (M - P_0)e^{-cM(t-t_0)}}$ $\quad t_0, P_0, c, M$ parameters

The exponential and logistic models were discussed in Section 2.2, and the linear model is based on the assumption that the size of the population is a linear function of time.

From the discussion in Section 2.2, we know that the logistic curve is such that M is the limiting size of the population, t_0 is the starting time for the study, P_0 is the original population size, and c is the growth parameter. We shall specify t_0 and P_0, and hence each of our three models will have two undetermined parameters. These parameters will be determined by the data of the study. We shall use the data for fruit fly growth shown in Table 2.18.

Using these data, we have $t_0 = 8$ and $P_0 = 10$ in the logistic model. We estimate the other two parameters for the logistic model and the two parameters in the linear and exponential models by the method of minimum discrepancy. Because we intend to use the least-squares measure to compare the models, we also use this measure to estimate the parameters for each model. In this way each parameter of each model will be assigned the value that makes the least-squares measure of the discrepancy of the model as small as possible.

We denote by $L_i, i = 1, 2, 3$, the least-squares measure of the discrepancy between the observed data and the data predicted by the function P_i. Then L_1 depends on the parameters

Table 2.18

t (in days)	Number of Observed Flies
8	10
16	50
24	160
32	300
40	330

m and b, L_2 depends on the parameters A and k, and L_3 depends on the parameters M and c. We first estimate the parameters of the function L_1. Our method requires computation of the partial derivatives of L_1 with respect to m and b, respectively. We have

$$L_1(m, b) = (8m + b - 10)^2 + (16m + b - 50)^2$$
$$+ (24m + b - 160)^2 + (32m + b - 300)^2 + (40m + b - 330)^2$$

and hence

$$\frac{\partial L_1}{\partial m} = 7040m + 240b - 55040$$

and

$$\frac{\partial L_1}{\partial b} = 240m + 10b - 1700$$

The solution of the system of equations $\partial L_1/\partial m = \partial L_1/\partial b = 0$ is given by $m = 11.12$ and $b = -97.00$ (accurate to two decimal places). Thus these values of m and b are candidates for the values that make L_1 as small as possible. It is easily shown that indeed these values do give a minimum for L_1 (Exercise 6). With this choice of the parameters m and b, the function L_1 has the value 3390.

We now turn to the exponential model and estimate the parameters A and k in the function L_2. Once again we compute the partial derivatives of the function L_2 and set these partial derivatives equal to 0. We have

$$L_2(k, A) = (Ae^{8k} - 10)^2 + (Ae^{16k} - 50)^2 + (Ae^{24k} - 160)^2$$
$$+ (Ae^{32k} - 300)^2 + (Ae^{40k} - 330)^2$$

Therefore,

$$\frac{\partial L_2}{\partial A} = 2e^{8k}\{A[e^{8k} + e^{24k} + e^{40k} + e^{56k} + e^{72k}]$$
$$- [10 + 50e^{8k} + 160e^{16k} + 300e^{24k} + 330e^{32k}]\}$$

and

$$\frac{\partial L_2}{\partial k} = 16Ae^{8k}\{A[e^{8k} + 2e^{24k} + 3e^{40k} + 4e^{56k} + 5e^{72k}]$$
$$- [10 + 100e^{8k} + 480e^{16k} + 1200e^{24k} + 1650e^{32k}]\}$$

Because inspection shows that L_2 does not have a minimum when A and k are such that $Ae^{8k} = 0$, the system of equations $\partial L_2/\partial A = \partial L_2/\partial k = 0$ can be solved by finding those values of A and k such that

$$A = \frac{10 + 50e^{8k} + 160e^{16k} + 300e^{24k} + 330e^{32k}}{e^{8k} + e^{24k} + e^{40k} + e^{56k} + e^{72k}}$$

and

$$A = \frac{10 + 100e^{8k} + 480e^{16k} + 1200e^{24k} + 1650e^{32k}}{e^{8k} + 2e^{24k} + 3e^{40k} + 4e^{56k} + 5e^{72k}}$$

If we let $x = e^{8k}$ and equate these expressions for A, we obtain the following algebraic equation for the variable x:

$$30x^{10} - x^9 - 15x^8 - 46x^7 - 20x^6 - 96x^5 - 55x^4 - 146x^3 - 90x^2 - 31x - 5 = 0$$

We solve this equation numerically (by computer) and obtain $x = 1.723$ as an approximate value for the single positive real solution of the equation (Exercise 9). This gives A the value 24.98, and for these values of x and A, the function L_2 has the value $L_2(k, A) = 11.515$. Note that we also have a value for k: $k = \frac{1}{8}\log(1.723) = 0.068$. At this stage we have not shown that these values of k and A are the ones that make L_2 as small as possible. We have only shown that they are values for which the partial derivatives of L_2 are 0. It is still necessary to show that a minimum exists at this point. The proof of this fact is somewhat involved, and we leave the details to Exercises 7 and 8.

We note that the linear model gives a much smaller least-squares value than the exponential model. However, neither model is able to predict the observed data with any great degree of accuracy. We next turn to a somewhat more sophisticated model, the logistic model. On the basis of the discussion of Section 2.2, we expect it to yield better results. The unspecified parameters for the logistic model are M and c, and the least-squares measure of the discrepancy between the actual and the predicted data is given by the function

$$L_3(M, c) = (10 - 10)^2 + \left(\frac{10M}{10 + (M - 10)e^{-8cM}} - 50\right)^2$$

$$+ \left(\frac{10M}{10 + (M - 10)e^{-16cM}} - 160\right)^2$$

$$+ \left(\frac{10M}{10 + (M - 10)e^{-24cM}} - 300\right)^2$$

$$+ \left(\frac{10M}{10 + (M - 10)e^{-32cM}} - 330\right)^2$$

In this case, and in most instances arising in practice, it is difficult to solve for the values of M and c that make L_3 as small as possible. The system of equations $\partial L_3/\partial M = \partial L_3/\partial c = 0$ is a complicated system of nonlinear equations and consequently is very difficult to solve. Moreover, it is difficult to show that L_3 has a minimum when evaluated at the solution of this system. In these cases the easiest method is often a search procedure that systematically tries different values of the parameters to find the best (or nearly best) ones. This procedure is almost always carried out on a computer, and there are many prepared

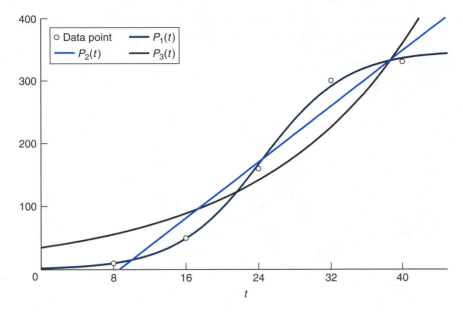

Figure 2.17

programs for such searches. We have used a very direct search procedure in this example. First a value of M is selected, and then the best value of c for that M is determined by comparing the values of L_3 for different choices of c and selecting the c that gives the smallest value of L_3. This process is carried out for a selection of values of M, and it yields a sequence of pairs (M, c), where the value of c is the best for that M. Finally, the values of L_3 are compared for these pairs to obtain the pair that gives the smallest value for L_3. The method is discussed in more detail in Exercise 9.

Using the search method discussed above, we obtain the pair $(M, c) = (347, 0.00062)$ as the pair that gives the smallest value for L_3. The L_3 value is

$$L_3(347, 0.00062) = 161.3$$

Thus we see that for these data, the logistic model is much better at "fitting" the observed growth of the population of fruit flies. The graphs of the function $P_1(t)$, $P_2(t)$, and $P_3(t)$ are shown in Figure 2.17, together with original data points.   ■

Exercises 2.8

1. Show that if any line is chosen that goes through two of the four points $(4, 15)$, $(6, 17)$, $(8, 29)$, and $(10, 34)$, then at least one of these four points is a greater distance from this line than it is from the line determined by these points using the least-squares method. Graph the seven lines.

2. Use the method of least squares to find the parabola that best fits the points $(1, 2.2)$, $(2, 4.1)$, $(3, 5.8)$, $(4, 4.4)$, and $(5, 2.4)$.

3. Use the method of least squares to estimate the parameters a and r in the function $P(t) = ae^{rt}$ so as to give the best fit to the following data:

t	Observed Value
5	10
10	12
15	30

4. Show that for $0 < r \leq 1$, the minimum of the function

$$\chi^2(r) = \frac{(33 - 50r)^2}{50r} + \frac{(30 - 50r)^2}{50r} + \frac{(27 - 50r)^2}{50r}$$

is obtained for

$$r = \left[\frac{(33)^2 + (30)^2 + (27)^2}{3(50)^2} \right]^{1/2}$$

5. Suppose that over a 10-week period, the demand for mobile homes is as follows:

Week	1	2	3	4	5	6	7	8	9	10
Demand	9	10	8	20	21	12	11	20	20	19

Use the chi-square measure to test the assumption that demand is constant on a week-to-week basis. Note that although the assumption of constant demand can be tested by assuming that demand is any natural number, there are relatively few natural numbers that are reasonable for these data. In particular, compute the chi-square measure of discrepancy for assumed demands for 14, 15, 16, 17, and 18 mobile homes per week. For ten degrees of freedom and the same 0.05 criterion used in the examples of this section, we have $\alpha = 18.3$. Which, if any, of the choices for a constant demand should be rejected and which accepted? Which choice minimizes the χ^2 measure of discrepancy?

6. Show that the minimum of the function

$$L_1(m, b) = (8m + b - 10)^2 + (16m + b - 50)^2 + (24m + b - 160)^2$$
$$+ (32m + b - 300)^2 + (40m + b - 330)^2$$

is obtained for values of m and b given by the solution of the system of equations $\partial L_1/\partial m = 0 = \partial L_1/\partial b$. Show that this system has a unique solution. *Hint:* Consider (in three dimensions) the graph of the function L_1 and examine the behavior of L_1 as $|m|$ and $|b|$ become very large.

7. Show that the minimum of the function

$$L_2(A, k) = (Ae^{8k} - 10)^2 + (Ae^{16k} - 50)^2 + (Ae^{24k} - 160)^2$$
$$+ (Ae^{32k} - 300)^2 + (Ae^{40k} - 330)^2$$

is obtained for values of A and k given by a solution of the system of equations

$$\frac{\partial L_2}{\partial A} = 0 = \frac{\partial L_2}{\partial k}$$

Hint: Consider the graph of the function L_2. In particular, note the behavior of L_2 for negative values of A and for large values of A and $|k|$. Finally, argue that given any large rectangle in the $A - k$ plane (with one side along the $A = 0$ axis, including the origin, and lying in the $A \geq 0$ half-plane), the minimum of L_2 will be taken at a point (A, k) in the interior of the rectangle.

8. For the function L_2 of Exercise 7, show that the system of equations $\partial L_2/\partial A = 0 = \partial L_2/\partial k$ has a unique solution. *Hint:* Let $x = e^{8k}$ and find the equation for x that must be satisfied if A and k are a solution of the system of equations. Then show that this equation in x has a single positive real root.

9. To find the minimum of the function $L_3(M, c)$ of Example 2.29, a direct computer search method was used. The method found a best value of c for each M by testing a range of values of c. Then the pairs (M, c) were compared to find a best pair. Construct a flowchart for this search procedure. Also discuss the problem of finding the proper step size to use in the search for the best c to go with a given M.

CHAPTER 3

Markov Chains and Related Stochastic Models

3.0   Introduction

Mathematical models are either deterministic or stochastic, and some settings can be represented with both deterministic and stochastic models. However, in many situations arising in the social and life sciences, there are phenomena for which stochastic models are the appropriate ones. In particular, in many circumstances the behavior of plants, animals, and people exhibits a degree of randomness that must be built into the models if predictions are to correspond with observations. There are a great variety of stochastic models—that is, sets of assumptions—that can be used to study these situations, and in this chapter we examine in detail only one rather special case. This special case, Markov chains, has proved to be widely applicable and a reasonably effective way of modeling many situations arising in the real world. Even in circumstances where detailed predictions based on the model differ from observations, we frequently gain insight into the process by studying these simple models.

We introduce the models studied in this chapter through a number of examples from a variety of settings. We then present the mathematical concepts and notation that we use in analyzing these situations. Finally, we develop parts of the theory of the models for two especially important subclasses of models. Throughout, we use the situations introduced in Section 3.1 to illustrate and apply our methods.

3.1   The Setting and Some Examples

A basic assumption we make throughout this chapter is that all situations we study have the property that we observe a system sequentially through time, and that at each observation the system can be determined to be in one of a finite number of states or to be satisfying a finite number of conditions. This is an assumption about our ability to classify circumstances or behaviors in useful ways. It is probably most effective to illustrate the notion through examples.

■ Animal Ranges

Consider a locale consisting of rocks, scrub brush, open meadow, and a stream (see Figure 3.1), and suppose that this locale is home for a small animal, say a marmot. We seek

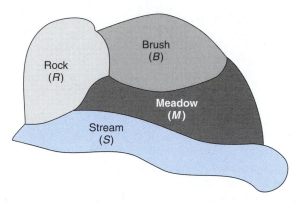

Figure 3.1

to model the movement of the marmot through time by noting its location at sequential observations and then forming a mathematical system that represents these movements in an appropriate way.

The use of a figure such as Figure 3.1 includes an assumption that the area can be partitioned meaningfully into the subareas shown, and we also assume that when an observation is made, we can determine which subarea contains the marmot. These may seem like natural and perhaps trivial assumptions, but in many experimental situations they are difficult to interpret or verify. For small animals, the task of keeping track of the creature may be a challenge, and for large animals, for which the use of tracking collars may help with the location problem, we may have the animal moving through many different subareas of its range. Also, if observations continue over an extended period of time, the nature of a specific area may change. Brush may become meadow, or during a very wet season, a part of the meadow may disappear into the stream. However, such issues are not a direct part of our current model, and we do not consider them further.

We suppose, therefore, that the location of the marmot can be tracked through time and that a sequence of observations can be represented as a sequence of locations and occupancy times. For instance, using the shorthand R, B, M, and S to denote the rocks, brush, meadow, and stream, respectively, then we might represent a particular sequence of observations by a sequence of letters (the locations) and numbers (the occupancy times):

$$R - 34.2 - B - 12.3 - R - 2.4 - B - 21.8 - M - \cdots$$

This sequence is to be interpreted as follows: When observations begin, the marmot is in the rocks, and it remains there for 34.2 minutes. It then moves to the brush, where it remains for 12.3 minutes, after which it again moves to the rocks, where it remains for 2.4 minutes, and so on. Although occupancy times play an important role in many models, we can illustrate the basic ideas of the model-building and analysis process by concentrating on the locations alone. Also, if (as is frequently the case) observations are made at discrete times, then the observational data consist solely of a sequence of locations. Depending on the criteria used to make observations, locations may or may not appear successively in the sequence. Thus the sequence of locations given above might be represented as $RBRBM \cdots$. In this representation, juxtaposition denotes the results of successive observations. If the definition of observation permits the marmot to be in the same location on successive observations, then a sequence of locations such as $RRRBMMS$ is possible. Here it might be suggestive to write the result of the seven observations as

$$R \to R \to R \to B \to M \to M \to S$$

When represented using juxtaposition or a diagram as above, a sequence of observations is frequently referred to as a *sample path*.

In this example, we identify a "state of the system" with a location of the marmot. From a biological perspective, the marmot may be sleeping or resting while in the rock pile, feeding while in the meadow, and simply in transit while in the brush. In a more complex situation where we have several marmots, we need a more elaborate definition of the state of the system. We must decide, for instance, whether we are going to keep track of individual marmots. If we decide to do so, then the state corresponding to marmot number 1 being in the rocks and marmot number 2 being in the brush is different from the state corresponding to marmot number 1 being in the brush and marmot number 2 being in the rocks. If we simply keep track of how many marmots are in the various locations, then these two situations correspond to the same state.

If observations are made periodically in time, then the marmot can be in any two locations on successive observations. If the time between successive observations is short compared with the time it takes the marmot to move through a location, then it is possible for the marmot to be observed either in the same location or in adjacent locations on successive observations. Sometimes the experiment is set up so that the marmot is observed each time it moves from one location to another. In such cases, two successive observations must have the marmot in different but adjacent locations. For each assumption about how observations are made, we have a diagram similar to that shown in Figure 3.2. In Figure 3.2a the arrows represent possible moves of the marmot under the assumption that the marmot is observed only when it moves from one area to another, and in Figure 3.2b the arrows represent possible moves when it is observed whenever it moves *and* at regularly spaced times— and can therefore be in the same area on successive observations. Figure 3.2 provides two examples of a **transition diagram.** It illustrates possible transitions between states of our process. Later we will add probabilities of the transitions to the diagram.

There are natural assumptions one can make about the movements of the marmot. The likelihood of the marmot moving from the brush to the rock may depend on several of the preceding moves, or it may depend only on the immediately preceding move, or it may be independent of preceding moves. Each of these assumptions leads to a mathematical model whose predictions can be compared with observations. Here, we consider in detail only one such assumption: We assume that the likelihoods of the various possible moves

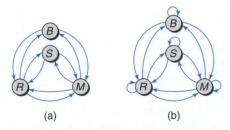

(a)   (b)

Figure 3.2

for the marmot depend only on the location of the marmot, not on how much time has elapsed since we began the observations and not on the previous moves of the marmot. It is this assumption (which we will make formal soon) that distinguishes Markov chains from other stochastic processes. In circumstances where the Markov assumption is appropriate, it is customary to arrange the likelihoods or probabilities of moves in a table or matrix. We illustrate this idea with two examples.

EXAMPLE 3.1 Assume that the marmot is observed only when it moves from one sub-area to another and that it is equally likely to make any move available to it. Then the probabilities of various moves are as shown in Table 3.1. ∎

Table 3.1

		Location after Move			
		R	B	M	S
Location before Move	R	0	$\frac{1}{3}$	$\frac{1}{3}$	$\frac{1}{3}$
	B	$\frac{1}{2}$	0	$\frac{1}{2}$	0
	M	$\frac{1}{3}$	$\frac{1}{3}$	0	$\frac{1}{3}$
	S	$\frac{1}{2}$	0	$\frac{1}{2}$	0

EXAMPLE 3.2 For this example, assume that the marmot is observed periodically and whenever it moves from one subarea to another. Assume that it is twice as likely to remain where it is as to move, and if it moves, then it is equally likely to make any move available to it. Under these assumptions, the probabilities of various moves are as shown in Table 3.2. ∎

Table 3.2

		Location after Move			
		R	B	M	S
Location before Move	R	$\frac{2}{3}$	$\frac{1}{9}$	$\frac{1}{9}$	$\frac{1}{9}$
	B	$\frac{1}{6}$	$\frac{2}{3}$	$\frac{1}{6}$	0
	M	$\frac{1}{9}$	$\frac{1}{9}$	$\frac{2}{3}$	$\frac{1}{9}$
	S	$\frac{1}{6}$	0	$\frac{1}{6}$	$\frac{2}{3}$

The tasks of verifying the entries in these tables and finding corresponding tables in other situations are topics of the exercises.

Table 3.2 lists the probabilities for single transitions by the marmot. In that table, a transition occurs whenever a fixed period of time has elapsed or the marmot has changed areas, whichever event occurs first. We are often interested in the results of multiple transitions, and we need the probabilities of each possible result. One method of computing these probabilities is by using a simple tree diagram. For example, suppose that the marmot is in the brush at a certain time, and we want to know how likely it is to be in each of the areas

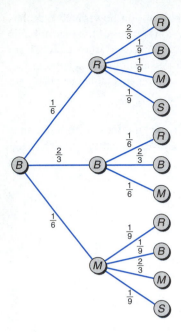

Figure 3.3

R, B, M, S after two transitions. The tree diagram in Figure 3.3 illustrates the computations needed to compute these four probabilities. We see that

$$\text{Probability}[B \to R \text{ in two transitions}] = \left(\frac{1}{6}\right)\left(\frac{2}{3}\right) + \left(\frac{2}{3}\right)\left(\frac{1}{6}\right) + \left(\frac{1}{6}\right)\left(\frac{1}{9}\right) = \frac{13}{54}$$

$$\text{Probability}[B \to B \text{ in two transitions}] = \left(\frac{1}{6}\right)\left(\frac{1}{9}\right) + \left(\frac{2}{3}\right)\left(\frac{2}{3}\right) + \left(\frac{1}{6}\right)\left(\frac{1}{9}\right) = \frac{26}{54}$$

$$\text{Probability}[B \to M \text{ in two transitions}] = \left(\frac{1}{6}\right)\left(\frac{1}{9}\right) + \left(\frac{2}{3}\right)\left(\frac{1}{6}\right) + \left(\frac{1}{6}\right)\left(\frac{2}{3}\right) = \frac{13}{54}$$

$$\text{Probability}[B \to S \text{ in two transitions}] = \left(\frac{1}{6}\right)\left(\frac{1}{9}\right) + \left(\frac{1}{6}\right)\left(\frac{1}{9}\right) = \frac{2}{54}$$

In this example it is relatively straightforward to determine the two-step transition probabilities using tree diagrams similar to Figure 3.3. In cases with more states or more steps, this method becomes unwieldy and we need an alternative.

■ The Effects of Group Structure on Small-Group Decision Making

In many group decision-making situations, we believe that in addition to the merits of the alternatives being considered, there are aspects of the dynamics of the group that influence the outcome. For instance, once a group of six people reaches a division of five to one in favor of some alternative, the mere fact of this division exerts some influence on the dissenting member. In this example we describe a model designed to test this conjecture in a setting in experimental psychology.

To isolate the possible influence of the group structure on decision making, we introduce an experiment designed to minimize other influences. In particular, care must be taken to ensure that the alternatives appear equally attractive and that no individual in the group assumes a leadership position. Suppose a group of people performs a sequence of trials. Each trial consists of the presentation of a stimulus—a set of alternatives—to be evaluated and a discussion of the merits of the alternatives. The discussion continues until consensus is reached. Suppose that a stimulus consists of a set of three geometrical designs that are to be evaluated according to some criteria. Each member of the group is able to convey a preference to the investigator without other members of the group knowing what that preference is, and preferences can be changed at will. The subjects are asked to express a preference as soon as the stimulus is shown to the group and then to begin discussions seeking to reach consensus. Each subject is to convey a preference change to the group right after it is conveyed to the investigator. After consensus is reached, the group is told which of the designs is "best" in terms of the criteria. The group is led to believe that there is a system behind the assignment of values to the designs in the set, but actually the designs in each set are ranked randomly. Consequently, as far as the group is concerned, each design is equally preferable. Also, techniques of selective reinforcement can be used to discourage the emergence of a group leader. For instance, selection of the best design can be manipulated so that each member of the group appears to have about the same percentage of "correct" initial selections.

Preference selections are monitored and recorded. The process continues either for a certain period of time or for a certain number of trials. In practice, it might be desirable to discard the first few trials because the subjects are becoming familiar with the experimental procedure during that period.

To make the discussion more specific, suppose that there are four individuals and that each stimulus consists of three designs. Each individual can select any of the three alternatives as the best, and consequently, there are $3^4 = 81$ possible distributions of preferences for the group. However, the experiment has been designed so that the alternatives appear equally attractive, and we ought not to distinguish among them. For example, if the alternatives are X, Y, and Z, and if the choices of the members are X, Y, Y, Y in one case and Y, X, X, X in another, then these choices should be viewed as equivalent from the standpoint of the structure of the group. Both represent a group structure in which three people vote for one alternative and a single individual votes for another. Also, there is no reason to distinguish among members of the group. That is, three votes for alternative X and one vote for Y should be considered the same regardless of which member votes for alternative Y.

It follows that the important information is the number of individuals who voted for the most popular alternative, the number who voted for the second most popular alternative, and the number who voted for the least popular alternative. That is, the relevant information is contained in a triple of integers (x, y, z), where x is the number of individuals voting for the most popular alternative, y is the number voting for the second most popular alternative, z is the number voting for the least popular alternative, and $x + y + z = 4$. The possible triples are $(4, 0, 0)$, $(3, 1, 0)$, $(2, 2, 0)$, and $(2, 1, 1)$. We refer to these triples as **group compositions** and we write them simply as xyz; thus 310 is the same as $(3, 1, 0)$.

Suppose that the group compositions are monitored continuously and every preference change is recorded. A change in group composition occurs whenever any subject changes a vote. Each preference change is equivalent to a change from one group composition to

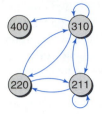

Figure 3.4

Table 3.3

		Composition after One Shift			
		400	310	220	211
Composition before Shift	310	$\frac{1}{8}$	$\frac{1}{8}$	$\frac{3}{8}$	$\frac{3}{8}$
	220	0	$\frac{1}{2}$	0	$\frac{1}{2}$
	211	0	$\frac{1}{4}$	$\frac{1}{4}$	$\frac{1}{2}$

another, possibly the same one. We assume that only one vote changes at a time. In the rare event that two people simultaneously indicate a change of vote, we arbitrarily select one to be changed first. It follows that preference changes by individuals are equivalent to shifts between group compositions that can be effected by the change of a single vote. For example, $211 \rightarrow 220$ is an admissible transition, but $211 \rightarrow 400$ is not. The possible shifts are shown by arrows in Figure 3.4.

Note that it is possible for a single vote to change and for a group with composition 211 to change to a group with the same composition. Likewise for a group with composition 310. Because a trial of the experiment ends when consensus is reached, there are no possible shifts from group composition 400.

The probabilities of various shifts can be conveniently summarized in a table. The entries in the table will depend, of course, on the assumptions about the voting behavior of the subjects. For example, if each subject is equally likely to change her or his vote and is equally likely to change to each of the other alternatives, then we have the information in Table 3.3. The task of verifying the entries in Table 3.3 is the topic of Exercise 5.

As a final comment in this section, we recall that an important part of a mathematical model is the assumptions. Our discussion of small-group decision making provided an example of one possible set of assumptions that can be made in that situation—of course, there are many alternatives. It is difficult (and frequently impossible) to directly test the validity of the assumptions. Instead, it is customary to compare the predictions based on the assumptions with observations. If the observations are consistent with the predictions, then one has reason to continue the study. If the observations are not consistent with the predictions, then the assumptions need to be reviewed and modified. In the following sections of this chapter, we develop a theory and techniques to make predictions, and in Chapter 4 we will develop techniques for making predictions based on simulation models. Making predictions based on assumptions and then comparing the predictions with observations are part of the cycle of model building described in Chapter 1.

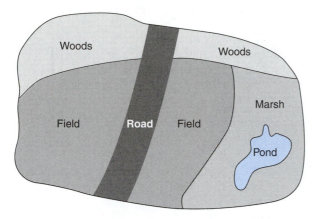

Figure 3.5

▉ Exercises 3.1

1. In the model of a marmot's range described in this section, verify the entries in Table 3.1 under the assumptions given in Example 3.1.

2. In the model of a marmot's range described in this section, verify the entries in Table 3.2 under the assumptions given in Example 3.2.

3. A deer has as its range the area diagrammed in Figure 3.5, and its movements are observed and recorded as follows: The location of the deer is noted every hour and every time it moves from one area of its range to another. For this purpose, the woods and the field to the east of the road are distinguished from the woods and the field to the west of the road. If the deer crosses the road, then it moves only from field to field; that is, it does not move to or from the woods when crossing the road. Suppose that the probabilities of moves depend only on its current location and not on what happened prior to its last move.

   Assume that the deer is twice as likely to remain where it is as to move and that every move that does not require it to cross the road is equally likely. Also, if it moves, each option that does not require it to cross the road is three times as likely to be selected as an option that involves crossing the road. Create a table similar to Table 3.1 for this situation.

4. A marmot lives in the region diagrammed in Figure 3.6. Suppose the marmot is observed every hour and each time it moves from one area to another. Suppose that the probabilities of moves depend only on its current location, and not on what happened prior to its last move. Also suppose it is equally likely to move and to remain where it is. If it moves, the probability of its moving to an adjoining area is proportional to the number of resources available to it in that area in comparison to the resources in all adjoining areas. The areas bordering the pond have water in addition to the other resources specified. Create a table similar to Table 3.1 for this situation.

5. Consider the small-group decision-making situation described in this section in which three alternatives are presented to four individuals. If each subject is equally likely to change her or his vote, and is equally likely to change to each of the other alternatives,

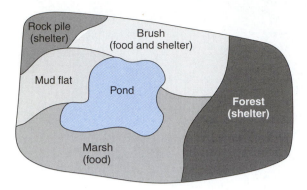

Figure 3.6

show that the probabilities of shifts between group compositions are as shown in Table 3.3.

6. Consider the small-group decision-making situation described in this section in which three alternatives are presented to four individuals. If from a specific group composition each possible shift to another group composition, possibly the same one, is equally likely, create a table similar to Table 3.3 for this case. How does the table change if only shifts to different group compositions are possible?

7. Consider a small-group decision-making situation similar to that described in this section with four alternatives presented to five individuals. What are the group compositions in this case? If each subject is equally likely to change her or his vote, and is equally likely to change to each of the other alternatives, find the probabilities of shifts between group compositions in this case and create a table similar to Table 3.3.

8. In the small-group decision-making experiment described in this section, define a shift toward consensus as one of the following: $211 \rightarrow 310, 220 \rightarrow 310, 310 \rightarrow 400$. Assume that a voter who can make a shift toward consensus is twice as likely to make a vote change as any other voter, and that if such a voter changes her or his vote, all changes are equally likely. Also, assume that all other voters are equally likely to change their votes and that all choices are equally likely. Create a table similar to Table 3.3 for this situation.

9. In the small-group decision-making experiment described in this section, define a shift toward consensus as one of the following: $211 \rightarrow 310, 220 \rightarrow 310, 310 \rightarrow 400$. Assume that a voter who can make a shift toward consensus is twice as likely to make a vote change as any other voter, and that if such a voter changes her or his vote, the change is twice as likely to be toward consensus as otherwise. Suppose that all other voters are equally likely to change their votes and that all vote changes are equally likely for these voters. Create a table similar to Table 3.3 for this situation.

10. Consider a small-group decision-making situation similar to the one described in this section, but with six individuals and three alternatives. Formulate a model similar to the one of this section under the following assumption: An individual who is the only person voting for an alternative is three times as likely to change her vote as an individual who

is one of two or more voting for an alternative. If an individual changes her vote, she is equally likely to change to any other alternative. Create a table similar to Table 3.3 for this case.

3.2 Basic Properties of Markov Chains

With the examples and discussion of Section 3.1 as a guide, we turn to a discussion of two more general settings that include the examples of Section 3.1. We consider a system that can be in any of N possible states, and we observe the system at n successive times. The concepts of *state* and *being in a state* or *occupying a state* are taken as undefined terms. When we construct logical models of systems in specific circumstances, we must assign meanings to these terms, but in this general discussion they are left undefined. We usually refer to the states simply by the integers $1, 2, \ldots, N$.

By the very nature of a Markov chain, it is likely that for most observations, the specific state occupied by the system cannot be determined in advance, and one knows only the probabilities that it will be in the various states. Consequently, the status of a system is usually given as a state vector.

Definition 3.1   A *state vector* $\mathbf{x}$ for a Markov chain with N states is an N-vector $\mathbf{x} = [x_1 \quad x_2 \quad \cdots \quad x_N]$, where x_i is the probability that the system is in state $i, i = 1, 2, \ldots, N$. The state vector on the mth observation will be denoted by $\mathbf{x}(m)$.

As an example, to say that the state of a four-state Markov chain is specified by the state vector $[0 \quad .25 \quad .5 \quad .25]$ means that the system is not in state 1, is in state 2 with probability .25, is in state 3 with probability .5, and is in state 4 with probability .25. If the system is known to be in a specific state, say state j, then the state vector has the jth coordinate equal to 1 and the remaining coordinates equal to zero. For instance, if a four-state Markov chain is known to be in state 2, then the state vector is $[0 \quad 1 \quad 0 \quad 0]$. In a Markov chain with N states, if the system is equally likely to be in any state, then the state vector has all coordinates equal to $1/N$.

If the system is in state i on the kth observation and in state j on the $(k + 1)$th observation, then we say that the system has made a *transition* from state i to state j at the kth *trial, step,* or *stage* of the process. We also say that the system has made a *move* from state i to state j.

It will be useful to work with another example, one that is somewhat simpler than those introduced in Section 3.1.

We use a setting that is familiar as a version of the classic maze of experimental psychology. The study of the behavior of mice (and rats) in mazes has been—and continues to be—important in generating and verifying hypotheses that lead to useful models for animal behavior.

EXAMPLE 3.3   Suppose that a mouse is released in the maze shown in Figure 3.7 and its behavior is observed. The illumination level in each compartment of the maze is maintained as shown in the figure. The system to be studied consists of the mouse and the maze, and we assume that the mouse is always in exactly one compartment and that it is possible to

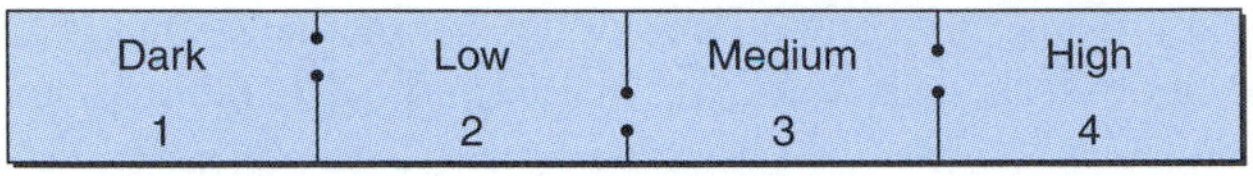

Dark	Low	Medium	High
1	2	3	4

Figure 3.7

determine that compartment. The system is said to be in state i if the mouse is in compartment i, $i = 1, 2, 3, 4$. Observations are to be made, and the state of the system recorded every 2 minutes and each time the mouse moves from one compartment to another, necessarily an adjacent one. To illustrate the possible transitions, suppose that on one observation the mouse is in compartment 2. On the next observation it may be in compartment 1, 2, or 3; it cannot be in compartment 4 according to our definition of observation. ■

In the situation of Example 3.3, we assume, as we would expect, that the mouse moves in unpredictable ways, and we describe its movements in probabilistic terms. Suppose that the mouse is in state i on observation k, and we wish to determine the probability that it is in state j on observation $k + 1$. In general, we might expect this probability to depend on the states i and j, the observation k, and the history of the movements of the mouse prior to its arrival in state i on the kth observation. There are, of course, many ways in which the transition could depend on the history of the process. For instance, we could assume that the mouse has complete recall of past movements and that its transition probabilities at the kth step depend on the total prior history of its movements. Although in some circumstances such an assumption may be appropriate, in many cases it leads to very complex models without yielding any improvements in predictions. A simpler assumption would be that the transition probabilities depend only on the most recent past, say the last 1, 2, or 3 moves. We investigate here the situation wherein transition probabilities depend only on the current state, not on the prior history of the process. This includes the assumption that they do not depend on k—the number of steps for which the process has been observed.

We summarize this discussion by giving the key assumption that distinguishes Markov chains from more general stochastic processes.

■ The Markov Assumption

A **Markov chain** is a stochastic process with a finite number of states and with the property that if it is in state i on one observation, then the probability that it will be in state j on the next observation depends on states i and j (which may be the same state) and not on the observation number or on the history of the process prior to the current observation.

It will be useful to introduce the following notation and terminology.

Definition 3.2 Let p_{ij} denote the conditional probability that if the system is in state i on one observation, then it will be in state j on the next observation, $1 \le i \le N$, $1 \le j \le N$. These probabilities are called *transition probabilities,* or, more precisely, one-step transition probabilities. For each Markov chain, the $N \times N$ matrix $\mathbf{P}$ whose ij-entry is p_{ij} is called the *transition matrix* for the Markov chain.

We have

$$
\mathbf{P} =
\begin{bmatrix}
p_{11} & p_{12} & p_{13} & \cdots & p_{1N} \\
p_{21} & p_{22} & p_{23} & \cdots & p_{2N} \\
\vdots & \vdots & \vdots & & \vdots \\
p_{N1} & p_{N2} & p_{N3} & \cdots & p_{NN}
\end{bmatrix}
$$

as the transition matrix for a Markov chain whose transition probabilities are p_{ij}, $1 \leq i \leq N, 1 \leq j \leq N$.

Remark   It is important to note that because of the Markov assumption, p_{ij} is the probability that if the system is in state i on observation k then it will be in state j on observation $k+1$, independent of k. Therefore, if one knows that (with states and observations specified) the transition probabilities do depend on the observation numbers, then the stochastic process is not a Markov chain according to our definition.

It follows from the definition of transition probability that the probability p_{ij} of making a transition from state i to state j at the kth step is the same as the probability of the system being in state j on the second observation given that it was in state i on the first observation. It is sometimes appropriate to maintain the assumption that the transition probabilities are independent of the history of the process but to permit them to depend on the time—that is, on how long the process has been observed. In this case we have a more general stochastic process usually called a nonhomogeneous Markov chain. We will not pursue this more general situation here.

Each entry in the ith row of the transition matrix is a probability, and if the system is in state i on one observation, then it must be in some state j, $1 \leq j \leq N$, on the next observation. Consequently, for each i we have $\sum_{j=1}^{N} p_{ij} = 1$, and the vectors $\mathbf{p}_i = [p_{i1} \quad p_{i2} \quad \cdots \quad p_{iN}]$, $i = 1, 2, \ldots, N$, are probability vectors. Each row of the transition matrix is a probability vector.

The transition matrix $\mathbf{P}$ has entries that are the probabilities of making transitions from one specified state to another in one step. There are corresponding matrices for multistep transition probabilities.

Definition 3.3   Let $\mathbf{P}(m) = [p_{ij}(m)]$ be the matrix for which the ij-entry is the probability of making a transition from state i to state j in m steps, $1 \leq i \leq N, 1 \leq j \leq N$, $m = 2, 3, \ldots$. Clearly, $\mathbf{P}(1) = \mathbf{P}$.

Remark   Note that we use the terms *step* and *move* in the same way as we use the term *transition*. It is common to talk about the probability of a transition from state i to state j in m steps, m moves, or m transitions.

EXAMPLE 3.4   Consider the situation described above in which a mouse moves in a maze with compartments illuminated at different levels, and formulate a Markov chain model under the following assumption: The mouse remains in the same compartment with probability .5, and the rest of the time it is equally likely to make any of the moves open to it.

We define the states as follows: The system is in state i if the mouse is in compartment i, $i = 1, 2, 3, 4$. Because the mouse remains in the same compartment half the time, $p_{11} = .5$, $p_{22} = .5$, $p_{33} = .5$, and $p_{44} = .5$. Next, if the mouse is in state 1 on one observation, then on the next observation it can be only in state 1 or 2. Consequently, using the assumption, we have $p_{12} = .5$. Likewise, $p_{43} = .5$. Also, if the mouse is in state 2 on one observation, then on the next observation it can be in state 1, 2, or 3. Consequently, because it is in state 2 with probability .5, it is in state 1 or 3 with probability .5, and because it is equally likely to be in either, we have $p_{21} = .25$ and $p_{23} = .25$. Similarly, $p_{32} = .25$, and $p_{34} = .25$. Finally, by the way observations are defined, p_{13}, p_{14}, p_{24}, p_{31}, p_{41}, and p_{42} all equal 0. Consequently, the transition matrix for this Markov chain is

$$
\mathbf{P} = \begin{bmatrix} p_{11} & p_{12} & p_{13} & p_{14} \\ p_{21} & p_{22} & p_{23} & p_{24} \\ p_{31} & p_{32} & p_{33} & p_{34} \\ p_{41} & p_{42} & p_{43} & p_{44} \end{bmatrix} = \begin{bmatrix} .5 & .5 & 0 & 0 \\ .25 & .5 & .25 & 0 \\ 0 & .25 & .5 & .25 \\ 0 & 0 & .5 & .5 \end{bmatrix} \qquad \blacksquare
$$

It is clear that the entries in the tables constructed in Section 3.1 are transition probabilities, and we will represent them in this way in the future.

■ State Vectors

A **state vector** is a probability vector that describes the status of a Markov chain at an observation, and the state vectors at two successive observations are related in a simple way. Indeed, if $\mathbf{x}(m)$ and $\mathbf{x}(m + 1)$ denote the state vectors at the mth and $(m + 1)$st observations, respectively, then $\mathbf{x}(m + 1) = \mathbf{x}(m)\mathbf{P}$. To verify this relationship, suppose that the state vector at a specific observation is $\mathbf{x} = [x_1 \quad x_2 \quad \cdots \quad x_N]$. (Here we suppress the dependence on the observation number m for notational convenience.) Then, using the definition of p_{j1} for $j = 1, 2, \ldots, N$, we can conclude that the probability it is in state 1 on the next observation is

$$
y_1 = x_1 p_{11} + x_2 p_{21} + \cdots + x_N p_{N1}
$$

Thus y_1 is the dot product of $\mathbf{x}$ and the first column of $\mathbf{P}$. Likewise, this time using the definition of p_{j2} for $j = 1, 2, \ldots, N$, we can conclude that the probability it is in state 2 on the next observation is

$$
y_2 = x_1 p_{12} + x_2 p_{22} + \cdots + x_N p_{N2}
$$

Continuing in this fashion, we find that the probability that it is in state N on the next observation is

$$
y_N = x_1 p_{1N} + x_2 p_{2N} + \cdots + x_N p_{NN}
$$

That is, if the state vector at one observation is $\mathbf{x} = [x_1 \quad x_2 \quad \cdots \quad x_N]$, then the state vector at the next observation is

$$
[y_1 \quad y_2 \quad \cdots \quad y_N] = \mathbf{xP}
$$

Thus, with $\mathbf{x}(m)$ and $\mathbf{x}(m + 1)$ as defined above, we have

(3.1) $$\mathbf{x}(m + 1) = \mathbf{x}(m)\mathbf{P}$$

■ Multistep Transitions and the Sequence of State Vectors

If the initial state vector is $\mathbf{x}$, then the state vector at the next observation is $\mathbf{xP}$, the state vector at the second observation is $(\mathbf{xP})\mathbf{P} = \mathbf{xPP} = \mathbf{xP}^2$, and so on. That is, the sequence of state vectors is

$$\mathbf{x}, \mathbf{xP}, \mathbf{xP}^2, \ldots, \mathbf{xP}^k, \ldots$$

A Markov chain is determined by the set of states, the transition matrix, and the initial state vector for the system. It is frequently useful to represent the information on the set of states and the transition probabilities in a *transition diagram*. The most common form of a transition diagram has the states represented by symbols (generally numbers or letters in small circles), an arrow directed from state i to state j if $p_{ij} > 0$, and a number near the arrow with the value of p_{ij}. The transition diagram for the model described in Example 3.3 is shown in Figure 3.8.

General stochastic processes can be studied using tree diagrams, and in particular, Markov chains can be studied in that way. There is, of course, a close connection between the information usually included on tree diagrams and the information included in a transition matrix. Given the initial state of the system, either tree diagrams or multistep transition matrices can be used to determine how the process evolves over time. Also, using a one-step transition matrix, it is always possible to determine the multistep matrix by using a tree diagram, and we illustrate this in the next example.

Figure 3.8

EXAMPLE 3.5 Consider the situation described in Example 3.4 and determine the 2-step transition matrix—that is, the matrix of two-step transition probabilities—by using tree diagrams.

Because the first row of the two-step transition matrix consists of the probabilities of making transitions from state 1 to states 1, 2, 3, 4 in two steps, we begin by constructing the tree diagram when the process begins in state 1. That tree diagram is shown in Figure 3.9(a). Using the information on the tree diagram, we find that

$$p_{11}(2) = .375, \quad p_{12}(2) = .5, \quad p_{13}(2) = .125, \quad p_{14}(2) = 0$$

To determine the entries in the second row of the two-step matrix, we use a tree diagram for which the process begins in state 2, as shown in Figure 3.9(b). Using the information on the tree diagram, we find that

$$p_{21}(2) = .25, \quad p_{22}(2) = .4375, \quad p_{23}(2) = .25, \quad p_{24}(2) = .0625$$

Similar arguments lead to the third and fourth rows of the two-step transition matrix. They are

$$p_{31}(2) = .0625, \quad p_{32}(2) = .25, \quad p_{33}(2) = .4375, \quad p_{34}(2) = .25$$
$$p_{41}(2) = 0, \quad p_{42}(2) = .125, \quad p_{43}(2) = .5, \quad p_{44}(2) = .375$$

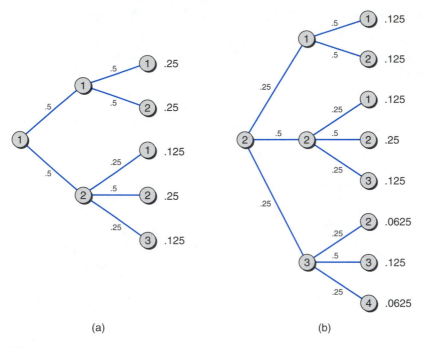

(a)   (b)

Figure 3.9

It follows that the two-step transition matrix $\mathbf{P}(2)$ is

$$\mathbf{P}(2) = \begin{bmatrix} .3750 & .5000 & .1250 & 0 \\ .2500 & .4375 & .2500 & .0625 \\ .0625 & .2500 & .4375 & .2500 \\ 0 & .1250 & .5000 & .3750 \end{bmatrix}$$ ∎

The technique illustrated in Example 3.5 is a very general one, and it can be used, in particular, to construct the m-step transition matrix for any integer m. However, it is clear that for Markov chains with a large number of states or a large number of steps m, the effort involved in using this method can be prohibitive. Indeed, one of the great benefits of using Markov chains in mathematical models is that there is a much simpler way to determine multistep transition matrices once you know the one-step transition matrix.

THEOREM 3.1   Let $\mathbf{P} = [p_{ij}]$ be the transition matrix of a Markov chain. Then the ij-entry of the m-step transition matrix $\mathbf{P}(m)$ is the ij-entry of $\mathbf{P}^m$, the mth power of the one-step transition matrix.

Proof.   The proof is a direct consequence of the definition of conditional probability and the Markov assumption, and it illustrates a useful approach to computing multistep transition probabilities. Indeed, the ij-entry of $\mathbf{P}(m)$ is the conditional probability that the system is in state j given that it began in state i and made m transitions. Consider each of the m-step sample paths from state i to state j as consisting of a path of length $m - 1$ followed by a single step. Then, after $m - 1$ steps the system must be in some state, say state k. The conditional probability that it is in state k given that it began in state i and made $m - 1$

transitions is the ik-entry of the $(m-1)$-step transition matrix, $\mathbf{P}(m-1)$. Therefore, the probability that the system is in state k after $m-1$ steps and in state j after m steps is equal to $p_{ik}(m-1)p_{kj}$. Finally, using the fact that each sample path is in some state k after $m-1$ transitions, we have

$$p_{ij}(m) = \sum_{k=1}^{N} p_{ik}(m-1)p_{kj}$$

This shows that $\mathbf{P}(m) = \mathbf{P}(m-1)\mathbf{P}$. Applying the same reasoning to $\mathbf{P}(m-1)$, we find that $\mathbf{P}(m-1) = \mathbf{P}(m-2)\mathbf{P}$, and consequently $\mathbf{P}(m) = \mathbf{P}(m-2)\mathbf{P}\mathbf{P} = \mathbf{P}(m-2)\mathbf{P}^2$. Continuing the argument, we find that

$$\mathbf{P}(m) = \mathbf{P}(m-1)\mathbf{P} = \mathbf{P}(m-2)\mathbf{P}^2 = \mathbf{P}(m-3)\mathbf{P}^3 = \cdots = \mathbf{P}(1)\mathbf{P}^{(m-1)}$$

But $\mathbf{P}(1) = \mathbf{P}$, and consequently $\mathbf{P}(m) = \mathbf{P}^m$.   ∎

EXAMPLE 3.6   Consider the situation described in Example 3.4, and compute the two-step transition matrix $\mathbf{P}(2)$ using Theorem 3.1.

In Example 3.4 we determined the transition matrix for the Markov chain to be

$$\mathbf{P} = \begin{bmatrix} .5 & .5 & 0 & 0 \\ .25 & .5 & .25 & 0 \\ 0 & .25 & .5 & .25 \\ 0 & 0 & .5 & .5 \end{bmatrix}$$

Using Theorem 3.1, we find that the two-step transition matrix $\mathbf{P}(2)$ is

$$\mathbf{P}(2) = \begin{bmatrix} .5 & .5 & 0 & 0 \\ .25 & .5 & .25 & 0 \\ 0 & .25 & .5 & .25 \\ 0 & 0 & .5 & .5 \end{bmatrix} \begin{bmatrix} .5 & .5 & 0 & 0 \\ .25 & .5 & .25 & 0 \\ 0 & .25 & .5 & .25 \\ 0 & 0 & .5 & .5 \end{bmatrix}$$

$$= \begin{bmatrix} .3750 & .5000 & .1250 & 0 \\ .2500 & .4375 & .2500 & .0625 \\ .0625 & .2500 & .4375 & .2500 \\ 0 & .1250 & .5000 & .3750 \end{bmatrix}$$

which is the same result we obtained in Example 3.5, as it must be.   ∎

EXAMPLE 3.7   Consider the process of group decision making described in Section 3.1. In particular, suppose we have three alternatives and a group of four individuals. We form a Markov chain model under the assumption that each individual is equally likely to change her or his vote and is equally likely to change to each of the other alternatives.

Table 3.3 contains the probabilities for shifts from the group compositions 310, 220, and 211 to the group compositions 400, 310, 220, and 211. To make use of the Markov chain concept, we need to include the group composition 400 as a state. The situation was originally described as an experiment that ended as soon as consensus was reached—that is, as soon as the group reached composition 400. However, for the purpose of our Markov chain concept, it is useful to view 400 just as we view any other state, but with the characteristic that the system never leaves that state. This can be accomplished by setting the transition probability from the state corresponding to group composition 400 to itself equal to 1, and the probability of making a transition from state 400 to any other state equal to 0.

For this example, we define the states of the system as the group compositions, and we define state 1 as group composition 400, and states 2, 3, and 4 as group compositions 310, 220, and 211, respectively. Then, with the understanding that once the system reaches group composition 400 it does not leave it, we have the transition matrix

$$\mathbf{P} = \begin{bmatrix} 1 & 0 & 0 & 0 \\ \frac{1}{8} & \frac{1}{8} & \frac{3}{8} & \frac{3}{8} \\ 0 & \frac{1}{2} & 0 & \frac{1}{2} \\ 0 & \frac{1}{4} & \frac{1}{4} & \frac{1}{2} \end{bmatrix}$$

Suppose the group is initially in state 4; that is, the group composition is 211. How many vote changes are required before the probability of the group reaching consensus first reaches .5? We answer this question by computing successive powers $\mathbf{P}^m$ of the transition matrix $\mathbf{P}$ and asking for the smallest integer m for which $p_{41}(m) \geq .5$. We have $p_{41}(k) < .5$ for $1 \leq k \leq 19$ and

$$\mathbf{P}(20) = \begin{bmatrix} 1 & 0 & 0 & 0 \\ .5536 & .1265 & .1049 & .2150 \\ .5064 & .1399 & .1160 & .2377 \\ .4941 & .1433 & .1189 & .2436 \end{bmatrix}$$

$$\mathbf{P}(21) = \begin{bmatrix} 1 & 0 & 0 & 0 \\ .5694 & .1220 & .1012 & .2074 \\ .5239 & .1349 & .1119 & .2293 \\ .5121 & .1383 & .1147 & .2350 \end{bmatrix}$$

from which we see that after 20 vote changes, the probability $p_{41}(20) = .494$, and after 21 vote changes, $p_{41}(21) = .512$. Thus 21 vote changes are required for the probability of the group reaching consensus to exceed .5. Recall that the system remains in state 1 once it arrives there. Consequently, the sequence of entries in the (4, 1) spot in the matrices $P(m)$ is a monotone nondecreasing sequence, and once an entry is greater than .5, all subsequent entries are also greater than .5. Of course, this answer depends heavily on the initial assumptions of the model. A different set of assumptions about the likelihood of vote changes would give a different matrix $\mathbf{P}$ and a different answer. ∎

Exercises 3.2

In these exercises, forming a Markov chain model requires that you identify the states and find the transition matrix.

1. Suppose that a mouse moves in the maze shown in Figure 3.7 and that observations are made every 5 minutes and every time the mouse changes compartments. Formulate a Markov chain model under the following assumptions: The mouse remains in the same compartment 40% of the time, and if it has a choice when it moves, it moves to a darker compartment twice as often as to a lighter one.

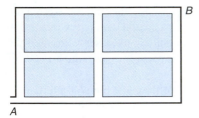

Figure 3.10

2. In the model formulated in Exercise 1, suppose the mouse is initially in the highly illuminated compartment.

   (a) Find the probability that it does not leave the highly illuminated compartment in the first five transitions.
   (b) Find the probability that it is in the same compartment after five observations.
   (c) Find the probability that it is not in the highly illuminated compartment on exactly one of the next five observations.

3. Suppose that a mouse moves in the maze shown in Figure 3.7 and that observations are made every time the mouse changes compartments. Formulate a Markov chain model under the following assumption: Whenever the mouse has a choice, it moves to a darker compartment three times as often as to a lighter one.

4. Suppose that we have the situation considered in Example 3.7. If the group is initially divided 220, how many vote changes are necessary before the probability of consensus first exceeds $\frac{1}{4}$?

5. In the group decision-making situation described in Section 3.1, define a shift toward consensus as one of the following: $211 \rightarrow 310$, $220 \rightarrow 310$, $310 \rightarrow 400$. Assume that a voter who can make a shift toward consensus is twice as likely to make a vote change as any other voter, and that if such a voter changes her or his vote, all changes are equally likely. Also, assume that all other voters are equally likely to change their votes and that all choices are equally likely.

   (a) Form a Markov chain model for this situation and find the transition matrix.
   (b) If the group is initially distributed 220, what is the most likely group composition after five vote changes? After ten vote changes?

6. An inebriated bicyclist cycles through the neighborhood shown in Figure 3.10. He begins at location A, and he traverses the streets at random. During each time interval he either rests at an intersection or pedals exactly one block.

   Suppose that at each intersection the bicyclist is three times as likely to pedal as to rest. If he pedals, he is equally likely to take any street open to him. Form a Markov chain model for this situation.

7. In the setting described in Exercise 6, suppose that the bicyclist never rests, that at any intersection he is equally likely to take any street available to him, and that once he reaches location B he stays there. Form a Markov chain model using these assumptions.

8. Consider a small-group decision-making situation similar to that described in Section 3.1 but with five individuals and three alternatives. Formulate a Markov chain

model under the following assumption: An individual who is the only person voting for an alternative is twice as likely to change her vote as a person who is one of a group of two or more voting for an alternative. If an individual changes her vote, then the probability of her changing to a particular alternative is proportional to the number of individuals voting for that alternative.

If the group initially has four individuals voting for the most popular alternative, find the probability that consensus is reached after at most three vote changes.

9. A Bloomington resident commutes to work in Indianapolis, and he encounters several traffic lights on the way to work each day. Over a period of time, the following pattern has emerged:

   - Each day the first light is green.
   - If a light is green, then the next one is always red.
   - If he encounters a green light and then a red one, then the next will be green with probability .6 and red with probability .4.
   - If he encounters two red lights in a row, then the next will be green with probability p and red with probability $1 - p$.

   Formulate a Markov chain model for this situation.

10. Consider a small-group decision-making situation similar to that described in Section 3.1 with five individuals and four alternatives. What are the group compositions in this case? Form a Markov chain model under the following assumption: Each subject is equally likely to change her or his vote and is equally likely to change to each of the other alternatives.

    If the group initially has four individuals voting for the most popular alternative, find the probability that the same holds after four vote changes.

11. A marmot lives in the region shown in Figure 3.11. Suppose that the marmot is observed every hour and each time it moves from one area to another. Formulate a Markov chain model under the following assumptions: The marmot is twice as likely to move as to remain where it is, and if it moves, the probability of its moving to a particular area is proportional to the number of resources available to it in that area in comparison to the number of resources available to it in the adjoining areas. The areas bordering the pond have water in addition to the resources specified.

    If the marmot begins in the rock pile, find the probability it is in the south meadow on the fifth observation.

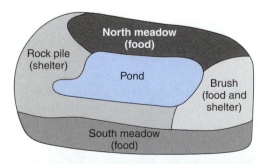

Figure 3.11

12. In the situation described in Exercise 11:

    (a) Suppose the marmot is initially in the north meadow, and find the probability that it is in the rocks after 10 transitions. Find the same probability after 20 and 40 transitions.
    (b) Suppose the marmot is initially in the brush, and determine the same probabilities as in part (a).

13. There are two coins, one fair and one biased with $\Pr[H] = .3$. A game is played by successively flipping the coins as follows:

    - The game begins with a flip of the fair coin, and the result, H or T, is noted.
    - If the result of a flip is H, then the other coin is used on the next flip, and the result is noted.
    - If the result of a flip is T, then the same coin is used on the next flip, and the result is noted.

    (a) Formulate a Markov chain model for this situation.
    (b) Find the probability that the fourth flip is a head.

14. There are six balls, two red and four green, distributed between boxes labeled 1 and 2, three balls in each box. When the game begins, there are two green balls and one red ball in box 1. The game is played as follows: A ball is selected at random from each box. The ball selected from box 1 is placed in box 2, the ball selected from box 2 is placed in box 1, and the colors of the balls in each box are noted. Then two more balls are selected, and play continues.

    (a) Formulate a Markov chain model for this situation. What are your states?
    (b) Find the probability that there are exactly two red balls in box 1 after three plays of the game.

15. Joe and Jess play a game as follows: An unfair coin with $\Pr[H] = .6$ is flipped, and the result is noted. If it comes up heads, then Jess pays Joe one dollar, and if it comes up tails, then Joe pays Jess one dollar. If each player has money, then the coin is flipped again. The game ends as soon as one player has all the money. When the game begins, Joe has one dollar and Jess has three dollars.

    (a) Formulate a Markov chain model for this game.
    (b) Find the probability that Jess has all the money after not more than four flips of the coin.

16. An experiment consists of flipping an unfair coin with $\Pr[H] = .6$ repeatedly, noting the result of each flip, until there are three consecutive heads. At that point the experiment ends.

    (a) Formulate a Markov chain model for this experiment.
    (b) Find the probability that the experiment ends after exactly six flips of the coin.

3.3   Classification of Markov Chains and the Long-Range Behavior of Regular Markov Chains

Markov chains, as examples of stochastic processes, can be used to yield information on the probabilities of events, events described in terms of states or sets of states. A key tool in studying Markov chains is the multistep transition matrix. In Section 3.2 we showed that for every Markov chain, the m-step transition matrix is the mth power of the one-step transition matrix. Beyond this common behavior, Markov chains are quite diverse. The goal of this section is to illustrate some of this diversity, to provide a useful way to classify Markov chains, and to study some selected classes in detail.

EXAMPLE 3.8 Consider two Markov chains with state space $\{1, 2, 3\}$; the first has transition matrix **P** and the second has transition matrix **T**.

$$\mathbf{P} = \begin{bmatrix} 0 & 1 & 0 \\ 0 & 0 & 1 \\ 1 & 0 & 0 \end{bmatrix} \quad \text{and} \quad \mathbf{T} = \begin{bmatrix} 0 & 1 & 0 \\ 0 & 0 & 1 \\ .5 & .5 & 0 \end{bmatrix}$$

The transition diagrams for these Markov chains are shown in Figure 3.12: Figure 3.12(a) shows the transition diagram for the Markov chain with transition matrix **P**, and Figure 3.12(b) shows the transition diagram for the Markov chain with transition matrix **T**.

First we study the Markov chain with transition matrix **P**. If the system begins in state 1, then the sequence of state vectors is $[1 \quad 0 \quad 0] \to [0 \quad 1 \quad 0] \to [0 \quad 0 \quad 1] \to [1 \quad 0 \quad 0] \to [0 \quad 1 \quad 0] \to \cdots$. That is, the system cycles repeatedly through states 1, 2, and 3 in that order. This can also be shown by examining the powers of the transition matrix **P**. Indeed, the third power of **P** is the identity matrix **I**, so if the system has state vector $\mathbf{x} = [x_1 \quad x_2 \quad x_3]$ on observation m, then it has the same state vector on observation $m + 3 \; (\mathbf{x}\mathbf{P}^3 = \mathbf{x})$.

The behavior of the Markov chain with transition matrix **T** is quite different. We have

$$\mathbf{T}^2 = \begin{bmatrix} 0 & 0 & 1 \\ .5 & .5 & 0 \\ 0 & .5 & .5 \end{bmatrix} \quad \mathbf{T}^3 = \begin{bmatrix} .5 & .5 & 0 \\ 0 & .5 & .5 \\ .25 & .25 & .5 \end{bmatrix} \quad \mathbf{T}^4 = \begin{bmatrix} 0 & .5 & .5 \\ .25 & .25 & .5 \\ .25 & .5 & .25 \end{bmatrix}$$

and

$$\mathbf{T}(30) = \mathbf{T}^{30} \begin{bmatrix} .2000 & .4000 & .4000 \\ .2000 & .4000 & .4000 \\ .2000 & .4000 & .4000 \end{bmatrix}$$

In fact, $\mathbf{T}(m)$ is the same as $\mathbf{T}(30)$—at least to the accuracy shown—for all $m > 30$. (See Exercise 5 for additional information on this situation.) We see that the rows of $\mathbf{T}(30)$ are all the same. One consequence of this is that for all observations numbered 30 and beyond, the state vector of the system is $[.2000 \quad .4000 \quad .4000]$ *independent of the initial*

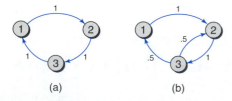

(a)   (b)

Figure 3.12

state. Indeed, for an initial state vector $[\,x \quad y \quad z\,]$, we have

$$
[\,x \quad y \quad z\,]\,\mathbf{T}(m) = [\,x \quad y \quad z\,]
\begin{bmatrix}
.2000 & .4000 & .4000 \\
.2000 & .4000 & .4000 \\
.2000 & .4000 & .4000
\end{bmatrix}
$$

$$
= [\,.2000x+.2000y+.2000z \quad .4000x+.4000y+.4000z \quad .4000x+.4000y+.4000z\,]
$$
$$
= [\,.2000(x+y+z) \quad .4000(x+y+z) \quad .4000(x+y+z)\,]
$$
$$
= [\,.2000 \quad .4000 \quad .4000\,]
$$

for any probability vector $[\,x \quad y \quad z\,]$.

That is, the system "forgets" the initial state, or the early history. The same conclusion holds for any matrix that differs from $\mathbf{T}$ only in the last row and that has a third row equal to $[\,p \quad 1-p \quad 0\,], 0 < p < 1$. In this case, the entries in the powers will be different (they will depend on the value of p), but the conclusion will be the same: As the number m of transitions increases, the m-step transition matrix $T(m)$ approaches a matrix with entries that are all positive and with rows that are all the same. ■

EXAMPLE 3.9   A Markov chain with state space $S = \{1, 2, 3, 4, 5\}$ has the transition diagram shown in Figure 3.13.

The transition matrix for this chain is

$$
\mathbf{P} =
\begin{bmatrix}
0 & 0 & .8 & .2 & 0 \\
0 & .3 & 0 & 0 & .7 \\
.1 & .6 & .3 & 0 & 0 \\
0 & .5 & .5 & 0 & 0 \\
0 & 1 & 0 & 0 & 0
\end{bmatrix}
$$

In this example, the probability of a direct transition from state 1 to state 2 is 0, but it is possible to go from state 1 to state 2 in more than one step, and in fact $p_{12}(2) > 0$. These facts are clear from Figure 3.13. However, we also see that it is impossible to go from state 2 to state 1 in any number of steps: $p_{21}(m) = 0$ for all values of m. It *is* possible to go from state 1 to states 3 and 4, and state 1 can be reached from states 3 and 4. Thus states 1, 3, and 4 are mutually accessible from each other. The set of all states can be partitioned using this "reachability" criterion, and we next turn to a systematic discussion of this idea. ■

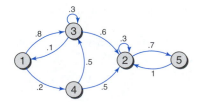

Figure 3.13

The idea introduced in Example 3.9 is helpful in classifying Markov chains, and we now show how to use it systematically. Suppose that we have a Markov chain with state space $S = \{1, 2, 3, \ldots, N\}$ and transition matrix $\mathbf{P} = (p_{ij})$. We say that state j is **accessible** from state i if there is an integer k such that $p_{ij}(k) > 0$, and we say that states i and j are **mutually accessible** if state i is accessible from state j and state j is accessible from state i. This concept can be used to partition the state space into classes of mutually accessible states. Begin with state 1, and let S_1 denote the set of all states that are mutually accessible from state 1. If $S_1 = S$, the entire set of states, then we are finished. If not, then there is a state, call it j, that is not in S_1. Let S_2 denote the set of all states that are mutually accessible with j. Continue in this way, and construct a collection of disjoint subsets of S whose union is S. Each subset consists of states that are mutually accessible, and no state belongs to more than one subset.

EXAMPLE 3.10 Applying our method of partitioning the states to Example 3.9, we have $S_1 = \{1, 3, 4\}$ and $S_2 = \{2, 5\}$. Note that $S_1 \cup S_2 = S$ and $S_1 \cap S_2 = \emptyset$. Now suppose that we form a transition matrix with the states reordered so that the states in the set S_2 are listed first and the states in the set S_1 are listed next. The order in which the states are listed within the sets S_1 and S_2 is unimportant, but the order must be the same in the rows and in the columns of the transition matrix. We have the new transition matrix

$$
\begin{array}{c c}
 & \overbrace{}^{S_2}\ \ \overbrace{}^{S_1} \\
 & \begin{array}{ccccc} 2 & 5 & 1 & 3 & 4 \end{array} \\
S_2 \left\{ \begin{array}{c} 2 \\ 5 \end{array} \right. & \left[\begin{array}{cc|ccc} .3 & .7 & 0 & 0 & 0 \\ 1 & 0 & 0 & 0 & 0 \\ \hline 0 & 0 & 0 & .8 & .2 \\ .6 & 0 & .1 & .3 & 0 \\ .5 & 0 & 0 & .5 & 0 \end{array} \right] \\
S_1 \left\{ \begin{array}{c} 1 \\ 3 \\ 4 \end{array} \right. &
\end{array}
$$

where the row and column labels denote the states. ∎

In Example 3.10 the transition matrix has two block matrices on the main diagonal. These are shown by the lines inside the transition matrix. The 2×2 block in the upper-left corner contains the transition probabilities for transitions between states in the set $\{2, 5\}$, and the 3×3 block in the lower-right corner contains the transition probabilities for the set of states $\{1, 3, 4\}$. There is a 2×3 matrix of zeros in the upper-right corner, a consequence of the fact that no state in the set $\{1, 3, 4\}$ is accessible from a state in the set $\{2, 5\}$. The 3×2 matrix in the lower-left corner contains transition probabilities for transitions from states in the set $\{1, 3, 4\}$ to states in the set $\{2, 5\}$.

The form of the transition matrix displayed in Example 3.10 can be achieved in the general case. That is, it is always possible to relabel the classes $S_1, S_2, \ldots, S_k$ as S_1', $S_3', \ldots, S_k'$ so that the resulting transition matrix has the form

$$
\begin{bmatrix}
A_1 & 0 & 0 & \cdots & 0 \\
X & A_2 & 0 & \cdots & 0 \\
X & X & A_3 & \cdots & 0 \\
\vdots & \vdots & \vdots & & \vdots \\
X & X & X & \cdots & A_k
\end{bmatrix}
$$

where each of the matrices A_j contains transition probabilities for transitions between states in the set S'_j, $j = 1, 2, \ldots, k$. In Example 3.10 we have $S'_1 = S_2$ and $S'_2 = S_1$. Entries in the transition matrix below these blocks, the blocks denoted by X's in the transition matrix, are transition probabilities between states that are not mutually accessible. Each "**0**" entry represents a block of zeros. A transition matrix written in this form is said to be in **canonical form.** One of the advantages of writing a matrix in canonical form is that the matrix of m-step transition probabilities has the same form. That is, the block submatrices on the diagonal contain the m-step transition probabilities within each class, there are zeros above these diagonal blocks, and the entries below the diagonal blocks are m-step transition probabilities between states in different classes.

Definition 3.4   A Markov chain for which every two states are mutually accessible is said to be **ergodic.**

We remark that the transition matrix of an ergodic Markov chain is in canonical form no matter how the states are ordered (of course, the order must be the same in the rows and columns).

Definition 3.5   Let j be a state. Then

The **index set** of the state j, denoted by $I(j)$, is the set of all integers m such that $p_{jj}(m) > 0$.
The **period** of state j, denoted by $d(j)$, is defined to be

(i)   0 if the index $I(j)$ is empty;
(ii)   the greatest common divisor of the integers in $I(j)$ if the index set is not empty.

Note that the index set of a state j consists of the set of all integers m such that there is a positive probability of making transitions from state j to state j in m steps.

EXAMPLE 3.11   The Markov chain with transition matrix **P** in Example 3.8 has index sets for states 1, 2, and 3 given by $I(1) = \{3, 6, 9, \ldots\}$, $I(2) = \{3, 6, 9, \ldots\}$, and $I(3) = \{3, 6, 9, \ldots\}$, respectively. It follows that $d(1) = 3$, $d(2) = 3$, and $d(3) = 3$.
     The Markov chain with transition matrix **T** in Example 3.8 has index sets for states 1, 2, and 3 given by $I(1) = \{3, 5, 6, \ldots\}$, $I(2) = \{2, 3, 4, \ldots\}$, and $I(3) = \{2, 3, 4, \ldots\}$, respectively. It follows that $d(1) = 1$, $d(2) = 1$, and $d(3) = 1$. Note that for matrix **T** the periods of all states are the same, but the index sets are not identical.   ■

EXAMPLE 3.12   Consider the Markov chain with the transition diagram given in Figure 3.14.

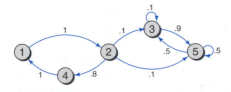

Figure 3.14

The transition matrix for this Markov chain, given in canonical form, is

$$
\begin{array}{c c}
& \begin{array}{c c c c c} 3 & 5 & 1 & 2 & 4 \end{array} \\
\begin{array}{c} 3 \\ 5 \\ 1 \\ 2 \\ 4 \end{array} &
\left[\begin{array}{c c | c c c}
.1 & .9 & 0 & 0 & 0 \\
.5 & .5 & 0 & 0 & 0 \\
\hline
0 & 0 & 0 & 1 & 0 \\
.1 & .1 & 0 & 0 & .8 \\
0 & 0 & 1 & 0 & 0
\end{array}\right]
\end{array}
$$

where the state numbers are listed to the left of the rows and above the columns. As we noted earlier (in Example 3.10), the order in which states 3 and 5 are listed is unimportant, and the order in which states 1, 2, and 4 are listed is unimportant. However, states 3 and 5 must be listed before states 1, 2, and 4, and the states must be listed in the same order in the rows as in the columns.

The index sets for states 1, 2, 3, 4, and 5 are given by $I(1) = \{3, 6, 9, \ldots\}$, $I(2) = \{3, 6, 9, \ldots\}$, $I(4) = \{3, 6, 9, \ldots\}$, $I(3) = \{1, 2, 3, \ldots\}$, and $I(5) = \{1, 2, 3, \ldots\}$. It follows that $d(1) = 3$, $d(2) = 3$, $d(4) = 3$, $d(3) = 1$, and $d(5) = 1$. ∎

We note from these examples that in each case, the periods of all states that are mutually accessible are the same. This is a general result.

THEOREM 3.2 If states i and j are mutually accessible, then $d(i) = d(j)$.

Proof. Let m and n be integers such that $p_{ij}(m) > 0$ and $p_{ji}(n) > 0$. Then

$$p_{ii}(m + n) \geq p_{ij}(m)p_{ji}(n) > 0 \qquad \text{and}$$
$$p_{jj}(m + n) \geq p_{ji}(n)p_{ij}(m) > 0$$

and it follows that $m + n \in I(i)$ and $m + n \in I(j)$.

Next, let k be any integer in $I(j)$ and h any divisor of the elements in $I(i)$. Then $p_{ii}(m + n + k) \geq p_{ij}(m)p_{jj}(k)p_{ji}(n) > 0$, and consequently $m + n + k \in I(i)$. Therefore, h divides $m + n + k$. However, h divides $m + n$, so h must divide k. But k was any element in $I(j)$, so h must divide every element of $I(j)$, and because $d(j)$ is the greatest common divisor of elements of $I(j)$, $h \leq d(j)$. Finally, because h was an arbitrary divisor of the elements of $I(i)$, we have $d(i) \leq d(j)$.

A similar argument shows that $d(j) \leq d(i)$, and consequently $d(i) = d(j)$. ∎

Definition 3.6 A Markov chain is a **regular Markov chain** if it is ergodic and the period of each state is 1.

The transition matrix **P** of Example 3.8 is not the transition matrix of a regular Markov chain because the period of each state is 3, but the transition matrix **T** is the transition matrix of a regular Markov chain. Indeed, for the Markov chain with matrix **T**, inspection of the transition diagram in Figure 3.12(b) shows that every two states are mutually accessible. Also, the index set for state 1 includes the integers 3 and 5, and consequently the period of state 1 is 1. As we will show later (in Theorem 3.4), the Markov chain with transition matrix **T** has the property that for every initial state vector $\mathbf{x_0}$, the state vector after m

transitions, $\mathbf{x}(m)$, tends to the state vector $[.2 \quad .4 \quad .4]$. This convergence of $\{\mathbf{x}(m)\}$ is a general property of regular Markov chains, and it is part of the content of Theorem 3.4.

Before turning to the main result of this section, Theorem 3.4, we remark that the definition of a regular Markov chain given in Definition 3.6 is equivalent to a condition on the powers of the transition matrix. This equivalent condition is frequently taken as the definition of a regular Markov chain. We state here a result that says our definition of *regular* implies the condition, and we provide a proof of the result in the Appendix to this chapter. The fact that the condition implies our definition of *regular* is the topic of Exercise 6.

THEOREM 3.3   If $\mathbf{P}$ is the transition matrix of a regular Markov chain, then there is an integer r such that $\mathbf{P}^n$ has only positive entries for all integers $n > r$. ■

Much of the usefulness of regular Markov chains as models rests on the fact that in the long run, the state vectors tend to a limit: $\lim_{m \to \infty} \mathbf{x}(m)$ exists. The limit is independent of the initial state, it has all positive coordinates, and it can be determined by solving a system of linear equations. These results are the content of Theorem 3.4, and the proofs are provided in the Chapter Appendix.

THEOREM 3.4   Let $\mathbf{P}$ be the transition matrix of a regular Markov chain. Then
   (i) The limit $\lim_{m \to \infty} \mathbf{P}^m$ exists and is a matrix $\mathbf{H}$ all of whose rows are the same vector $\mathbf{s}$ (called the steady-state vector for $\mathbf{P}$). The coordinates in $\mathbf{s}$ are all positive.
  (ii) The vector $\mathbf{s}$ is a probability vector that satisfies the equation $\mathbf{s} = \mathbf{sP}$.
 (iii) If $\mathbf{x}$ is any probability vector that satisfies the equation $\mathbf{x} = \mathbf{xP}$, then $\mathbf{x} = \mathbf{s}$. ■

We note that the convergence of the m-step transition matrices $\mathbf{P}(m)$ to a limit whose rows are all the same means that the m-step state vectors $\mathbf{x}(m)$ converge to a limit and that the limit is the common row of the limit of the transition matrices. The limit of the state vectors is independent of the initial state vector. This result can be interpreted as meaning that as the number of transitions increases, the system "forgets" the initial state. The long-term behavior of the state vector in a regular Markov chain does not depend on the initial state. The same conclusion need not hold for other ergodic Markov chains.

EXAMPLE 3.13   Consider the mouse moving in a maze as described in Examples 3.3 and 3.4, and determine the probability that the mouse will be in the dark compartment in the long run.

When phrased in this way, the problem asks for the coordinate of the limiting state vector (if the limit exists) corresponding to the dark compartment. First, the Markov chain is regular—that is, it is ergodic and each state is of period 1—so the state vectors tend to a limit. Next, the limit vector can be obtained by finding a probability vector $\mathbf{x}$ that satisfies the system of equations $\mathbf{x} = \mathbf{xP}$, where $\mathbf{P}$ is the transition matrix of Example 3.3:

$$\mathbf{P} = \begin{bmatrix} .5 & .5 & 0 & 0 \\ .25 & .5 & .25 & 0 \\ 0 & .25 & .5 & .25 \\ 0 & 0 & .5 & .5 \end{bmatrix}$$

We write the system $\mathbf{x} = \mathbf{x}\mathbf{P}$ as $\mathbf{x}(\mathbf{I} - \mathbf{P}) = 0$; if we set $\mathbf{x} = [\,x_1 \quad x_2 \quad x_3 \quad x_4\,]$, the system of equations becomes

(3.2) $[\,x_1 \quad x_2 \quad x_3 \quad x_4\,]$
$$\begin{bmatrix} .5 & -.5 & 0 & 0 \\ -.25 & .5 & -.25 & 0 \\ 0 & -.25 & .5 & -.25 \\ 0 & 0 & -.5 & .5 \end{bmatrix} = [\,0 \quad 0 \quad 0 \quad 0\,]$$

The system of equations (3.2) has infinitely many solutions. However, we are interested only in solutions that are probability vectors, and adding that requirement, namely

(3.3) $x_1 + x_2 + x_3 + x_4 = 1$

gives a unique solution.

The system of equations that consists of the four equations (3.2) and Equation (3.3) consists of five equations in four variables, and a unique solution is determined by any three of the equations of (3.2) together with Eqation (3.3). Equation (3.3) must be retained, and we can select any three of the four equations in (3.2). We choose the first three equations of (3.2) and Equation (3.3). We have

$$\begin{array}{rrrrl} x_1 + & x_2 + & x_3 + & x_4 & = 1 \\ .5x_1 - & .25x_2 & & & = 0 \\ -.5x_1 + & .5x_2 - & .25x_3 & & = 0 \\ & -.25x_2 + & .5x_3 - & .25x_4 & = 0 \end{array}$$

Solving this system, we find

$$x_1 = \frac{1}{6}, \quad x_2 = \frac{1}{3}, \quad x_3 = \frac{1}{3}, \quad x_4 = \frac{1}{6}$$

Using this information, we can answer the original question. The mouse will be in the dark compartment, compartment 1, about $\frac{1}{6}$ of the time in the long run.   ∎

Remark   The system of equations $\mathbf{x} = \mathbf{x}\mathbf{P}$, or $\mathbf{x}(\mathbf{I} - \mathbf{P}) = 0$, has the unknown vector $\mathbf{x}$ on the left and the matrix on the right. This is a departure from the common notation for systems of equations, and it arises from the way we define transition probabilities. It is common for systems of linear equations to be written with the variable $\mathbf{x}$ on the right—that is, in the form $\mathbf{Ax} = \mathbf{b}$, where $\mathbf{A}$ is the coefficient matrix for the system.

Exercises 3.3

1. Transition matrices for ten Markov chains are shown below. In each case, write the transition matrix in canonical form and find the period of each state. If the chain is

regular, find the steady-state vector. Which chains are ergodic but not regular?

a. $\begin{bmatrix} 0 & 0 & 1 & 0 & 0 \\ 0 & 0 & 1 & 0 & 0 \\ 0 & .8 & 0 & 0 & .2 \\ .4 & 0 & 0 & 0 & .6 \\ 0 & 0 & 0 & 1 & 0 \end{bmatrix}$
b. $\begin{bmatrix} 0 & 0 & 1 & 0 & 0 \\ 0 & .4 & 0 & .6 & 0 \\ .2 & .3 & 0 & .5 & 0 \\ 0 & .5 & 0 & .1 & .4 \\ 0 & 0 & 0 & 1 & 0 \end{bmatrix}$
c. $\begin{bmatrix} .2 & 0 & .8 & 0 & 0 \\ 0 & 0 & 1 & 0 & 0 \\ 0 & .8 & 0 & 0 & .2 \\ 0 & 0 & .4 & 0 & .6 \\ 0 & 0 & 0 & 1 & 0 \end{bmatrix}$

d. $\begin{bmatrix} 0 & 1 & 0 & 0 & 0 \\ 0 & 0 & 1 & 0 & 0 \\ .5 & 0 & 0 & .5 & 0 \\ 0 & 0 & 0 & 0 & 1 \\ 0 & 0 & 1 & 0 & 0 \end{bmatrix}$
e. $\begin{bmatrix} 0 & 1 & 0 & 0 & 0 \\ 0 & 0 & 0 & 0 & 1 \\ 0 & 0 & 0 & 1 & 0 \\ 0 & 0 & .4 & 0 & .6 \\ .2 & 0 & 0 & .8 & 0 \end{bmatrix}$
f. $\begin{bmatrix} 0 & 0 & .2 & 0 & .8 \\ 0 & 0 & 0 & 0 & 1 \\ 1 & 0 & 0 & 0 & 0 \\ 0 & 1 & 0 & 0 & 0 \\ 0 & 0 & 0 & 1 & 0 \end{bmatrix}$

g. $\begin{bmatrix} 0 & 0 & 0 & 1 & 0 \\ 0 & 0 & 1 & 0 & 0 \\ 0 & .5 & 0 & .5 & 0 \\ 0 & 0 & .8 & 0 & .2 \\ 1 & 0 & 0 & 0 & 0 \end{bmatrix}$
h. $\begin{bmatrix} 0 & 0 & 1 & 0 & 0 \\ 0 & .5 & 0 & 0 & .5 \\ .8 & 0 & 0 & .2 & 0 \\ 0 & 0 & 0 & 1 & 0 \\ 0 & 1 & 0 & 0 & 0 \end{bmatrix}$
i. $\begin{bmatrix} 0 & 1 & 0 & 0 & 0 \\ .8 & 0 & 0 & .2 & 0 \\ 0 & 0 & 0 & .6 & .4 \\ 0 & 0 & 1 & 0 & 0 \\ 0 & .8 & 0 & 0 & .2 \end{bmatrix}$

j. $\begin{bmatrix} .5 & 0 & .5 & 0 & 0 \\ .7 & 0 & .1 & 0 & .2 \\ .5 & 0 & .1 & .4 & 0 \\ 0 & 0 & 1 & 0 & 0 \\ 0 & 1 & 0 & 0 & 0 \end{bmatrix}$

2. A transition matrix for a Markov chain is shown below. Write this matrix in canonical form.

$$\begin{bmatrix} 0 & 0 & 1 & 0 & 0 & 0 & 0 & 0 \\ .6 & .2 & 0 & .2 & 0 & 0 & 0 & 0 \\ 0 & 1 & 0 & 0 & 0 & 0 & 0 & 0 \\ 0 & 0 & 0 & 1 & 0 & 0 & 0 & 0 \\ 0 & 0 & 0 & 0 & .2 & .5 & .3 & 0 \\ 0 & 0 & 0 & .1 & .6 & 0 & 0 & .3 \\ 0 & 0 & 0 & 0 & 0 & 0 & 1 & 0 \\ 0 & 0 & 0 & 0 & 1 & 0 & 0 & 0 \end{bmatrix}$$

3. In the setting described in Exercise 11 of Section 3.2, find the long-range probability of the marmot being in each of the areas.

4. In the setting described in Exercise 13 of Section 3.2, find the long-range probability that the flip is made with the biased coin.

5. Show that if **P** is the transition matrix for a Markov chain and **H** is a matrix with all rows equal to the same probability vector **w**, then **PH** is a matrix all of whose rows are the vector **w**.

6. Show that if $\mathbf{P}$ is the transition matrix of a Markov chain for which there is an integer r such that $\mathbf{P}^r$ has only positive entries, then the Markov chain is ergodic and of period 1. That is, the Markov chain is regular according to Definition 3.6.

7. Let $\mathbf{P}$ be the transition matrix for a Markov chain, and suppose there is an integer r such that all entries of $\mathbf{P}^r$ are greater than $h > 0$. Show that for all integers $m > r$, the entries of $\mathbf{P}^m$ are also greater than h.

8. Amy is studying the feeding habits of a certain bird. She observes that the bird always comes the first day she makes food available. After that, however, whenever food is available the pattern of feeding is as follows:

   - If the bird feeds one day, then it never feeds the next day.
   - If the bird feeds on day $n - 1$ and does not feed on day n, then it feeds on day $n + 1$ with probability .75 and does not feed on day $n + 1$ with probability .25.
   - If the bird feeds neither on day $n - 1$ nor on day n, then it feeds on day $n + 1$ with probability .85 and does not feed on day $n + 1$ with probability .15.

   (a) Formulate a Markov chain model for this situation.
   (b) In the long run, on what fraction of the days does the bird feed?

9. In the situation of Exercise 8, suppose the first two parts of the feeding pattern remain the same but the third part is replaced by the following:

   - If the bird feeds neither on day $n - 1$ nor on day n, then it feeds on day $n + 1$ with probability p and does not feed on day $n + 1$ with probability $1 - p$.

   (a) Formulate a Markov chain for this situation.
   (b) After an extended period of feeding, the probability that the bird feeds on a particular day is a function of p. Find this function and graph it for $0 < p < 1$.

10. Suppose that a mouse moves in the maze shown in Figure 3.7 and that observations are made every 5 minutes and every time the mouse moves from one compartment to another. Assume that the mouse remains where it is with probability .4 and that whenever it has a choice, it is three times as likely to move to a darker compartment as to a lighter one. In the long run, what is the probability that it is in the compartment with low illumination?

11. A computer consultant allocates her time in one-week blocks among two employers and vacations. She is very well paid by employer A, but she dislikes the work. She enjoys working for employer B, but the pay is poor. She always takes a week of vacation when she shifts from one employer to the other, and she never takes more than one week of vacation at a time. If she is on vacation, then she selects an employer at random, and A is selected with probability .6. If this is her first week working for employer A, then she will take a vacation next week with probability .2, and if she has worked for employer A for two weeks (or more), then she will take a vacation next week with probability .5. If this is her first week working for employer B, then she will take a vacation next week with probability .1, and if she has worked for employer B for two weeks (or more), then she will take a vacation next week with probability .3. Suppose she starts by taking a vacation and then beginning to work for employer A.

   (a) Formulate a Markov chain model for this situation and find the transition matrix.
   (b) In the long run, how much time does she spend on vacation?

3.4 Absorbing Chains and Applications to Ergodic Chains

In Section 3.3 we introduced a classification of Markov chains, and we considered the special case of chains (ergodic and regular Markov chains) with the property that any two states are mutually accessible—that is, the states form a single equivalence class. Those chains that have at least one pair of states that are not mutually accessible are more complex, and the behavior of systems modeled with such chains shows a great deal of variety. Rather than conducting a systematic study of these more general chains, we turn to another special case: Markov chains in which one or more of the equivalence classes of states consist of a single state. As we shall see, these chains are also very useful as models.

> **Definition 3.7**   A state i of a Markov chain is an **absorbing state** if $p_{ii} = 1$. A Markov chain is said to be an **absorbing Markov chain** if
>
> 1. There is at least one absorbing state, and
> 2. For each nonabsorbing state j, there is an absorbing state k and a number m of steps such that the probability of making a transition from j to k in m steps is positive—that is, $p_{jk}(m) > 0$.

The definition of an absorbing state means that state i is absorbing if the ith row of the transition matrix has a 1 in the ith column, and then, necessarily all other entries in the ith row are 0. Entries in the ith row and the ith column are said to be on the *main diagonal* of the transition matrix. It is important to note that a state j with a 1 in the jth row of the transition matrix in a position other than on the main diagonal is not an absorbing state.

Condition 2 of Definition 3.7 can be stated as follows: For each nonabsorbing state, it is possible to make a transition to some absorbing state in some number of steps. It is not necessary that each absorbing state be accessible from each nonabsorbing state, only that *some* absorbing state be accessible.

It is conventional to write the canonical form of the transition matrix for an absorbing Markov chain with the absorbing states listed first.

EXAMPLE 3.14   Let **P** and **T** be the transition matrices for Markov chains with five states.

$$
\mathbf{P} = \begin{bmatrix} 0 & 0 & 1 & 0 & 0 \\ 0 & 1 & 0 & 0 & 0 \\ 0 & .8 & 0 & 0 & .2 \\ .4 & 0 & 0 & 0 & .6 \\ 0 & 0 & 0 & 1 & 0 \end{bmatrix} \qquad
\mathbf{T} = \begin{bmatrix} .5 & 0 & .3 & 0 & .2 \\ .6 & .2 & 0 & .2 & 0 \\ .3 & 0 & .6 & 0 & .1 \\ 0 & 0 & 0 & 1 & 0 \\ .6 & 0 & .4 & 0 & 0 \end{bmatrix}
$$

For the matrix **P**, state 2 is an absorbing state and all other states are nonabsorbing. Note that state 1 is nonabsorbing even though the first row has one entry equal to 1 and all other entries equal to zero; the entry 1 is not on the main diagonal. A similar comment holds for state 5. The absorbing state can be reached in one step from state 3 and in two or more steps from each of the other nonabsorbing states. Therefore, both conditions are satisfied and **P** is the transition matrix of an absorbing Markov chain. A canonical form for the matrix **P**, with states as shown, is

$$
\begin{array}{c c c c c c}
 & 2 & 1 & 3 & 4 & 5 \\
\begin{matrix} 2 \\ 1 \\ 3 \\ 4 \\ 5 \end{matrix} &
\begin{bmatrix}
1 & 0 & 0 & 0 & 0 \\
0 & 0 & 1 & 0 & 0 \\
.8 & 0 & 0 & 0 & .2 \\
0 & .4 & 0 & 0 & .6 \\
0 & 0 & 0 & 1 & 0
\end{bmatrix}
\end{array}
$$

For the transition matrix **T**, state 4 is an absorbing state and all other states are non-absorbing. It is possible to reach state 4 in a single step from state 2. However, there is a single absorbing state, and it is not possible to reach this state from state 1, 3, or 5. Consequently, **T** is not the transition matrix of an absorbing Markov chain. A canonical form for the matrix **T**, with states listed as shown, is

$$
\begin{array}{c c c c c c}
 & 4 & 1 & 3 & 5 & 2 \\
\begin{matrix} 4 \\ 1 \\ 3 \\ 5 \\ 2 \end{matrix} &
\begin{bmatrix}
1 & 0 & 0 & 0 & 0 \\
0 & .5 & .3 & .2 & 0 \\
0 & .3 & .6 & .1 & 0 \\
0 & .6 & .4 & 0 & 0 \\
.2 & .6 & 0 & 0 & .2
\end{bmatrix}
\end{array}
$$

Note that the equivalence class of states $\{1, 3, 5\}$ has the property that once the system enters the class, it never leaves it. ■

Our convention is that the canonical form of the transition matrix of an absorbing Markov chain has states ordered so that absorbing states are listed first. That is, if there are N states and k of them are absorbing, then we suppose that the states have been relabeled so that states $1, 2, \ldots, k$ are absorbing and states $k + 1, k + 2, \ldots, N$ are nonabsorbing. With this convention, the transition matrix **P** has the form

$$
(3.4) \qquad\qquad \mathbf{P} = \begin{bmatrix} \mathbf{I} & \mathbf{0} \\ \mathbf{R} & \mathbf{Q} \end{bmatrix}
$$

where **I** is a $k \times k$ identity matrix whose row and column labels correspond to absorbing states; **0** is a matrix with all entries equal to zero; **R** is a $(N - k) \times k$ matrix in which the row labels correspond to nonabsorbing states and the column labels correspond to absorbing states; the entries of **R** give the probabilities of direct transitions from nonabsorbing states to absorbing states; and **Q** is an $(N - k) \times (N - k)$ matrix whose entries give the probabilities of transitions between nonabsorbing states. As usual, the states must be listed in the same order in the rows and in the columns. When we refer to a transition matrix of an absorbing Markov chain being written in canonical form, we mean the form (3.4).

One consequence of writing the transition matrix in this form is that the multistep transition matrices have a particularly simple form—for example,

$$
\mathbf{P}(2) = \mathbf{P}^2 = \begin{bmatrix} \mathbf{I} & \mathbf{0} \\ \mathbf{R}_2 & \mathbf{Q}^2 \end{bmatrix}
$$

where $\mathbf{R}_2 = \mathbf{R} + \mathbf{QR}$. In general, for any integer m, $m \geq 2$,

$$(3.5) \qquad \mathbf{P}(m) = \mathbf{P}^m = \begin{bmatrix} \mathbf{I} & \mathbf{0} \\ \mathbf{R}_m & \mathbf{Q}^m \end{bmatrix}$$

where $\mathbf{R}_m$ can be computed successively as

$$(3.6) \qquad \mathbf{R}_m = \mathbf{R} + \mathbf{QR}_{m-1}, \qquad \mathbf{R}_1 = \mathbf{R}$$

or as

$$(3.7) \qquad \mathbf{R}_m = \mathbf{R}_{m-1} + \mathbf{Q}^{m-1}\mathbf{R}, \qquad \mathbf{R}_1 = \mathbf{R}$$

Both of these equations for $\mathbf{R}_m$ will be useful, as we shall see. The matrix $\mathbf{P}(m)$ contains the m-step transition probabilities, and therefore the entries of $\mathbf{Q}^m$ are the m-step transition probabilities from one nonabsorbing state to another, and the entries of $\mathbf{R}_m$ are the m-step transition probabilities from nonabsorbing states to absorbing states.

We note for emphasis that the state labels in the transition matrix in canonical form may differ from the original state labels. In situations where questions about the original setting are to be answered, care must be used in keeping track of the changes in state labels.

Several properties of the matrices $\mathbf{Q}^m$ and $\mathbf{R}_m$ are useful and will enable us to develop techniques for answering many interesting questions. We begin with an examination of the matrix $\mathbf{Q}$.

Suppose we have an absorbing Markov chain with a transition matrix $\mathbf{P}$ in canonical form, and suppose the system begins in the nonabsorbing state i. If j is another nonabsorbing state, $i \neq j$, then the probability that the system is in state j on the first (subsequent) observation is q_{ij}. The probability that it is in state j on the second observation is $q_{ij}(2)$, and so on. Let $E(i, j; m)$ be the expected number of times the system is in state j given that it started in state i and continued for m transitions. We develop an expression for $E(i, j; m)$.

For fixed states i and j, define a random variable X_k by

$$X_k = \begin{cases} 1 & \text{if the system began in state } i \text{ and is in state } j \text{ after the } k\text{th transition} \\ 0 & \text{if the system began in state } i \text{ and is not in state } j \text{ after the } k\text{th transition} \end{cases}$$

It follows from the definition of the expected value of a random variable that $E[X_k] = q_{ij}(k)$, for $k = 1, 2, \ldots, m$, and

$$(3.8) \qquad \begin{aligned} E(i, j; m) &= E[X_1] + E[X_2] + \cdots + E[X_m] \\ &= q_{ij}(1) + q_{ij}(2) + \cdots + q_{ij}(m) \end{aligned}$$

If states i and j are the same, then the expression has an additional term as a result of the fact that the system began in state j:

$$(3.9) \qquad E(j, j; m) = 1 + q_{jj}(1) + q_{jj}(2) + \cdots + q_{jj}(m)$$

Because i and j could be any nonabsorbing states, we have shown that the ij-entry in the matrix

$$(3.10) \qquad \mathbf{I} + \mathbf{Q} + \mathbf{Q}^2 + \cdots + \mathbf{Q}^m$$

is the expected number of times the system is in state j given that it started in state i and made m transitions. Note that the matrix I in (3.10) has the same dimensions as Q.

If the system begins in a nonabsorbing state, there is a positive probability that it will reach some absorbing state, and after doing so, the system remains there for all subsequent observations. In fact, as the number of observations increases, the probability of finding the system in a nonabsorbing state becomes arbitrarily small, and the probability that the system is in an absorbing state approaches 1. Indeed, the probability that the system is in a nonabsorbing state after m transitions becomes small sufficiently fast (as m increases) that the series

$$(3.11) \qquad \mathbf{I} + \mathbf{Q} + \mathbf{Q}^2 + \mathbf{Q}^3 + \cdots$$

converges. The convergence of the series (3.11), together with the meanings of the partial sums (3.10) as given in (3.8) and (3.9), gives a highly useful result.

THEOREM 3.5 If the transition matrix of an absorbing Markov chain is written in canonical form as

$$\mathbf{P} = \begin{bmatrix} \mathbf{I} & \mathbf{0} \\ \mathbf{R} & \mathbf{Q} \end{bmatrix}$$

then the matrix $\mathbf{I} - \mathbf{Q}$ has an inverse, and the series $\mathbf{I} + \mathbf{Q} + \mathbf{Q}^2 + \mathbf{Q}^3 + \cdots$ converges to the inverse of $\mathbf{I} - \mathbf{Q}$.

The matrix $\mathbf{N} = (\mathbf{I} - \mathbf{Q})^{-1}$ is called the **fundamental matrix** of the Markov chain, and the ij-entry in $\mathbf{N}$ is the expected number of visits to nonabsorbing state j given that the system began in nonabsorbing state i and continued until an absorbing state was reached. The sum of the entries in the ith row of $\mathbf{N}$ is the expected number of transitions before an absorbing state is reached. The state labels are those of the transition matrix in canonical form. ■

The details of arguments justifying this theorem are included in the Chapter Appendix.

EXAMPLE 3.15 An absorbing Markov chain has the transition matrix

$$\begin{bmatrix} 0 & 1 & 0 & 0 & 0 \\ .2 & 0 & .1 & .1 & .6 \\ 0 & 0 & 1 & 0 & 0 \\ 0 & 0 & 0 & 1 & 0 \\ .2 & .2 & .3 & .1 & .2 \end{bmatrix}$$

(a) If the system begins in state 2, find the expected number of visits to state 5 before an absorbing state is reached.
(b) If the system begins in state 2, find the expected number of transitions before an absorbing state is reached.

We begin by writing the transition matrix in canonical form.

$$
\begin{array}{c c c c c c}
 & 3 & 4 & 1 & 2 & 5 \\
3 & \begin{bmatrix} 1 \\ 0 \\ 0 \\ .1 \\ .3 \end{bmatrix} & \begin{matrix} 0 \\ 1 \\ 0 \\ .1 \\ .1 \end{matrix} & \begin{matrix} 0 \\ 0 \\ 0 \\ .2 \\ .2 \end{matrix} & \begin{matrix} 0 \\ 0 \\ 1 \\ 0 \\ .2 \end{matrix} & \begin{matrix} 0 \\ 0 \\ 0 \\ .6 \\ .2 \end{matrix} \end{bmatrix}
\end{array}
$$

Because the question is posed in terms of the original state labels, we retain those labels above and to the left of the new matrix. The matrices $Q, I - Q$, and N are

$$
Q = \begin{bmatrix} 0 & 1 & 0 \\ .2 & 0 & .6 \\ .2 & .2 & .2 \end{bmatrix} \quad I - Q = \begin{bmatrix} 1 & -1 & 0 \\ -.2 & 1 & -.6 \\ -.2 & -.2 & .8 \end{bmatrix} \quad N = \begin{bmatrix} 1.7 & 2 & 1.5 \\ .7 & 2 & 1.5 \\ .6 & 1 & 2 \end{bmatrix}
$$

Recall that the state labels for the first, second, and third rows of $Q, I - Q$, and N, are 1, 2, and 5, respectively.

Using the fundamental matrix N, we can answer the questions. The answer to question (a) is the entry in the second row and third column of N: 1.5. If the system begins in state 2, then the expected number of visits to state 5 before an absorbing state is reached is 1.5.

The answer to question (b) is the sum of the entries in the second row of N: $.7 + 2 + 1.5 = 4.2$. If the system begins in state 2, then the expected number of transitions before an absorbing state is reached is 4.2. Of course, it may reach an absorbing state in one transition: There is a positive probability that it moves directly from state 2 to state 3 (or 4). However, it may also take more than one transition, and we now know that the expected number is 4.2. ∎

EXAMPLE 3.16   Consider the small-group decision-making situation described in Section 3.1, and suppose we have a group of four individuals and three alternatives. Define a vote change as a "shift toward consensus" if it results in one of the group composition shifts: $310 \rightarrow 400$, $220 \rightarrow 310$, $211 \rightarrow 310$. Assume that an individual who can effect a shift toward consensus is twice as likely to change a vote as one who cannot. Also assume that the probability of changing a vote to another alternative is proportional to the number of individuals voting for that alternative. Suppose the group initially has group composition 211. Find the expected number of vote changes before consensus is reached.

Denote the group compositions 400, 310, 220, 211 as states 1, 2, 3, and 4, respectively. Then a Markov chain model for this situation under these assumptions has transition matrix (see Exercise 6)

$$
\begin{bmatrix}
1 & 0 & 0 & 0 \\
\frac{2}{5} & 0 & \frac{3}{5} & 0 \\
0 & 1 & 0 & 0 \\
0 & \frac{4}{9} & \frac{2}{9} & \frac{3}{9}
\end{bmatrix}
$$

This is a transition matrix for an absorbing chain, and the matrix is already written in canonical form. The matrix $\mathbf{Q}$ and the fundamental matrix $\mathbf{N}$ are

$$
\mathbf{Q} = \begin{bmatrix} 0 & \frac{3}{5} & 0 \\ 1 & 0 & 0 \\ \frac{4}{9} & \frac{2}{9} & \frac{3}{9} \end{bmatrix} \qquad \mathbf{N} = \begin{bmatrix} \frac{5}{2} & \frac{3}{2} & 0 \\ \frac{5}{2} & \frac{5}{2} & 0 \\ \frac{15}{6} & \frac{11}{6} & \frac{9}{6} \end{bmatrix}
$$

The third row of $\mathbf{N}$ is associated with state 4, group composition 211. Therefore, the expected number of transitions before reaching state 1 is $\frac{15}{6} + \frac{11}{6} + \frac{9}{6} = \frac{35}{6}$. ∎

The matrix $\mathbf{N}$ can also be used to determine the probabilities of absorption in the various absorbing states. To determine how, we return to Equations (3.6) and (3.7),

$$\mathbf{R}_m = \mathbf{R}_{m-1} + \mathbf{Q}^{m-1}\mathbf{R} \qquad \text{and} \qquad \mathbf{R}_m = \mathbf{R} + \mathbf{Q}\mathbf{R}_{m-1}, \ \mathbf{R}_1 = \mathbf{R}$$

Suppose that the system begins in the ith nonabsorbing state. The probability that it is in the jth absorbing state after the first transition is r_{ij}. The probability that it is in the jth absorbing state after the second transition is the ij-entry of the matrix $\mathbf{R}_2$, and in general, the probability that it is in the jth absorbing state after the mth transition is the ij-entry of the matrix $\mathbf{R}_m$.

Next, from the expression $\mathbf{R}_m = \mathbf{R}_{m-1} + \mathbf{Q}^{m-1}\mathbf{R}$ we see that $\mathbf{R}_m \geq \mathbf{R}_{m-1}$, where the symbol $\geq$ means entrywise inequality. That is, each of the sequences of numbers obtained by fixing i and j and taking the ij-entry in the matrix $\mathbf{R}_m$ as the mth entry in the sequence is a monotone nondecreasing sequence of numbers. Moreover, each of these numbers must be less than or equal to 1 because the rows of the m-step transition matrix are probability vectors. Therefore, each of the sequences of numbers converges, and so the sequence of matrices $\{\mathbf{R}_m\}$ converges. Define the matrix $\mathbf{A}$ to be the limit of the sequence $\{\mathbf{R}_m\}$:

$$\mathbf{A} = \lim_{m \to \infty} \mathbf{R}_m$$

It follows from the discussion of the meaning of the entries in $\mathbf{R}_m$ that the ij-entry of the matrix $\mathbf{A}$ has the following interpretation:

> If an absorbing Markov chain is initially in state i, then the probability that it is absorbed in nonabsorbing state j is the ij-entry of the matrix $\mathbf{A}$.

Here as elsewhere in the discussion, it is important to remember that the references to states i and j refer to the states of the matrix in canonical form, and references to the original state labels must be translated into the new state labels.

The definition of the matrix $\mathbf{A}$ given above is as a limit—not particularly well suited for computation—and it is useful to have an alternative means of determining $\mathbf{A}$. There is an expression for the matrix $\mathbf{A}$ that involves only the matrices $\mathbf{R}$ and $\mathbf{N}$. To determine the expression, we recall that $\mathbf{R}_m = \mathbf{R} + \mathbf{Q}\mathbf{R}_{m-1}$, and if we take the limit of both sides as $m \to \infty$, we have

$$\mathbf{A} = \mathbf{R} + \mathbf{Q}\mathbf{A}$$

From this we have

$$\mathbf{A} - \mathbf{Q}\mathbf{A} = (\mathbf{I} - \mathbf{Q})\mathbf{A} = \mathbf{R}$$

Finally, because the inverse of $\mathbf{I} - \mathbf{Q}$ exists and is equal to $\mathbf{N}$, if we multiply both sides of the expression $(\mathbf{I} - \mathbf{Q})\mathbf{A} = \mathbf{R}$ on the left by $\mathbf{N}$, we have $\mathbf{A} = \mathbf{NR}$. We summarize this result as a theorem.

THEOREM 3.6 Suppose that the transition matrix of an absorbing Markov chain is written in canonical form as

$$\mathbf{P} = \begin{bmatrix} \mathbf{I} & \mathbf{0} \\ \mathbf{R} & \mathbf{Q} \end{bmatrix}$$

and that $\mathbf{N}$ is the fundamental matrix for $\mathbf{P}$. Then the entry in the ith row and jth column of the matrix $\mathbf{A} = \mathbf{NR}$ is the probability that the system is absorbed in the jth absorbing state given that it began in the ith nonabsorbing state. ∎

EXAMPLE 3.17 An absorbing Markov chain has the following transition matrix.

$$\begin{bmatrix} 0 & 0 & .5 & 0 & .5 \\ 0 & 1 & 0 & 0 & 0 \\ .2 & .1 & .2 & .1 & .4 \\ 0 & 0 & 0 & 1 & 0 \\ .1 & .1 & 0 & .6 & .2 \end{bmatrix}$$

If the system is initially in state 3, find the probability that it is absorbed in state 4.

We rewrite the transition matrix with the states listed in the order 2, 4, 1, 3, 5. Then the matrices $\mathbf{Q}$, $\mathbf{R}$, and $\mathbf{N}$ are

$$\mathbf{Q} = \begin{bmatrix} 0 & .5 & .5 \\ .2 & .2 & .4 \\ .1 & 0 & .2 \end{bmatrix} \quad \mathbf{R} = \begin{bmatrix} 0 & 0 \\ .1 & .1 \\ .1 & .6 \end{bmatrix} \quad \mathbf{N} = \begin{bmatrix} 1.28 & .80 & 1.20 \\ .40 & 1.50 & 1.00 \\ .16 & .10 & 1.40 \end{bmatrix}$$

Therefore, the matrix $\mathbf{A} = \mathbf{NR}$ is

$$\mathbf{A} = \begin{bmatrix} .20 & .80 \\ .25 & .75 \\ .15 & .85 \end{bmatrix}$$

To complete the example, we need to identify the entry of $\mathbf{A}$ that gives the probability of absorption in state 4 given a start in state 3. The states have been relabeled in the order 2, 4, 1, 3, 5. Therefore, state 3 corresponds to the second row of the matrix $\mathbf{A}$, and state 4 corresponds to the second column. It follows that the desired probability is .75. ∎

■ Applications of Absorbing Chains to Ergodic Chains

One of the common uses of the ideas and techniques introduced here for absorbing chains is to determine useful information about ergodic chains. In many situations we can do so by constructing an absorbing Markov chain that is based on the ergodic chain and the information desired.

Consider an ergodic chain with transition matrix $\mathbf{P}$. By the definition of an ergodic chain, for any states i and j, there is an integer m such that the probability of making a transition from state i to state j in m steps is positive. What is the expected number of transitions to first reach state j given a start in state i? To answer the question, we construct an absorbing chain with the same states, with transition matrix $\mathbf{P}'$ that has the same rows as $\mathbf{P}$ with the exception of the jth row. The jth row of $\mathbf{P}'$ has a 1 on the main diagonal and zeros in all other entries. That is, we have replaced state j by an absorbing state. Transition probabilities among all other states are as in the original chain, and, in particular, transition probabilities into the new state j are the same as those into the old state j. The absorbing chain behaves as follows:

- If the system begins in state j, then it remains there
- If the system begins in state i, $i \neq j$, then it proceeds just as in the original ergodic system until it reaches state j for the first time. Once in state j, it remains there.

Because the original system was ergodic, it is possible to reach state j from every other state, and the new process satisfies the conditions for an absorbing Markov chain.

Next, write the transition matrix for the new chain in canonical form and find the fundamental matrix $\mathbf{N}$. By Theorem 3.5, the entries in $\mathbf{N}$ give the expected number of times the system is in each nonabsorbing state prior to reaching the absorbing state (there is a single absorbing state in this case). Interpreting this in terms of the original process, we see that these numbers give the expected number of times the system is in each state i, $i \neq j$, before it first reaches state j.

EXAMPLE 3.18   A regular Markov chain has the transition matrix

$$\mathbf{P} = \begin{bmatrix} .25 & .25 & .5 & 0 \\ 0 & .25 & .5 & .25 \\ .25 & .25 & .5 & 0 \\ .25 & 0 & .5 & .25 \end{bmatrix}$$

If the system is initially in state 2, find (a) the expected number of visits to state 3 before it first reaches state 4, and (b) the expected number of transitions before it first reaches state 4.

Because this is an ergodic Markov chain (every regular chain is ergodic), we can answer the question by constructing an absorbing chain. We are interested in what happens before the system first reaches state 4, and we construct an absorbing chain by replacing state 4 with an absorbing state. The transition matrix for the new absorbing chain is

$$\begin{bmatrix} .25 & .25 & .5 & 0 \\ 0 & .25 & .5 & .25 \\ .25 & .25 & .5 & 0 \\ 0 & 0 & 0 & 1 \end{bmatrix}$$

Writing this matrix in canonical form yields

$$\begin{array}{c c} & \begin{array}{cccc} 4 & 2 & 1 & 3 \end{array} \\ \begin{array}{c} 4 \\ 2 \\ 1 \\ 3 \end{array} & \begin{bmatrix} 1 & 0 & 0 & 0 \\ .25 & .25 & 0 & .5 \\ 0 & .25 & .25 & .5 \\ 0 & .25 & .25 & .5 \end{bmatrix} \end{array}$$

where the original state labels are as shown. The matrix $\mathbf{Q}$ and the fundamental matrix $\mathbf{N}$ are

$$\mathbf{Q} = \begin{bmatrix} .25 & 0 & .5 \\ .25 & .25 & .5 \\ .25 & .25 & .5 \end{bmatrix} \qquad \mathbf{N} = \begin{bmatrix} 4 & 2 & 6 \\ 4 & 4 & 8 \\ 4 & 3 & 9 \end{bmatrix}$$

The rows and columns of the matrix $\mathbf{N}$ correspond to states 2, 1, 3, in that order. Consequently, we conclude that if the system begins in state 2, then the expected number of visits to state 3 before it reaches an absorbing state is 6. Thus, in the original setting the expected number of visits to state 3 before first reaching state 4 is 6.

Also, if the absorbing chain begins in state 2, then the expected number of transitions before it is absorbed is $4 + 2 + 6 = 12$. Consequently, in the original setting the expected number of transitions before the system first reaches state 4 is 12. ∎

EXAMPLE 3.19 Consider the situation described in Example 3.3 of Section 3.2, in which a mouse moves in a maze with compartments illuminated at different levels. Assume that half the time the mouse is in the same compartment on successive observations, and half the time it moves from its starting compartment to an adjacent compartment between observations. If it moves, then it is equally likely to move to any compartment open to it. If the mouse begins in compartment 1 (see Figure 3.7), find the expected number of transitions before it first reaches compartment 4.

The transition matrix for this system, determined in Example 3.4 of Section 3.2, is

$$\mathbf{P} = \begin{bmatrix} .5 & .5 & 0 & 0 \\ .25 & .5 & .25 & 0 \\ 0 & .25 & .5 & .25 \\ 0 & 0 & .5 & .5 \end{bmatrix}$$

Because the task is to find the number of transitions before the system reaches state 4, we construct an absorbing Markov chain with state 4 replaced by an absorbing state. The transition matrix for the new chain is

$$\mathbf{P'} = \begin{bmatrix} .5 & .5 & 0 & 0 \\ .25 & .5 & .25 & 0 \\ 0 & .25 & .5 & .25 \\ 0 & 0 & 0 & 1 \end{bmatrix}$$

We write the new transition matrix in canonical form, and we label the rows and columns with the original state labels to retain that information. We have

$$\mathbf{P} = \begin{array}{c} \\ 4 \\ 1 \\ 2 \\ 3 \end{array} \begin{array}{cccc} 4 & 1 & 2 & 3 \\ \begin{bmatrix} 1 & 0 & 0 & 0 \\ 0 & .5 & .5 & 0 \\ 0 & .25 & .5 & .25 \\ .25 & 0 & .25 & .5 \end{bmatrix} \end{array}$$

The fundamental matrix for this chain is

$$N = \begin{bmatrix} 3 & 2 & 1 \\ 2 & 4 & 2 \\ 1 & 2 & 3 \end{bmatrix}$$

The system began in state 1, and state 1 corresponds to the first row of this fundamental matrix. The sum of the entries in the first row is 6, and consequently, the expected number of transitions before the system first reaches state 4, given that it begins in state 1, is 6. ∎

In Examples 3.18 and 3.19 we have used the technique of creating a new Markov chain with an absorbing state to determine the expected number of transitions before the system first reaches a specified state. The same technique can be used to find the probability that the system visits state i before state j. For example, given an ergodic Markov chain and a specified starting state (different from states i and j), we can find the probability that state i is visited before state j by creating a new Markov chain with states i and j absorbing—as in Examples 3.18 and 3.19—and then using the associated matrix $\mathbf{A}$ as in Example 3.17.

Exercises 3.4

1. Transition matrices for six Markov chains are shown below. In each case determine whether the Markov chain is absorbing, and write the transition matrix in canonical form. If the chain is absorbing, find the fundamental matrix $\mathbf{N}$.

a. $\begin{bmatrix} 0 & 1 & 0 & 0 & 0 \\ 0 & 0 & .5 & 0 & .5 \\ 0 & 0 & 1 & 0 & 0 \\ .8 & 0 & .2 & 0 & 0 \\ 0 & 1 & 0 & 0 & 0 \end{bmatrix}$ b. $\begin{bmatrix} 0 & 0 & 0 & 1 & 0 \\ 0 & .1 & .3 & .1 & .5 \\ 0 & 0 & 1 & 0 & 0 \\ .4 & 0 & 0 & .2 & .4 \\ 0 & 0 & 0 & .8 & .2 \end{bmatrix}$

c. $\begin{bmatrix} 1 & 0 & 0 & 0 & 0 \\ 0 & 0 & 0 & .5 & .5 \\ 0 & 0 & 0 & 0 & 1 \\ .5 & .5 & 0 & 0 & 0 \\ 0 & .5 & .5 & 0 & 0 \end{bmatrix}$ d. $\begin{bmatrix} 0 & 1 & 0 & 0 & 0 \\ 0 & 1 & 0 & 0 & 0 \\ 0 & 0 & 0 & .4 & .6 \\ 0 & 0 & 0 & 1 & 0 \\ .5 & 0 & .5 & 0 & 0 \end{bmatrix}$

e. $\begin{bmatrix} 0 & 0 & 1 & 0 & 0 \\ 0 & .4 & 0 & .6 & 0 \\ .1 & 0 & .1 & 0 & .8 \\ 0 & 0 & 0 & 1 & 0 \\ 0 & .5 & 0 & 0 & .5 \end{bmatrix}$ f. $\begin{bmatrix} 0 & 0 & 0 & 1 & 0 \\ 0 & 1 & 0 & 0 & 0 \\ 1 & 0 & 0 & 0 & 0 \\ 0 & 1 & 0 & 0 & 0 \\ .2 & 0 & 0 & 0 & .8 \end{bmatrix}$

2. The transition diagram for a Markov chain is shown in Figure 3.15.
   (a) Find the transition matrix and write it in canonical form.
   (b) If the system is initially in state 2, find the expected number of transitions before it first reaches an absorbing state.

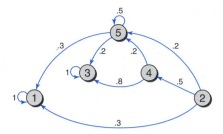

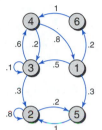

Figure 3.15

Figure 3.16

3. A Markov chain has the transition diagram shown in Figure 3.16. Describe the behavior of the Markov chain in as much detail as you can, including (to the extent possible) information on the long-run behavior of the state vector. To what extent does the long-run behavior depend on the initial state vector?

4. A transition matrix for a Markov chain is shown below.

$$\begin{bmatrix} 0 & 1 & 0 & 0 & 0 & 0 & 0 & 0 \\ .5 & .2 & 0 & .3 & 0 & 0 & 0 & 0 \\ 1 & 0 & 0 & 0 & 0 & 0 & 0 & 0 \\ 0 & 0 & 0 & 1 & 0 & 0 & 0 & 0 \\ 0 & 0 & 0 & 0 & .3 & .3 & .4 & 0 \\ 0 & 0 & 0 & .2 & .5 & 0 & 0 & .3 \\ 0 & 0 & 0 & 0 & 0 & 0 & 1 & 0 \\ 0 & 0 & 0 & 0 & 0 & 1 & 0 & 0 \end{bmatrix}$$

(a) Write this matrix in canonical form.
(b) Find the fundamental matrix for your canonical form.
(c) If the process is initially in state 2, find the expected number of transitions before it first reaches an absorbing state.
(d) If the process begins in state 8, find the probability that it is absorbed in state 4.
(e) If the process begins in state 8, find the probability that it is in state 3 before it is absorbed.
(f) If the process is initially in state 3, find the expected number of visits to state 1 before it is absorbed.

5. A transition matrix for a Markov chain is shown below.

$$\begin{bmatrix} 0 & .5 & 0 & 0 & .5 & 0 & 0 & 0 & 0 & 0 \\ .2 & .2 & 0 & 0 & 0 & .6 & 0 & 0 & 0 & 0 \\ 0 & 1 & 0 & 0 & 0 & 0 & 0 & 0 & 0 & 0 \\ 0 & 0 & 0 & 1 & 0 & 0 & 0 & 0 & 0 & 0 \\ 0 & 0 & 0 & 0 & 0 & .2 & .2 & .6 & 0 & 0 \\ 0 & 0 & 0 & .3 & 0 & .5 & 0 & 0 & .2 & 0 \\ 0 & 0 & 0 & 0 & 0 & 0 & 1 & 0 & 0 & 0 \\ 0 & 0 & 0 & 0 & 0 & 0 & .5 & .2 & .3 & 0 \\ 0 & 0 & 0 & 0 & 0 & 0 & 0 & 0 & 1 & 0 \\ 0 & 0 & 0 & 0 & 0 & .4 & 0 & 0 & .2 & .4 \end{bmatrix}$$

(a) Write this matrix in canonical form.
(b) Find the fundamental matrix for your canonical form.
(c) If the process is initially in state 5, find the expected number of transitions before it is absorbed.
(d) If the process begins in state 2, find the probability that it is absorbed in state 9.
(e) If the process begins in state 2, find the probability that it is in state 8 before it is absorbed.
(f) If the process is initially in state 1, find the expected number of visits to state 5 before it is absorbed.

6. Consider the small-group decision-making situation described in Section 3.1, and suppose we have a group of four individuals and three alternatives. Define a vote change as a "shift toward consensus" if it results in one of the group composition shifts: $310 \rightarrow 400$, $220 \rightarrow 310$, $211 \rightarrow 310$. Assume that an individual who can effect a shift toward consensus is twice as likely to change a vote than one who cannot. Also assume that the probability of changing a vote to another alternative is proportional to the number of individuals voting for that alternative. Suppose that the group initially has group composition 211. Formulate a Markov chain model for this situation, find the transition matrix, and find the fundamental matrix.

7. Consider a small-group decision-making situation similar to that described in Section 3.1 with six individuals and three alternatives. Formulate a Markov chain model under the following assumption: An individual who is the only person voting for an alternative is twice as likely to change her vote as a person who is one of a group of two or more voting for an alternative. If an individual changes her vote, then the probability of changing to an alternative is proportional to the number of individuals voting for that alternative.

(a) If the group is initially divided 3, 2, and 1, find the expected number of vote changes before consensus is reached.
(b) If the group is initially divided 2, 2, and 2, find the probability that at some point it is divided 3, 3, and 0 before consensus is reached.

8. From one academic year to another, each student at Gigantic State University moves on to the next class, flunks out, or remains in the same class with probabilities p, q, and r, respectively. A student is said to be in state 1 if graduated, 2 if flunked out, 3 if a

senior, 4 if a junior, 5 if a sophomore, and 6 if a freshman. Formulate a Markov chain model for this situation, and then answer the following:

(a) Find the transition matrix.
(b) If $p = .7, q = .2$, and $r = .1$, find the fundamental matrix.
(c) With p, q, and r as in (b), how long does a student need to be in college before there is a better than even chance of his or her graduating?

9. A marmot lives in the region shown in Figure 3.6. Suppose the marmot is observed every hour and each time it moves from one area to another. Also, suppose that it will be observed in the same location on successive observations with probability .2. If it moves, the probability of its moving to an area is proportional to the number of resources available to it in that area in comparison to the number of resources available to it in the adjoining areas. The areas bordering the pond have water in addition to the other resources specified, and the marmot is never actually in the pond.

(a) Formulate a Markov chain model for this situation.
(b) If the marmot begins at its den in the rock pile, find the expected number of observations before it first reaches the marsh.
(c) If the marmot begins at its den in the rock pile, find the probability that it reaches the forest without going through the marsh.

10. Joe and Jess play a game as follows: An unfair coin with $\Pr[H] = .6$ is flipped and the result is noted. If it comes up heads, then Jess pays Joe one dollar, and if it comes up tails, then Joe pays Jess one dollar. If each player has money, then the coin is flipped again. The game ends as soon as one player has all the money. When the game begins, Joe has one dollar and Jess has three dollars. (See also Exercise 15 of Section 3.2.)

(a) Find the expected number of plays before the game ends.
(b) Find the probability that Jess wins the game.

11. There are six balls, three red and three green, distributed between boxes labeled 1 and 2, with three balls in each box. A game is played as follows: A ball is selected at random from each box. The ball selected from box 1 is placed in box 2, the ball selected from box 2 is placed in box 1, and the colors of the balls in each box are noted. Then two more balls are selected, and play continues. When the game begins there are two green balls and one red ball in box 2. Find the probability that all balls in box 1 are green before all balls in box 2 are green. (See also Exercise 14 of Section 3.2.)

12. An unfair coin with $\Pr[H] = .4$ is flipped repeatedly, and the result of each flip is noted. Find the expected number of flips before there are three consecutive heads for the first time.

13. Each morning Reba decides how to get from her apartment to the university. She can walk, bike, take the bus, or ride with a friend who also has an 8:00 a.m. class. It is too much work to plan in advance, so each morning she makes a random choice subject to the following conditions:

- If she walks on one day, then the next day she walks with probability .4; all other choices are equally likely.
- If she rides the bus one day, then the next day she is twice as likely to ride the bus as not; all other choices are equally likely.

- If she rides with a friend one day, then she does not ride with a friend next day; all other choices are equally likely.
- If she rides her bike one day, then she always walks the next day.

If she rides with a friend on Monday, find the expected number of days before she first rides the bus.

Chapter Appendix: Mathematical Details

The primary focus of this book is on model building. However, there are times when a slightly deeper look into the mathematical aspects of the models yields interesting and useful information. Thus, in this Appendix we consider in greater detail the following topics that were introduced but not explored in Chapter 3.

1. The relation between Markov chains that are ergodic and of period 1 and Markov chains with a transition matrix $\mathbf{P}$ for which some power $\mathbf{P}^r$ has only positive entries.
2. The long-run behavior of the powers of the transition matrix of a regular Markov chain. This result can be used to give information on the long-run behavior of the state vector for a regular Markov chain.
3. The behavior of the powers $\mathbf{Q}^m$, where $\mathbf{Q}$ is the matrix of transition probabilities among the nonabsorbing states in an absorbing Markov chain, and the invertibility of the matrix $(\mathbf{I} - \mathbf{Q})$.

It is the details of the second and third topics that are most likely to be useful, and we begin with them. (The details of the proof of Theorem 3.3 are included primarily for completeness, although they do provide some insights into the structure of Markov chains with period 1.)

Proof of Theorem 3.4   The hypothesis of the theorem is that we have a Markov chain with a transition matrix $\mathbf{P}$ and an integer r such that $\mathbf{P}^r$ has only positive entries. We show that there is a probability vector $\mathbf{s}$ such that

$$(3.12) \qquad \lim_{r \to \infty} \mathbf{P}^r = \begin{bmatrix} \mathbf{s} \\ \mathbf{s} \\ \vdots \\ \mathbf{s} \\ \mathbf{s} \end{bmatrix}.$$

To this end, let r be an integer such that all entries in $\mathbf{P}^r$ are positive—say all are greater than $h > 0$. It follows that all entries of $\mathbf{P}^m$ are greater than h for all $m \geq r$ (Exercise 7 of Section 3.3). Consider the first column of $\mathbf{P}^m$ and denote it by $\mathbf{p}^1(m)$. Set

$$s(m) = \min\{p_{i1}(m), \ i = 1, 2, \ldots, N\} \quad \text{and}$$
$$t(m) = \max\{p_{i1}(m), \ i = 1, 2, \ldots, N\}$$

Thus $s(m)$ is the smallest probability that the system is in state 1 on the mth observation given a start in state $1, 2, \ldots, N$, and $t(m)$ is the largest such probability. The following is a useful fact about Markov chains, and it will be helpful in verifying our result.

> The sequences $\{s(m)\}$ and $\{t(m)\}$ are monotone increasing and monotone decreasing, respectively. That is, $s(m + 1) \geq s(m)$ and $t(m + 1) \leq t(m)$ for $m = 1, 2, 3, \ldots$.

To confirm this monotonicity, we denote by $\mathbf{p}_i$ the ith row of the matrix $\mathbf{P}$, and we note that

$$s(m+1) = \min\{p_{i1}(m+1), \ i = 1, 2, \dots, N\}$$
$$= \min\{\mathbf{p}_i \cdot \mathbf{p}^1(m), \ i = 1, 2, \dots, N\}$$

Also, since $\mathbf{p}_i$ is a probability vector,

$$\mathbf{p}_i \cdot \mathbf{p}^1(m) = \sum_{j=1}^{N} p_{ij} p_{j1}(m) \geq s(m) \sum_{j=1}^{N} p_{ij} = s(m)$$

and consequently the minimum of these inner products—which is $s(m+1)$—must also be greater than or equal to $s(m)$. The proof that $t(m+1) \leq t(m)$ is similar.

Returning to the proof of the Theorem, we have $s(m) \leq t(m) \leq t(1)$ for all m, and consequently $\{s(m)\}$ is an increasing sequence of real numbers that is bounded above. By a fundamental property of real numbers, the limit $\lim_{m\to\infty} s(m)$ exists, and we denote the limit by s. Also, the sequence $\{t(m)\}$ is monotone decreasing and $t(m) \geq s(m) \geq s(1)$, and consequently $\lim_{m\to\infty} t(m) = t$ exists. Because $s(m) \leq t(m)$ for all m, we conclude that $s \leq t$. If $s = t$, then we have shown that the entries in the first column of $\{\mathbf{P}^r\}$ all converge to a common limit as m increases. We show that the remaining possibility, $s < t$, is impossible. In this case, the sequences $\{s(m)\}$ and $\{t(m)\}$ and the limits s and t are as shown in Figure 3.17.

Now set $d = t - s$, and note that we are assuming $d > 0$. We complete the proof by showing that $s(m)$ increases by more than a fixed amount—an amount depending on h and d—in each sequence of r transitions. In fact, we show that if

1. $p_{i1}(m) \geq h$ for $1 \leq i \leq N$ and $m \geq r$,
2. $s(m) \leq s < t \leq t(m)$ for all m, and
3. $t - s = d$,

then

$$s(m+r) \geq s(m) + hd \qquad \text{for } m = 1, 2, 3 \dots$$

To verify this, we let k be an index for which $s(m+r) = p_{k1}(m+r)$. We let $\mathbf{u}$ be an N-vector all of whose coordinates are 1. Then for $m = 1, 2, \dots$, we have

$$s(m+r) = \mathbf{p}_k(r) \cdot \mathbf{p}^1(m)$$
$$= \mathbf{p}_k(r) \cdot [\mathbf{p}^1(m) - s(m)\mathbf{u}] + s(m)(\mathbf{p}_k(r) \cdot \mathbf{u})$$
$$\geq s(m) + \max\{p_{kj}(r)[p_{j1}(m) - s(m)], \ j = 1, 2, \dots\}$$
$$\geq s(m) + hd$$

The second inequality results from the fact that at least one of the terms $p_{j1}(m) - s(m)$ is as large as d and all of the $p_{kj}(r)$ are at least as large as h. To see that $s < t$ is

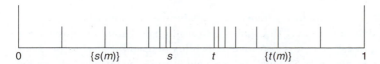

0   {s(m)}   s   t   {t(m)}   1

Figure 3.17

impossible, we argue as follows: Because $s(m) \to s$ as $m \to \infty$, there is an integer m_0 such that $s(m) > s - hd/2$ for all $m > m_0$. However, the above argument shows that $s(m + r) \geq s(m) + hd \geq s + hd/2$ for all $m > m_0$. But this is impossible by the meaning of s.

The proof that all entries in the first column of $\{\mathbf{P}^r\}$ tend to the same limit is complete. Also, because $s(r) > 0$, that limit must be positive. The same proof can be applied to entries in the second, ..., Nth columns of $\{\mathbf{P}^r\}$. The limit for each column will, in general, depend on the column index. The verification of Equation (3.12) is complete, and this is part (i) of Theorem 3.4.

For part (ii) of Theorem 3.4, we let $\mathbf{H}$ denote the matrix $\lim_{m \to \infty} \mathbf{P}^m$, a matrix all of whose rows are the same vector $\mathbf{s}$. We write $\mathbf{P}^{m+1} = \mathbf{P}^m \mathbf{P}$. Next, let m tend to infinity in this expression, and use Equation (3.12). Then $\mathbf{H} = \mathbf{HP}$, and the first row of each side of this expression gives $\mathbf{s} = \mathbf{sP}$.

For part (iii) of Theorem 3.4, let $\mathbf{x}$ be a probability vector that satisfies $\mathbf{x} = \mathbf{xP}$. If we multiply each side of this equation on the right by the matrix $\mathbf{P}$, we have $\mathbf{xP} = \mathbf{xPP} = \mathbf{xP}^2$, and because $\mathbf{x} = \mathbf{xP}$, we have $\mathbf{x} = \mathbf{xP}^2$. Continuing in this way, we have $\mathbf{x} = \mathbf{xP}^m$ for $m = 2, 3, 4, \ldots$. Letting m tend to infinity and using Equation (3.12), we have

$$\mathbf{x} = \mathbf{x} \begin{bmatrix} \mathbf{s} \\ \mathbf{s} \\ \vdots \\ \mathbf{s} \\ \mathbf{s} \end{bmatrix}$$

Finally, because $\mathbf{x}$ is a probability vector, the sum of the coordinates in $\mathbf{x}$ is 1, and consequently the right-hand side of the expression just above is $\mathbf{s}$. ∎

Proof of Theorem 3.5 We show first that if $\mathbf{P}$ is the transition matrix of an absorbing Markov chain written in canonical form,

$$\mathbf{P} = \begin{bmatrix} \mathbf{I} & \mathbf{0} \\ \mathbf{R} & \mathbf{Q} \end{bmatrix}$$

then the matrix $\mathbf{I} - \mathbf{Q}$ is invertible. We do so by showing that the infinite series in expression (3.11)

$$\mathbf{I} + \mathbf{Q} + \mathbf{Q}^2 + \mathbf{Q}^3 + \cdots$$

converges and that the matrix $\mathbf{N}$, defined as the sum of the series, is the inverse of $\mathbf{I} - \mathbf{Q}$.

First we show that $\lim_{m \to \infty} \mathbf{Q}^m = \mathbf{0}$. Begin by noting that the multistep transition matrices have a particularly simple form. Indeed, as shown in Equation (3.5), we have

$$\mathbf{P}^m = \begin{bmatrix} \mathbf{I} & \mathbf{0} \\ \mathbf{R}_m & \mathbf{Q}^m \end{bmatrix}$$

where the entries in $\mathbf{R}_m$ are the multistep transition probabilities from nonabsorbing states to absorbing states, and the entries in $\mathbf{Q}^m$ are multistep transition probabilities within the set of nonabsorbing states. Because there is a positive probability of reaching an absorbing state from each nonabsorbing state, for each nonabsorbing state i there is an integer m depending on i such that the row in $\mathbf{R}_m$ corresponding to state i has a positive entry. Also, because the columns of $\mathbf{R}_m$ correspond to absorbing states, once an entry in $\mathbf{R}_m$ is positive, it remains positive for all larger integers m. For each state i, let $m(i)$ be an integer m determined as above, and let $M = \max\{m(i)$ for all nonabsorbing states $i\}$. Then each row of $\mathbf{R}_M$ has at least one positive entry. Let h be the smallest positive entry in $\mathbf{R}_M$. Then each row in $\mathbf{Q}^M$ has coordinates whose sum is no larger than $1 - h < 1$, and, in particular, each entry is no larger than $1 - h$.

It follows that each entry in $\mathbf{Q}^M\mathbf{Q}^M = \mathbf{Q}^{2M}$ is no larger than $(1 - h)^2$, and, in general, each entry in $\mathbf{Q}^{kM}$ is no larger than $(1 - h)^k$. Next, consider the sequence $\{\mathbf{Q}^m\}$. For each column in $\mathbf{Q}^m$, the sequence of largest entries forms a monotone decreasing sequence, and consequently the entries in each column of $\mathbf{Q}^m$ tend to zero, and therefore $\lim_{m\to\infty} \mathbf{Q}^m = \mathbf{0}$.

The estimates on the size of the entries in $\mathbf{Q}^m$ derived just above enable us to show that the series $\mathbf{I} + \mathbf{Q} + \mathbf{Q}^2 + \mathbf{Q}^3 + \cdots$ converges. Indeed, for each position in the sum of matrices (3.11), the associated series of numbers consists of M consecutive entries each less than $1 - h$ (which is less than 1), followed by another M consecutive entries each less than $(1 - h)^2$, followed by another M consecutive entries each less than $(1 - h)^3, \ldots$; and the resulting series of numbers converges. The entries in each position in the series $\mathbf{I} + \mathbf{Q} + \mathbf{Q}^2 + \mathbf{Q}^3 + \cdots$ converge, and therefore the infinite series of matrices converge. Denote the sum by $\mathbf{N}$. Finally,

$$(\mathbf{I} - \mathbf{Q})(\mathbf{I} + \mathbf{Q} + \mathbf{Q}^2 + \mathbf{Q}^3 + \cdots + \mathbf{Q}^m) = \mathbf{I} - \mathbf{Q}^{m+1}$$

Because $\lim_{m\to\infty} \mathbf{Q}^m = \mathbf{0}$, it follows that

$$\lim_{m\to\infty} (\mathbf{I} - \mathbf{Q})(\mathbf{I} + \mathbf{Q} + \mathbf{Q}^2 + \mathbf{Q}^3 + \cdots + \mathbf{Q}^m) = \lim_{m\to\infty} (\mathbf{I} - \mathbf{Q}^{m+1}) = \mathbf{I}$$

and therefore,

$$(\mathbf{I} - \mathbf{Q}) \lim_{m\to\infty} (\mathbf{I} + \mathbf{Q} + \mathbf{Q}^2 + \mathbf{Q}^3 + \cdots + \mathbf{Q}^m) = (\mathbf{I} - \mathbf{Q})\mathbf{N} = \mathbf{I}$$

A similar argument shows that $\mathbf{N}(\mathbf{I} - \mathbf{Q}) = \mathbf{I}$. Therefore, $(\mathbf{I} - \mathbf{Q})$ is invertible and $(\mathbf{I} - \mathbf{Q})^{-1} = \mathbf{N}$.   ∎

Proof of Theorem 3.3   The goal is to show that if $\mathbf{P}$ is the transition matrix of an ergodic Markov chain with period 1, then there is a power of $\mathbf{P}$ all of whose entries are strictly positive. The proof is based on a result from elementary number theory concerning the greatest common divisor of a set of integers:

> If d is the greatest common divisor of a set of integers $\{n_1, n_2, n_3, \ldots, n_k\}$,
> then there are integers $x_1, x_2, x_3, \ldots, x_k$ such that

(3.13) $$d = n_1 x_1 + n_2 x_2 + \cdots + n_k x_k$$

Suppose that the integers n_j are labeled so that all the positive x_j occur before any negative ones. Then d can be written as the sum $N_1 - N_2$, where N_1 is the portion of the sum (3.13) that includes all positive x_j, and $-N_2$ is the portion of the sum that includes all negative x_j. If the greatest common divisor is 1, then $1 = N_1 - N_2$, with N_1 and N_2 defined as above.

Set $N = N_2^2$. Any integer $n \geq N$ can be written as $n = N + k = N_2^2 + k$, with k a nonnegative integer. We write $k = aN_2 + b$, with $0 \leq b < N_2$ and a equal to the integer j

that satisfies $jN_2 \leq k < (j+1)N_2$. With a and b defined in this way, we have

$$n = N_2^2 + k = N_2^2 + aN_2 + b = N_2^2 + aN_2 + b(N_1 - N_2) = (N_2 + a - b)N_2 + bN_1$$

which gives us a representation of n as a linear combination of N_1 and N_2 with *positive* coefficients. Thus, for any integer $n > N$, we have a representation of n as a linear combination of the integers $\{n_1, n_2, n_3, \ldots, n_k\}$ with positive coefficients.

Now, for a specific state i with period 1, let $\{n_1, n_2, n_3, \ldots, n_k\}$ be integers such that $p_{ii}(n_j) > 0$ for $j = 1, 2, \ldots, k$, and the greatest common divisor of $\{n_1, n_2, n_3, \ldots, n_k\}$ is 1. Then for every sufficiently large n, there are positive integers y_j, $j = 1, 2, \ldots, k$, for which we have

$$p_{ii}(n) = p_{ii}(y_1 n_1 + y_2 n_2 + \cdots + y_k n_k) \geq \prod_{j=1}^{k} p_{ii}(y_j n_j) \geq \prod_{j=1}^{k} p_{ii}(n_j)^{y_j} > 0$$

It follows that for any states i and j and any integer m such that $p_{ij}(m) > 0$, we have $p_{ij}(m+n) \geq p_{ii}(n)p_{ij}(m) > 0$ for all sufficiently large integers n.

We now have enough information to complete the proof of Theorem 3.3. For every pair of states i and j, there is an integer m depending on i and j and denoted by $m(i, j)$ such that $p_{ij}(m(i, j)) > 0$. By the result just above, $p_{ij}(m(i, j) + n) > 0$ for all sufficiently large n. Let M be the maximum of $m(i, j)$ over all pairs i and j, let N be the maximum of the integers N_2^2 (there is a value of N_2 for each state), and define $M^* = M + N$. Then

$$p_{ij}(M^*) = p_{ij}(m(i, j) + M^* - m(i, j)) \geq p_{ij}(m(i, j))p_{ii}(M^* - m(i, j)).$$

Now $p_{ij}(m(i, j)) > 0$ and $M^* - m(i, j) = N + [M - m(i, j)] > N$, and consequently both factors on the right-hand side are positive. This shows that $p_{ij}(M^*) > 0$ for all states i and j. ∎

CHAPTER 4

Simulation Models

4.0   Introduction

In this chapter we discuss in more detail the computer implementation of models used to study activities or processes by imitating or simulating their actual behavior. We introduced the broad ideas and provided examples in Section 2.6. In this chapter we continue the discussion in greater depth. Our view, however, is that simulation is just one of a number of modeling techniques, and therefore we consider only the basic aspects.

Simulation models are widely used and their popularity is increasing. Many systems of current interest are large and complex, and simulation models may be a very effective way to gain insight. Indeed, software that facilitates simulation has become increasingly common, and the cost of computing continues to decline. Consequently, simulation models have become a very attractive option as an aid to understanding complex systems. Because the development of simulation models for complex systems is a time-consuming task, a number of special-purpose simulation languages have been created to reduce the effort. Often these languages are designed to handle a relatively restricted class of situations, but to do so in a convenient and efficient way. Our goal is to illustrate the fundamental concepts, so we use only widely available computer software: the scientific software package MAPLE and the spreadsheet EXCEL; there are many alternatives to both.

4.1   The Simulation Process

The simulation process is intended to help us understand the behavior of a system by using a computer to imitate its behavior or certain aspects of its behavior. Although the term *simulation* is sometimes used to refer to the use of a computer in models for completely deterministic situations—that is, models in which a specific set of inputs always yields the same set of outputs—we will use the term *numerical model* or *computational model* for such situations. We will reserve the term *simulation model* for situations in which some of the quantities or aspects of the system being studied are described in probabilistic terms. In such situations, the output data are themselves random, and consequently we can obtain only estimates of the actual behavior of the system. For instance, some of the population models studied in Section 2.2 are numerical models in the sense that the model is deterministic and we used a computer to generate the data on which the predictions were based. In contrast, the population model of Section 2.6 has stochastic components and the model is a simulation model. Of course, deterministic models are special cases of stochastic models, and the nuances in terminology should not be overemphasized.

Our goal is to gain understanding of, or answer questions about, the system being studied. As one approach to attaining that goal, we consider certain attributes of the system and seek information about the changes in the attributes; usually we are interested in changes as time passes. The values of the attributes at a specific time are used to define the state of the system at that time. Appropriate attributes and the associated set of states are determined by the questions being asked. For example, if we have a model for a homogeneous population, then the set of states might be the set of population sizes. If, on the other hand, we have a population that has been divided into subgroups, then the attributes of interest might be the sizes of the populations in each of the subgroups, and we could define a state as a vector whose coordinates are the population sizes of the subgroups. We suppose that the state of a system is determined by a finite number of attributes, and we call these attributes the state variables.

In most cases, we will be concerned with computer-implemented models for situations in which the state variables change only at the times in a discrete (finite or countable) set. It is convenient to think of the times in this set as the times at which the system is observed, and we frequently refer to this set as the set of observation times. The state variables may change between two successive observations, or they may be the same on two successive observations; they are monitored only at the observation times. Observations may occur at times specified in advance, and regularly spaced times are common. Another alternative is that the observations may be made at times defined as values of one of the random variables of the system. An example of a situation where the latter view is appropriate is a phone call center where the interval between incoming calls is given by a random variable, the duration of the calls is given by another random variable, and the system is observed each time a call is received or terminated.

A third alternative is that the observation times may be described by events whose actual times are either unknown or irrelevant to the study. Examples of this alternative include the ecological situations described in Chapter 3, in which an animal is moving from one subarea of its range to another and is monitored whenever it changes subareas. Another example is the activity (described in Section 1.5) of collecting tokens from bags of cat food. We will use this example later in this section, and we restate it here for convenience: Suppose that a cat food producer includes a plastic letter in each bag of cat food. The letters are C, A, T, and S, and suppose that every customer who collects all four letters receives a free bag of cat food. Given the distribution of letters in bags, estimate the number of bags of cat food that a customer needs to buy to collect all four letters. In these examples the relation between observations and actual time is unimportant. In such circumstances, it is common to think of the observations as taking place at an ordered set of times denoted by the positive integers.

Although simulation models have been applied in a variety of actual situations, there is a common set of activities that are part of most simulations. A simple and useful view of the simulation process is one in which the process consists of three parts:

- Setup and initialization
- Main simulation
- Collection and presentation of results

In the setup and initialization phase, the task is to decide which data generated in the simulation should be collected and how to collect them. In some cases the data are obvious, and in other cases it requires thought and creativity to identify appropriate data. For instance, consider a study of a homogeneous population with a random reproduction rate, where the goal is to project the population size at each time in some set. In this case the input data would be the values and probabilities of the reproduction rate, the initial population size, the time interval over which the population is to be modeled, and the times at which the population size is to be recorded.

As another example, consider the token collection activity described above. In this situation, the goal is to collect all four tokens. It is important to know whether a C has been obtained, but it is less important to know how many C's have been collected. As far as the task of estimating the total number of bags purchased is concerned, it is unimportant whether, at the time all tokens are in hand, you have purchased four bags that contained a C, two that contained a T, and one each with an A and an S, or two bags that contained a C, four that contained a T, and one each with an A and an S. In each situation you have purchased a total of eight bags to obtain all tokens. The simulation should be designed so that the selection process continues until all four tokens have been selected and stops as soon as that condition is fulfilled. The goal is to estimate the total number of bags of cat food purchased, and we can do so by recording the total number of tokens selected when the simulation stops.

The main activity in a simulation is that part of the process where the behavior of the system is actually simulated. This may be relatively straightforward, as with simulations of coin tossing and other situations involving combinatorial probability, or it may be quite complex, as with the simulation of the spread of an actual disease. Frequently it is useful to describe the main simulation process with a version of a flow diagram that shows how the logic of the process is structured. Such a diagram shows the steps, or in some cases the main steps, of the simulation in sequential order. This makes possible an analysis of the simulation that is independent of the details of the implementation. Also, this approach facilitates additional subdivision of the process into independent segments or the combination of several segments into a single block.

The final part of the simulation is the important task of extracting information from the results and presenting that information in a way that is helpful in responding to the original questions. In some cases the most useful information can be displayed in a single table or graph, and in other cases—especially situations in which alternative scenarios are considered—it may be necessary to prepare a number of tables, graphs, and associated narrative to convey what has been learned from the simulation.

It is customary to refer to a single simulation of the activity being studied—a basketball game, the evolution of a population, a customer selecting bags of cat food until all four tokens are collected—as one *run* of the simulation. Because a simulation run is one realization of a sequence of outcomes, each outcome selected from a set (when a coin is flipped, it can land with heads or tails up; the weather can be clear, cloudy, or rainy; an individual can be susceptible to a disease, infected, or immune) and each of which occurs with some likelihood, there is no assurance that the results of one run are representative of even the broad features of the process. That is, the results of one run may share features with the results of many other runs, or they may not. Looking at the output of a single simulation run usually gives little information about the situation, and the analysis of a simulation is almost always based on multiple runs.

Rather than continuing to discuss the process in general terms, let's consider examples to illustrate the concepts. We use relatively simple examples to introduce and illustrate the basic concepts. More realistic and complex examples are developed in later sections. Our first example is a stochastic model for the time evolution of the size of a population, the situation introduced in Section 2.6.

EXAMPLE 4.1   In this situation we have a homogeneous population, and we suppose that the size is noted at equally spaced times and that the rate of population change between observations is a random variable. We begin each run with a population of size 100, and we simulate the population growth (which can be an increase, no change, or a decline) through 20 units of time. We write the rate of change from one observation to the next in the form $1 + r$, where r is a random variable with values -0.01, 0, 0.01, or 0.02 and probabilities given in Table 4.1.

Table 4.1

Value of r	Probability
-0.01	.3
0	.1
0.01	.4
0.02	.2

In the first part of the simulation process, one must determine what variables are needed and how they should be defined. We are interested in the size of the population after 20 units of time, and for convenience, we denote this by *finalpop* (for "final population size"). For each run of the simulation, we obtain finalpop by sampling successively 20 times—one for each observation—from the values of r, using the probabilities of Table 4.1, and tracking the time evolution of the population size from its initial value of 100 to its final value, finalpop. At each step we denote the population size by *popsize*. We begin with popsize equal to 100.

The main simulation consists of 20 steps. In each step we begin with a population size (popsize) and end with a new population size obtained by multiplying popsize by $1 + r$, where the value of r is one of the values in Table 4.1 and the probability of selecting r is also given in Table 4.1. For example, if the size of the population is 104 at the start of a step and if the random variable has value 0.01 at that step, then the size of the population at the end of the step is $(1.01) \times (104) = 105.4$. At the end of this step we have a new value of the population size, a value that we again call popsize. After 20 repetitions, the value of popsize is used as the value of finalpop for this simulation run, and a single repetition of the simulation is complete. The logic for the main simulation for this situation is shown in Figure 4.1. A complete simulation consists of a specified number of runs, and for each run we record the value of the final population size. These data are usually presented in the form of a table or graph. For instance, if the number of runs is 100,000 and the data intervals are to be of length 5, then a specific simulation has the outcome shown in Table 4.2 and the graph shown in Figure 4.2. These are the results reported in Section 2.6.   ■

For our next example, we consider a stochastic model for a basketball player in a situation similar to that described in Section 2.6.

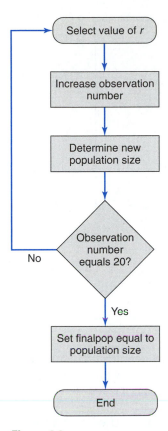

Figure 4.1

Table 4.2

Interval	Number of Values of finalpop	Probability
[90, 95)	81	.00081
[95, 100)	2949	.02949
[100, 105)	14234	.14234
[105, 110)	28622	.28622
[110, 115)	35621	.35621
[115, 120)	14218	.14218
[120, 125)	3775	.03775
[125, 130)	477	.00477
[130, 135)	23	.00023

EXAMPLE 4.2   Suppose that a basketball player is an inconsistent free throw shooter, and the likelihood of his making a free throw depends on his results earlier in the game. In particular, suppose that he is equally likely to make or to miss his first free throw. At any time during a game, if he makes a free throw, then he is equally likely to make or to

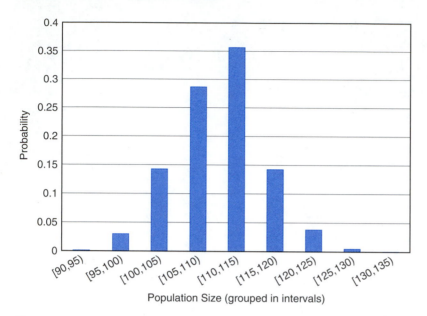

Figure 4.2

miss the next free throw. On the other hand, if he misses a free throw, then he increases his concentration, and he makes the next with probability .7 and misses it with a probability .3. If the player usually has about 10 free throws in each game, what is the average number of points per game that he scores by making free throws?

One run of the simulation will represent a typical game—that is, 10 free throw attempts. The data collection part of this simulation is straightforward. In each run of the simulation, we are interested in the total number of free throws made in the game. To determine the total number, we note the result of each attempt and then increment the number of successes each time a free throw is made.

The logic of the simulation is shown in the diagram in Figure 4.3. Note that on the first free throw and after every successful attempt, the probability of a success is .5, whereas after each unsuccessful attempt, the probability of success is .7. After each attempt we record the result—that is, update the number of free throws made if the attempt is a success—and we update the number of attempts. After 10 attempts the simulation run is complete.

For a simulation consisting of multiple runs, a single run is embedded as part of a more complex program consisting of several runs, together with the associated initialization and data collection activities. It is important to reset all counters before each run and to be certain that the data are collected in appropriate ways. In the current example, the points scored must be reset to zero before each run, and to determine the average number of points scored in a simulation consisting of N runs, we add the points for each of the runs and divide by N.

Data for simulations consisting of 100, 1000, and 10,000 runs are shown in Table 4.3 and are presented in graphical form in Figure 4.4 on page 160. Note that in the column corresponding to 100 runs in Table 4.3, there are no runs in which fewer than 3 points are scored. In contrast, in the column corresponding to 10,000 runs, there are 41 runs in which 2 points are scored and 1 run in which 1 point is scored. In the figure, the vertical axis gives the fraction of

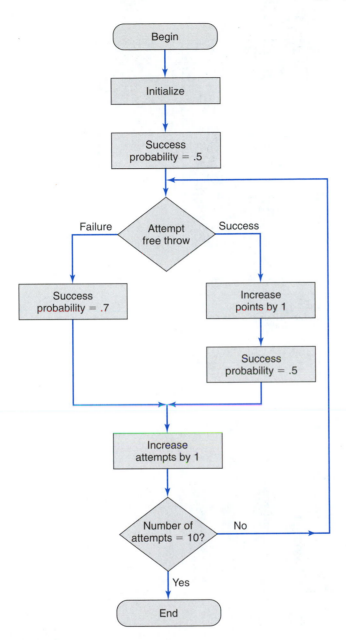

Figure 4.3

the total number of runs yielding a specified number of points. In general, the information provided by an appropriately designed simulation is expected to become more reliable as the number of runs increases. For instance, in the present situation, the estimate of the probability that the player scores 6 points provided by a simulation with 10,000 runs is more accurate (in a sense made precise in Section 4.5) than an estimate based on a simulation with 100 runs.   ∎

Table 4.3

Points	Number of Runs		
	100	1000	10,000
0	0	0	0
1	0	0	1
2	0	8	41
3	6	29	360
4	14	132	1168
5	23	249	2580
6	36	298	3077
7	15	186	1972
8	5	81	667
9	0	17	125
10	1	0	9
Mean	5.600	5.767	5.755
Standard deviation	1.305	1.332	1.283

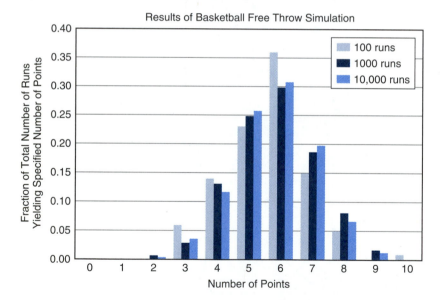

Figure 4.4

For the next example, we return to the token collection situation introduced in Section 2.6 and restated earlier in this section.

EXAMPLE 4.3 The flow of activities in the actual situation is as follows: A customer purchases a bag of cat food, examines the token inside, and notes whether it is a C, an A, a T, or an S. We are interested in the total number of bags of cat food purchased to obtain all four letters, and the record of the total number of bags purchased must be increased by one. If the customer already has the token in the bag, then the set of tokens that the customer has collected does not change. If the token in the bag is not one the customer already has collected, then the customer notes the new token acquired. If the customer now has all four tokens, then the process ends. If not, then the customer purchases another bag of cat food.

From this description, it is clear that to respond to the question asked, it is necessary to keep track of two pieces of information after each purchase: the total number of bags of cat food purchased and which tokens the customer has collected. Actually, what matters is whether or not the customer has all tokens after each purchase. In terms of the three parts of the process described above, the setup and initialization phase involve defining variables that enable us to keep track of the total number of bags purchased (or, equivalently, the total number of tokens selected) and whether or not all four tokens have been selected. We can handle the latter task by introducing four variables each of which has the value 1 or 0 according to whether the corresponding token has or has not been collected. The initialization step is to set each of the variables and the total number of tokens selected equal to zero.

The main simulation must mimic the purchase activity. As in the example of Section 2.6, we assume that the probabilities that the specific tokens are found in randomly selected bags of cat food are as shown in Table 4.4. The purchase process can be simulated by sampling from the set $\{C, A, T, S\}$ successively with the selection probabilities given by Table 4.4. That is, we simulate the process by beginning with the first selection, noting which token has been selected, increasing the total number of tokens selected to the value 1, and changing the value of the appropriate token variable from 0 to 1. For the next purchase, the total number of tokens selected is increased by 1, and the value of the token variable is incremented from 0 to 1 if it has not been selected and remains 1 if it has been selected previously. To check whether all tokens have been selected, we ask whether all token variables have the value 1, or, equivalently, whether the sum of the values of all token variables is 4. As soon as all token variables have the value 1, the simulation ends. The logic of the main simulation for this situation is depicted in flowchart form in Figure 4.5.

For a single run of the simulation, the total number of tokens selected—call it N—is the relevant information to be collected. The number N must be at least 4, but there is no fixed upper bound on N. If we collect data for a series of runs, then for each run we have a number, and the data can be displayed in a table or in a graph. For a large number of simulations, each consisting of 1000 runs, the number of tokens selected was found to be no more than 75. Data for two simulations each consisting of 1000 runs are shown in Figure 4.6. The average number of tokens selected for the runs in simulation 1 is 14.01, and the average for simulation 2 is 14.36. ∎

Table 4.4

Token	Probability of Selection
C	.3
A	.1
T	.3
S	.3

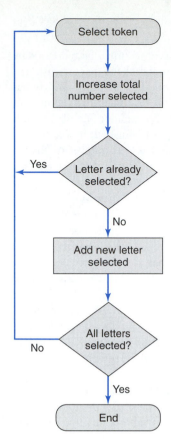

Figure 4.5

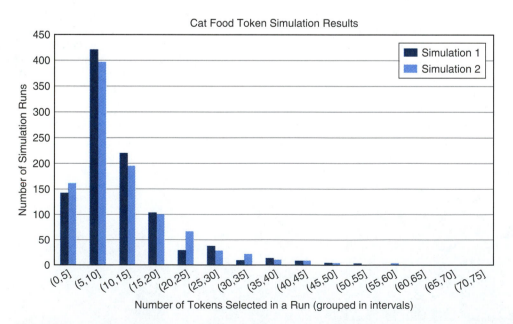

Figure 4.6

In these examples we have outlined in abbreviated form the three parts of the simulation process. There is detail in the main simulation step that has been embedded in a narrative description that must be expanded with greater detail for the simulation to be completely described. This detail includes using a random number generator to select values of the relevant random variable that are consistent with the probabilities given as part of the setting of the situation. This step is crucial to the accuracy of the simulation, and it is the topic of the next section.

Exercises 4.1

1. Given an unfair coin with $\Pr[H] = .3$, describe how you would conduct a simulation to estimate the number of flips before you first have three heads.

2. Given an unfair coin with $\Pr[H] = .3$, describe how you would conduct a simulation to estimate the number of flips before you first have both heads and tails.

3. In the setting of Exercise 2, use an analytical argument to determine the expected number of flips before both a head and a tail occur.

4. Given an unfair coin with $\Pr[H] = .3$, describe how you would conduct a simulation to estimate the number of flips before you first have three consecutive heads.

5. In the situation of Example 2.21, suppose that if Hack makes his first free throw, then the probability of his making any subsequent free throw is .5, and that if he does not make the first free throw, then the probability of his making any subsequent free throw is .3. In each game the probability of his making the first free throw is .4. The task is to estimate the average number of points he will score from free throws in a game where he shoots 25 free throws. Describe how you would conduct a simulation for this situation.

6. In the setting of Exercise 5, use a tree diagram and a probability argument to determine the expected number of points he will score on free throws.

7. In the situation of Example 4.2 of this section, suppose that a basketball player's free throw shooting can be described as follows:

    - If he makes a free throw, then the probability of his making the next free throw is .5.
    - If he makes a free throw and then misses one, or if he misses the first free throw, then the probability of his making the next free throw is .6.
    - If he misses two consecutive free throws, then the probability of his making the next free throw is .8.
    - In each game the probability of his making the first free throw is .4.

    The task is to estimate the average number of points he will score from free throws in a game where he shoots 20 free throws. Describe how you would conduct a simulation for this situation.

8. In the situation described in Example 4.3, suppose that if the customer does not have all four letters after buying ten bags, she becomes discouraged and stops purchasing cat food. Describe how you would conduct a simulation for this situation.

9. In the situation described in Example 4.3, suppose the customer becomes discouraged and stops purchasing cat food once five tokens with the same letter have been collected. Describe how you would conduct a simulation for this situation.

10. A marketing strategist plans to use tokens as described in Example 4.3 to stimulate sales. Suppose that the goal is to assign probabilities p, p, p, and q to the letters C, T, S, and A, respectively, $3p + q = 1$, so that the expected number of bags of cat food required to obtain all four tokens is no smaller than 15. Describe a simulation to estimate the largest value of q for which this goal is achieved.

11. Describe how you would conduct a simulation of a basketball player shooting 3-point shots in a game. Assume that the player is equally likely to shoot 2, 3, 4, or 5 such shots per game and that the probability of making the shot depends on how many such shots have already been attempted. For the first two shots the probability of success is .25, for the third .3, for the fourth .35, and for the fifth .40. The simulation is to find the average number of points scored per game on 3-point shots and the standard deviation.

12. Use analytical methods to find the expected number of points scored on 3-point shots per game in the setting of Exercise 11.

13. Describe how you would simulate the following situation. A player shoots a certain number of free throws, 2-point shots, and 3-point shots each game. The associated probabilities for the number of shots taken are as follows:

number of free throws:	0, 1, 2, 3, or 4	all equally likely
number of 2-point shots:	0, 1, 2, 3, ..., 8	all equally likely
number of 3-point shots:	0, 1, 2, 3	all equally likely

The probabilities of making these kinds of shots are as follows:

free-throws:	.8 per shot
2-point shot:	.4 per shot
3-point shot:	.3 per shot

The goal of the simulation is to find both the average number of points per game and the standard deviation of the number of points per game for this player.

14. Use analytical methods to find the average number of points per game for the player in the setting of Exercise 13.

4.2 Generating Values of Discrete Random Variables

At some point in the simulations described earlier in this chapter, and also in the examples of Chapter 2, it is necessary to generate values of a random variable with a known probability distribution. This step is therefore an essential one in the simulation process, and it is the topic of this section and of Section 4.4. Our approach is to assume that we have a method of generating values of a random variable that is uniformly distributed in the interval $[0, 1)$, and we use these values to generate the values of more general random variables. We begin with a brief discussion of the task of generating values of a random variable uniformly distributed on $[0, 1)$, and then we turn to the general situation.

It is helpful to begin with an example. Consider an experiment that consists of selecting at random a number from the set $\{0, 1, 2, \ldots, 999\}$. There are 1000 numbers in the set, and the phrase *selected at random* means that the probability distribution for the experiment assigns probability $\frac{1}{1000}$ to each of the numbers in the set. If the experiment is repeated a number of times, then the resulting collection of numbers will yield a set of one-digit, two-digit, or three-digit random numbers in the interval $[0, 999)$. These integers can be converted to numbers in the interval $[0, 1)$ by multiplying each by .001. For instance, multiplying by .001 converts the integer 317 to .317. This example can be extended in an obvious way to generate a set of k-digit random numbers in the interval $[0, 1)$.

Historically, the production of a large table of random numbers has not been an easy task. Early efforts in simulation used numbers generated by physical processes that were viewed as random. For instance, flipping coins, rolling dice, spinning roulette wheels, and sampling electronic noise were used to generate tables of numbers that were accepted as random. A particularly noteworthy example, because of its widespread use in the 1950s and 1960s, is the RAND table of one million random digits generated through the use of electronic noise. With the availability of computers, it was natural to turn to computation to generate numbers that would serve the purpose. It is common to use the term *pseudo-random* for numbers generated in this way. More precisely, a sequence of pseudo-random numbers $\{u_i\}$ is a sequence of numbers in $[0, 1)$ that is generated using a deterministic process and has the relevant statistical properties. Although the topic of "relevant statistical properties" is a complex one, it is enough for us to note that first, the distribution of numbers in the sequence must be uniform. That is, in a set containing k of the numbers u_i, $\{u_i\}_{i=1}^{k}$, the fraction of the numbers in any interval $[a, b)$, $0 \le a < b \le 1$ approaches the length of the interval $b - a$ as k becomes large. For example, one tenth of the numbers should be in the interval $[.2, .3)$. Second, the entries must be independent. The independence question can be studied by considering the distribution of points (u_j, u_{j+1}) in the plane or by considering a similar question in m-space for small values of m (say, $m \le 6$). For example, 25% of the points (u_i, u_{i+1}) should be in each of the four quadrants of the unit square in the plane.

A common way to generate a sequence of pseudo-random numbers is to begin with an integer x_0, called the seed, and compute x_i, $i \ge 1$, recursively by setting

$$x_i = ax_{i-1} \ (\mathrm{mod} \ m), \quad i = 1, 2, \ldots,$$

where a and m are given positive integers. The expression "mod m" means that the left-hand side, x_i, is defined as the result of dividing the right-hand side, ax_{i-1}, by m and taking the remainder. This definition means that each x_i is an integer between 0 and $m - 1$. Then define u_i as x_i/m. This approach to generating pseudo-random numbers is called the *multiplicative congruential method*. Because each of the numbers x_i is one of the integers $0, 1, \ldots, m - 1$, it follows that after $m + 1$ of the x_i have been generated, there must be duplication, and after that point, the values in the sequence $\{u_i\}$ will repeat. Therefore, one of the goals must be to select values of a and m such that a large number of the values of x_i can be generated before repetition occurs. Although pseudo-random numbers generated by this method, and by related methods, are obtained deterministically, they have been used as reasonable approximations to random numbers, and from now on we will omit the prefix *pseudo-* when referring to numbers generated in this way and will simply refer to them as random numbers.

Most computer languages, including popular spreadsheets and mathematical software packages such as MAPLE and Mathematica, have built-in random number generators— generally quite sophisticated—that can be used to generate random numbers. We do not pursue how these numbers are generated or the statistical properties of the sequence generated. Rather, we take the output as random numbers distributed uniformly on [0, 1) and proceed.

■ Random Numbers Distributed as a Discrete Random Variable

We continue the discussion by using random numbers uniformly distributed on [0,1) to generate random numbers distributed according to a given discrete random variable. To that end, suppose X is a discrete random variable whose values are $x_1, x_2, x_3, \ldots, x_n$, where $x_1 < x_2 < x_3 < \cdots < x_n$, and the associated probabilities are $p_1, p_2, p_3, \ldots, p_n$. That is, $\Pr[X = x_i] = p_i$, for $i = 1, 2, \ldots, n$, and $\sum_{i=1}^{n} p_i = 1$. The density function for X is the function that assigns the probability p_i to the value $x_i, i = 1, 2, \ldots, n$. The distribution function for X (sometimes called the cumulative distribution function) is a function of a real variable t, whose value at t is denoted by $F_X(t)$, defined for all t by $F_X(t) = \Pr[X \le t]$. Because the value of the distribution function at a specific t is the probability that X takes a value in the interval $(-\infty, t]$, it follows that $F_X(t)$ does not decrease as t increases.

EXAMPLE 4.4   Consider a random variable X that has five possible values and whose density function is given in Table 4.5. Find the graphs of the density and distribution functions of X.

The density function of X is defined by the Table 4.5, and its graph is shown in Figure 4.7(a). The distribution function F_X is defined as follows:

$$
F_X(t) = \begin{cases}
0, & t < -2 \\
.1, & -2 \le t < 2 \\
.6, & 2 \le t < 3 \\
.8, & 3 \le t < 5 \\
.9, & 5 \le t < 8 \\
1.0, & 8 \le t
\end{cases}
$$

and the graph of F_X is shown in Figure 4.7(b). Note that the size of the jump in the graph of F_X at the value $t = -2$ is .1, which is the probability that X takes the value -2. Similarly, the size of the jump of the graph of F_X at the value $t = 2$ is .5, and the probability that X takes the value 2 is .5; and so on. In each case, the size of the jump of F_X at the value of t equal to one of the x_i is the associated probability p_i.   ■

The task of determining a sequence of numbers $\{v_i\}$, which is distributed as a random variable X, can be viewed geometrically as follows: Suppose X has the distribution function

Table 4.5

Value of X	Probability
$x_1 = -2$	.1
$x_2 = 2$	.5
$x_3 = 3$	.2
$x_4 = 5$	.1
$x_5 = 8$	.1

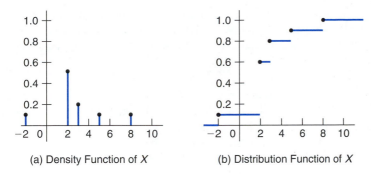

(a) Density Function of X (b) Distribution Function of X

Figure 4.7

shown in Figure 4.8, and let U denote a random variable that is uniformly distributed on $[0, 1)$. Select a value u of U and view this as a number on the vertical axis in Figure 4.8. If $u < p_1$, set $v = x_1$. If $u \geq p_1$, then, find the corresponding value x_i of X for which the index i satisfies

$$p_1 + p_2 + \cdots + p_{i-1} \leq u < p_1 + p_2 + \cdots + p_i$$

and set $v = x_i$. In Figure 4.8 the value of X associated with the u shown is x_3 because $p_1 + p_2$ is less than u and $p_1 + p_2 + p_3$ is larger than u. For any sequence $\{u_j\}$ uniformly distributed on $[0, 1)$, the sequence $\{v_i\}$ determined this way is distributed as the random variable X.

For $n = 5$ this process can be described explicitly in the following way:

Generate $\{u_j\}$ uniformly on $[0, 1)$, and define $\{v_j\}$ by

(4.1)

if $u_j < p_1$	then set $v_j = x_1$
if $p_1 \leq u_j < p_1 + p_2$	then set $v_j = x_2$
$\vdots$	
if $p_1 + p_2 + p_3 + p_4 \leq u_j < 1$	then set $v_j = x_5$

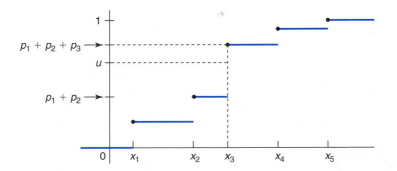

Figure 4.8

EXAMPLE 4.5   Consider a random variable X that records the outcome of flipping an unfair coin with probability of a head (H) equal to .6. Denote the outcome "head" by 1 and the outcome "tail" by 2. Then $p_1 = \Pr[H] = .6$ and $p_2 = .4$. This is a simple application of the relations (4.1) with two values of X. Suppose that values of a uniformly distributed random variable are 0.7191, 0.9057, 0.3424, 0.8488, and 0.9620. Then the associated values of X are 2, 2, 1, 2, and 2. These values correspond to the results of flips: tail, tail, head, tail, and tail.   ∎

EXAMPLE 4.6 Consider a random variable X with density function shown in the first two columns of Table 4.6 (See Example 4.1). Table 4.6 also includes, in additional columns, data that help illustrate the connection between the density function and the process described in the relations (4.1). To illustrate the process, suppose that we are given five values of a uniformly distributed random variable: 0.7215, 0.3994, 0.6278, 0.8222, and 0.5019. Beginning with the first value $u = 0.7215$, we note that the value is in the interval [0.4, 0.8). Using the process described in (4.1) and the information in Table 4.6, we see that the associated value of X is 0.1. The details are similar for the other values, and for the remaining values of U we have the associated values of the random variable X to be 0, 0.1, 0.2, and 0.1. Another set of five values of a uniformly distributed random variable is {0.0841, 0.6633, 0.8403, 0.5571, 0.1883}. For these values of U, the associated values of X are −0.01, 0.01, 0.02, 0.01, and −0.01.   ∎

Table 4.6

Value of X (x_i)	Probability ($\Pr[X = x_i]$)	Cumulative Probability ($\Pr[X \leq x_i]$)	u-Interval Where Associated Value of X is x_i
−0.01	.3	.3	[0, 0.3)
0	.1	.4	[0.3, 0.4)
0.01	.4	.8	[0.4, 0.8)
0.02	.2	1.0	[0.8, 1.0)

EXAMPLE 4.7 Each October, Rolf's Pretty Good Convenience Store sells Halloween party kits. Rolf must order these kits in May without knowing what the demand will be in October. However, judging on the basis of sales from earlier years and sales for similar products, Rolf estimates the possible demands and associated probabilities, and these estimates are shown in Table 4.7. The party kits can be ordered only in blocks of 25, and each year, on November 1, the kits are essentially worthless and are discarded. The kits cost Rolf $10 each, and he sells them for $20 each. Thus he makes a profit of $10 on each one he sells, and he has a loss of $10 on each kit he does not sell during October. Rolf must

Table 4.7

Demand	Probability
100	.20
125	.20
150	.30
175	.20
200	.05
225	.05

decide how many kits to order. On the basis of the data shown in Table 4.7, he assumes that he can always sell 100 kits, so that is a safe order. If he orders 100 kits, then his profit is $1000 ($10 × 100). However, Rolf also knows that he will make more money if he orders more than 100 kits and sells them. For example, if he orders 200 kits and sells them all, then he makes $2000 ($10 × 200). Of course, if he orders 200 kits and sells only 100, then he makes no money at all [($10 × 100) − ($10 × 100)]. Rolf wants to pick an order size that maximizes his expected (average) profit.

Remark This problem is simple enough to be solved without using a simulation (Exercise 10). However, it is useful to simulate this simple setting as a first step in solving similar, more realistic problems that are difficult to solve without simulation. We consider such a modified example in Section 4.5.

■ A Simulation

The first step of the simulation is to pick an order amount for Rolf from the set of possible orders {100, 125, 150, 175, 200, 225}. To illustrate the method, let's assume an order amount of 150. Next we generate a demand by using the discrete random variable given by Table 4.7, together with a random number generator that gives values u uniformly distributed on $[0, 1)$. Suppose we have $u = 0.35$, which gives a demand of 125. Then we know that Rolf ordered 150 but sold only 125, so his profit for this one value of u is $(125 \times \$10) - (25 \times \$10) = \$1000$.

In the same way as the special case $u = 0.35$, we can generate thousands of other values of u, and for each one we compute the profit to Rolf, given that Rolf ordered 150 kits. Finally, for all these values we can compute the average profit and the standard deviation. Using 2000 values of u, we obtain

$$\text{Average profit} = 1198.5$$

$$\text{Standard deviation} = 402.7$$

At this point in our simulation we have considered only the order value of 150, so we need to repeat the simulation for each of the other possible values: 100, 125, 175, 200, and 225. The results of these simulations are shown in Table 4.8, and they show that Rolf should order 150 items if his goal is to maximize his expected profit.

It is interesting to note the large values of the standard deviations in Table 4.8 relative to the averages. These values are a consequence of the high level of uncertainty in the demand

Table 4.8

Number of Kits Ordered	Average Profit	Standard Deviation
100	1000.0	0
125	1156.5	195.0
150	1198.5	402.7
175	1087.0	551.2
200	893.0	618.0
225	677.0	669.0

for Halloween kits (Table 4.7), and they are a measure of the risk to Rolf in ordering various numbers of kits. This risk is inherent in the situation, and it may affect the decision Rolf makes when he orders kits. If he is risk adverse, then he may choose to order only 125 kits. This lowers his expected profit from 1198.5 to 1156.5, but it also lowers the standard deviation (and the associated risk) from 402.7 to 195. In percentage terms, the reduction in expected profit is about 3.5%, and the reduction in standard deviation is about 51.5%. ∎

Exercises 4.2

1. Suppose that X is the random variable of Example 4.4, and suppose that the numbers 0.8095, 0.1017, 0.4789, 0.2448 and 0.9447 are values of U, a random variable uniformly distributed on $[0, 1)$. Use these values to find five numbers distributed as X. Use a random number generator, and find 20 additional numbers distributed as X.

2. Suppose that X is the random variable of Example 4.5, and suppose that the numbers 0.5375, 0.0812, 0.4529, 0.3524 and 0.9823 are values of U, a random variable uniformly distributed on $[0, 1)$. Use these values to find five numbers distributed as X. Use a random number generator, and find 20 additional numbers distributed as X and the associated results of the coin flip.

3. A basketball player makes each of her free throw shots with probability .4. Assume that she attempts three shots and that the results of all shots are independent. Let X be a random variable whose values are the number of successful shots.

   (a) Generate ten values of U, a random variable uniformly distributed on $[0, 1)$, and use them to generate ten numbers distributed as X. Find the mean of these numbers.
   (b) Generate 100 values distributed as X. Find the mean and standard deviation.

4. There are two red balls and five green balls in a box. Three balls are selected simultaneously and at random. A random variable X is defined as the number of red balls selected.

   (a) Find the density and distribution functions of X.
   (b) Given a random variable U uniformly distributed on $[0, 1)$, describe how to construct a sequence $\{v_i\}$ distributed as X.
   (c) Simulate 100 draws of three balls from the box, and find the average number of red balls drawn.

5. In a box containing ten ZIP disks, six are full of data and four are not. A sample of three disks is selected at random, and a random variable X is defined as the number of disks that are not full.

   (a) Find the density and distribution functions of X.
   (b) Given a random variable U uniformly distributed on $[0, 1)$, describe how to construct a sequence $\{v_i\}$ distributed as X.
   (c) Use a random number generator to generate 100 values of U. Find the corresponding values of X and the mean of these values.

6. You have two coins; one is fair and the other is unfair with $\Pr[H] = .6$. A coin is selected at random and flipped twice. A random variable X is defined as the number of heads.

Table 4.9

Demand	Probability
100	.1
125	.1
150	.2
175	.3
200	.2
225	.1

   (a) Find the density and distribution functions of X.
   (b) Simulate this experiment 1000 times, and find the mean and standard deviation of the resulting values.

7. There are three red balls and two blue balls in box 1, and there are four red balls and one blue ball in box 2. A box is selected at random and two balls are drawn. A random variable is defined as the number of blue balls selected.

   (a) Find the density and distribution functions of X.
   (b) Simulate this experiment 1000 times, and find the mean and standard deviation of the resulting values.

8. There are ten balls labeled 1 through 10 in a box. A ball with an even number is worth $1, and a ball with an odd number is worth $2. Three balls are selected simultaneously and at random. A random variable X is defined as the total value (in dollars) of the three balls selected.

   (a) Find the density and distribution functions of X.
   (b) Given a random variable U uniformly distributed on $[0, 1)$, describe how to construct a sequence $\{v_i\}$ distributed as X.
   (c) Simulate the experiment 1000 times, and find the mean and standard deviation of the resulting values.

9. Repeat the simulation of Example 4.7 using the data in Table 4.9 (in place of Table 4.7). Use the results from your simulation to prepare a table similar to Table 4.8. How many kits should Rolf order in this situation?

10. Use analytical methods to determine the expected profit for each of the possible order sizes in Example 4.7. Compare those values with those obtained from the simulations.

4.3   Discrete Event Simulation

In this section we illustrate the concepts introduced in Sections 4.1 and 4.2 with several examples. We consider relatively simple situations to constrain the amount of technical detail, and because of the simplicity, some of the examples can be studied using analytical methods. In such cases, comparing the results of simulation models with the analytical results illustrates the variation between predictions based on the two models. The last part of the section discusses the use of simulation techniques in the study of Markov chain models.

Our first example is one wherein the time period over which a situation is being modeled is specified in advance and does not depend on the behavior of the system. In contrast to this situation, we later consider examples wherein the period over which a situation is modeled is not specified in advance and does depend on the behavior of the system.

EXAMPLE 4.8   A manager of a small business notes that sales volume appears to vary with the weather. Suppose that sunny, cloudy, and rainy days occur at random and that the probability of a sunny day is .5, the probability of a cloudy day is .3, and the probability of a rainy day is .2. Also, suppose that good and bad sales days occur at random but with frequencies as follows:

- On a sunny day, the probability of good sales is .5.
- On a cloudy day, the probability of good sales is .8.
- On a rainy day, the probability of good sales is .3.

Use a simulation model to provide an estimate of the expected number of days with good sales in a month of 30 days.

The goal of the simulation is to determine the number of days on which good sales will occur when the weather and the sales behave randomly with the given probabilities. We let the number of days of good sales in a month be a random variable X—an integer random variable that takes values between 0 and 30—and we seek the expected value of X.

The core step in the simulation is what happens on a single day. Because the likelihood of a good sales day depends on the weather, the first step is to determine the weather. Once the weather is determined, we determine whether the sales are good or bad. The data on sales are recorded, and we proceed to the next day.

These steps are carried out using the methods of Section 4.2. For instance, to determine the weather, we can introduce a random variable W with values 1, 2, and 3, where 1 corresponds to sunny weather, 2 to cloudy weather, and 3 to rainy weather. The probability density function for W is given by the data of the problem, as shown in Table 4.10. Once values distributed as W have been generated, the next step is to determine whether sales are good or bad. We introduce three random variables, one for each type of weather. Each of these random variables takes values 1 and 2, where 1 corresponds to good sales and 2 corresponds to bad sales. Again we use the techniques of Section 4.2. For instance, if X_C is a random variable whose values describe sales on a cloudy day, then $\Pr[X_C = 1] = .8$ and $\Pr[X_C = 2] = .2$.

The logic for a simulation model for one month is shown in the portion of Figure 4.9 that starts with the *Begin month* symbol and ends with the *End month* symbol.

One run of this segment of the simulation gives one value of X, the number of days with good sales in that month, and to estimate the expected value of X, we need many runs. The expanded flowchart for a complete simulation consists of taking the flowchart for one

Table 4.10

Value of W	Probability
1	.5
2	.3
3	.2

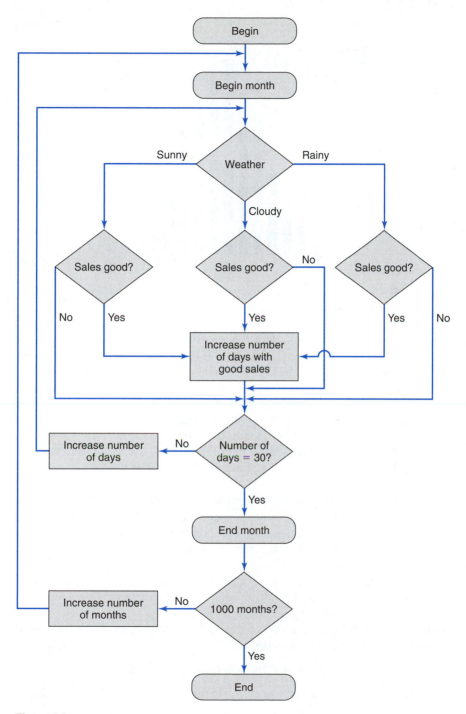

Figure 4.9

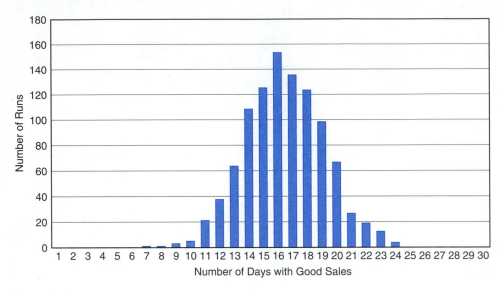

Figure 4.10

month and embedding it in a larger flowchart that repeats the activities of one month many times. Figure 4.9 shows this expansion for 1000 repetitions of one month each. The details of the initialization and the nature of the data to be accumulated vary from one situation to another. To estimate the expected value of the number of days with good sales, we record the value of X for each of 1000 months and take the average of this set of numbers. Using these data, we can also investigate other properties of the distribution of X, such as the standard deviation of the number of good days per month. One simulation consisting of 1000 runs of one month each resulted in the data graphed in Figure 4.10. Using the data from this simulation, we find that the average number of days with good sales is 16.48. ■

As we noted earlier, in some cases our examples are simple enough that information about the situation can be obtained by analytical methods. Example 4.8 is such a case. In the analytical approach to this problem, we first compute the probability that any given day in the month is a good sales day, and then, using that probability, we compute the average number of good sales days per month. The key here is that we assume that the sales on any given day (good or bad) are independent of the sales on any other day. Then, in a month of 30 days, we have 30 Bernoulli trials with the same probability p of good sales on each day. Thus the average number of days with good sales is $E[X] = 30p$.

To compute p, we use the information given about the weather and the probability of good sales for each type of weather. We obtain

$$p = (.5)(.5) + (.3)(.8) + (.2)(.3) = .55$$

Thus $E[X] = 30(.55) = 16.5$, and the standard deviation of X is $\sigma[X] = \sqrt{30(.55)(.45)} = 2.7249$.

Recall that the mean or expected value of a random variable X is in general a centrally located number relative to the probability density function for X. The variance (and standard deviation) of X is a measure of dispersion of the values of X. If X assumes values far from the mean with relatively large probabilities, then the variance will be large, whereas if X assumes values near the mean with relatively large probabilities, then the variance will be small.

The next example is of a game whose duration cannot be predicted prior to the play but, rather, depends on the outcomes of events that take place during the play of the game.

EXAMPLE 4.9   Jessica and Joe play a game with the following rules: A die is rolled, and if it comes up 1 or 2, then Jessica pays Joe $1, and if it comes up 3, 4, 5, or 6, then Joe pays Jessica $1. When play begins, Joe has $5 and Jessica has $2. The game ends when one of the players has all the money, $7 in this case. We use a simulation model to estimate the expected number of times the die is rolled before the game ends and to estimate the probability that Jessica wins the game.

The logic of one play of the game is shown in Figure 4.11. For each play of the game, we record the number of times the die was rolled and whether Jess won. A simulation consisting of 10,000 repetitions of the game yields the data summarized in Table 4.11. The data give 9.70 as an estimate of the expected number of rolls of the die in each play of the game. To estimate the probability that Jess wins, we use the ratio of the number of times

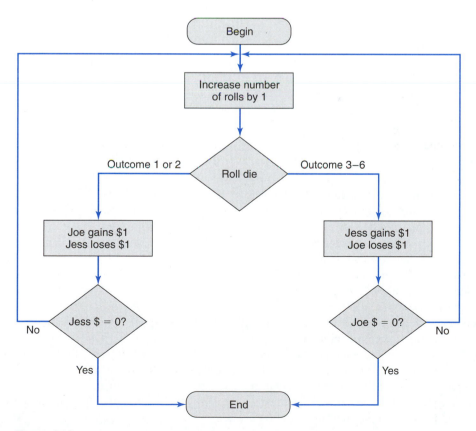

Figure 4.11

Table 4.11

Interval	Number of Runs with Game Length in the Interval	Number of Runs with Game Length in the Interval: Jess Wins
[0, 5)	1631	0
[5, 10)	4390	3955
[10, 15)	1899	1645
[15, 20)	1269	1192
[20, 25)	391	347
[25, 30)	258	244
[30, 35)	77	71
[35, 40)	55	53
[40, 45)	15	12
[45, 50)	10	8
[50, 55)	4	3
[55, 60)	1	1
[60, 65)	0	0
[65, 70)	0	0
[70, 75)	0	0

Jess wins to the total number of plays of the game. On the basis of the results of these 10,000 runs of the simulation, an estimate for the probability that Jess wins is .75. The expected duration of the game and the probability that Jess wins can also be determined by formulating a Markov chain model for the situation (Exercise 4.13). ∎

For the next example, we return to a setting introduced in Section 2.6, a setting where the free-throw-shooting skill of a basketball player is to be studied.

EXAMPLE 4.10 A basketball player makes her first free throw with probability .2. After the first free throw, she makes a free throw with probability .6 if she made the preceding one, and she makes a free throw with probability .3 if she missed the preceding one. Use a simulation to estimate the expected number of points she makes from free throws in a game during which she shoots ten free throws.

The logic for one game is shown in Figure 4.12. The decision process represented by the decision box labeled *Make next shot?* is more complicated than the decision process in Examples 4.8 and 4.9. In this case the probability of making a shot depends on whether the preceding shot was made or missed. The result of the preceding shot must be retained in order to determine the probability that the present shot is made.

To estimate the expected number of points, we would repeat the simulation for one game many times, and the flowchart would be enlarged in a way similar to that described in Example 4.9. Data for 1000 repetitions of a simulated game are shown in Table 4.12. Using these data, we find that the expected number of free throw points scored by the player in a game is 3.98 and that the standard deviation is 1.95. ∎

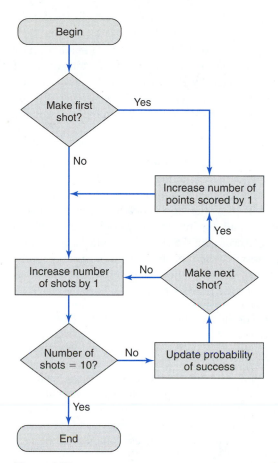

Figure 4.12

Table 4.12

Number of Points	Number of Runs with This Number of Points
0	28
1	84
2	124
3	168
4	201
5	188
6	104
7	66
8	25
9	10
10	2

■ Simulating Markov Chains

We next consider the simulation of Markov chains. Recall (see Chapter 3) that a Markov chain is determined by a set of states, a transition matrix, and an initial state vector. If the transition matrix for a Markov chain is denoted by $\mathbf{P}$ and the initial state vector by $\mathbf{x}_0$, then the relation between successive state vectors is given by the recurrence relation $\mathbf{x}_{n+1} = \mathbf{x}_n\mathbf{P}, n \geq 0$, and the state vector at the nth observation can be determined analytically by $\mathbf{x}_n = \mathbf{x}_0\mathbf{P^n}$.

As an alternative to determining the coordinates of the state vector at the nth observation using analytical techniques, we can estimate these coordinates using a simulation. This approach has the advantage of providing a technique that applies to stochastic processes that are more general than Markov chains—to situations in which there may be no simple analytical techniques. Also, the results of a simulation contain information about the distribution of the values being estimated, and this information is often very useful in studying a stochastic process. We begin by determining the state vector after n transitions for a specific initial state, say state i. To determine the probability that the system will be in state j on observation n, we begin in state i and track the system through n transitions using data from the transition matrix. We repeat the process K times and use the fraction

(number of times the system is in state j on observation n)$/K$

as an estimate of the n-step transition probability $p_{ij}(n)$, the probability that the system is in state j after n transitions given that it began in state i.

The process of tracking the system through n transitions consists of n steps, each step as follows: Suppose the system is in state m. Use the mth row of the transition matrix as a probability density function for a random variable whose values are the states occupied by the system at the next observation. For instance, if there are four states and the system is in state 1, then a random variable X whose values are the states occupied on the next observation has the probability density function shown in Table 4.13.

Given the initial state of the system, we determine the state after one transition by using the methods described in Section 4.2 with random variables whose density functions are the rows of the transition matrix. The process used on the first step is repeated for each step, a total of n steps if we are interested in the state vector on the nth observation. Note that the density function used to determine the next state depends on the current state. Each repetition of this set of n transitions constitutes a simulation run. Because we are dealing with a Markov chain, the transition probabilities depend only on the current state, and consequently the random variables have probability distributions that do not depend on the transition number. We provide more details in Example 4.11. The logic of a single run is shown in Figure 4.13.

Repeating the simulation many times provides data that can be used to estimate the n-step transition probabilities and the coordinates of the state vector at the nth observation. The details are illustrated in Examples 4.11 and 4.12.

Table 4.13

State	1	2	3	4
Probability	p_{11}	p_{12}	p_{13}	p_{14}

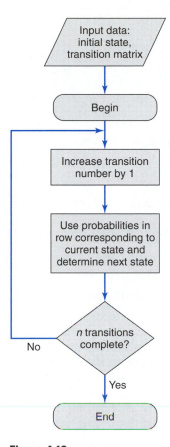

Figure 4.13

EXAMPLE 4.11   Consider a Markov chain with four states and transition matrix **P**. Use simulation to estimate the five-step transition matrix.

$$\mathbf{P} = \begin{bmatrix} .6 & .3 & 0 & .1 \\ .5 & 0 & .5 & 0 \\ 0 & .2 & .2 & .6 \\ .2 & 0 & .8 & 0 \end{bmatrix}$$

The five-step transition matrix is a matrix in which the four entries in the first row are the probabilities that the system is in states 1, 2, 3, and 4, respectively, given that it was intially in state 1 and there were five transitions. To estimate these probabilities, we follow the outline described just above, and we let X_1 denote a random variable with values 1, 2, 3, and 4 (the state labels) and with the probability density function given by the first row of the matrix **P** and shown in Table 4.14. We define X_2, X_3, and X_4 similarly.

Table 4.14

State	1	2	3	4
Probability	.6	.3	0	.1

Table 4.15

State	1	2	3	4
Fraction of 10,000 Times That the System Is in the State after five Transitions	.3489	.1727	.2681	.2103

A simulation using the flowchart of Figure 4.13 for a single run consisting of five transitions was carried out for 10,000 runs. On each of the 10,000 repetitions, the system is in a specific state after five transitions. We use the ratio

(number of times the system is in state 1 on observation 5)/10,000

as an estimate for the probability that the system makes a transition from state 1 to state 1 in five steps. Similarly, we use

(number of times the system is in state 2 on observation 5)/10,000,

(number of times the system is in state 3 on observation 5)/10,000, and

(number of times the system is in state 4 on observation 5)/10,000

as estimates for the probabilities that the system will be in states 2, 3, and 4, respectively, after five transitions, given that it began in state 1. For a specific simulation, these results are summarized in Table 4.15. The entries in Table 4.15 provide an estimate for the first row of the five-step transition matrix.

If we repeat this process beginning in state 2, then we have estimates for the entries in the second row of the five-step transition matrix; if we repeat the process beginning in state 3, then we have estimates for the entries in the third row of the five-step transition matrix; and if we repeat the process beginning in state 4, then we have estimates for the entries in the fourth row of the five-step transition matrix. Using the results of a specific simulation consisting of 10,000 runs, we have the matrix

$$
\begin{bmatrix}
.3489 & .1727 & .2681 & .2103 \\
.3365 & .1321 & .3831 & .1483 \\
.2460 & .1779 & .2679 & .3082 \\
.2929 & .1111 & .4356 & .1604
\end{bmatrix}
$$

as an estimate for the five-step transition matrix. ∎

In this example we used a simulation to obtain an estimate of the five-step transitions for a given transition matrix. It is interesting to compute the exact value of the five-step transition matrix using the methods developed in Chapter 3. We find that

$$
P^5 = \begin{bmatrix}
.3475 & .1762 & .2617 & .2145 \\
.3360 & .1318 & .3751 & .1570 \\
.2504 & .1755 & .2713 & .3029 \\
.3021 & .1132 & .4250 & .1598
\end{bmatrix}
$$

Table 4.16

State	1	2	3	4
Probability	.1	.6	.1	.2

Remark   We note that the approximation of $\mathbf{P}^5$ given in Example 4.11 has entries that are close to those of $\mathbf{P}^5$, but they are not the same, and they often differ by about 0.01 from the exact value. This shows that even after 10,000 runs, because of the stochastic nature of the Markov chain, the average value in each entry is not the theoretical value. In Section 4.5 we discuss the problem of deciding how many runs are needed to provide an estimate of a certain accuracy.

EXAMPLE 4.12   Suppose we have a Markov chain with transition matrix $\mathbf{P}$ of Example 4.11 and initial state vector $\mathbf{x}_0 = [.1 \quad .6 \quad .1 \quad .2]$. Use a simulation to estimate the state vector after five transitions.

   We can accomplish the task of this problem by using the flowchart of Figure 4.13 and adding a preliminary step. Indeed, we use the information about the initial state vector to define a random variable X_0 with the probability density function shown in Table 4.16.

   The first step of the simulation is to generate values (which will serve as initial states) distributed as X_0. To do so, we select a random number u uniformly distributed on $[0, 1)$. If $u < 0.1$, then the initial state is state 1; if $0.1 \leq u < 0.7$, then the initial state is state 2; if $0.7 \leq u < 0.8$, then the initial state is state 3; and if $0.8 \leq u$, then the initial state is state 4. After the initial state is selected, the simulation proceeds as in Example 4.11. Note that the probabilities must be updated each time a transition takes place. A simulation of 10,000 runs gives the vector $[.3259 \quad .1348 \quad .3582 \quad .1811]$ as an estimate for the state vector after five transitions.   ∎

EXAMPLE 4.13   Consider an absorbing Markov chain with the transition matrix $\mathbf{P}$:

$$\mathbf{P} = \begin{bmatrix} 1 & 0 & 0 & 0 \\ .5 & 0 & .5 & 0 \\ 0 & .2 & .2 & .6 \\ .2 & 0 & .8 & 0 \end{bmatrix}.$$

Use simulation to estimate the expected number of transitions before the system first reaches an absorbing state, given that the initial state is state 3.

   We again use a flowchart built on that shown in Figure 4.13. In this example, state 1 is the only absorbing state, and we modify the flowchart to show that the simulation stops once state 1 is reached and to show that the number of transitions when the system first reaches state 1 is recorded. A flowchart with these revisions is shown in Figure 4.14.

   A simulation consisting of 1,000,000 runs generated data summarized in Table 4.17 on page 183. The mean value of the data—that is, the expected number of transitions before the system first reaches state 1—is 8.1873.   ∎

   This example provides another instance where we can compare the results obtained by simulation with those obtained using analytical methods. Using results for absorbing

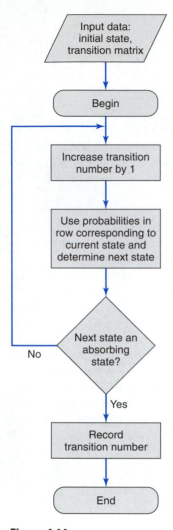

Figure 4.14

Markov chains developed in Chapter 3, we compute the fundamental matrix **N** for the matrix **P**.

$$\mathbf{N} = \begin{bmatrix} 1.4545 & 2.2727 & 1.3636 \\ 0.9091 & 4.4545 & 2.7273 \\ 0.7273 & 3.6364 & 3.1818 \end{bmatrix}$$

The entries in the second row of **N** are the expected numbers of visits to states 2, 3, and 4, respectively, given that the system is initially in state 3. Therefore, if the system is initially in state 3, then the expected number of transitions before it first reaches an absorbing state is $0.9091 + 4.4545 + 2.7273 = 8.1819$, a result that agrees well with the results of the simulation.

Table 4.17

Interval	Number of Runs with Steps to Absorption in This Interval
[0, 10)	694, 497
[10, 20)	231, 634
[20, 30)	55, 818
[30, 40)	13, 663
[40, 50)	3, 251
[50, 60)	846
[60, 70)	212
[70, 80)	64
[80, 90)	11
[90, 100)	3
[100, 110)	1

Exercises 4.3

1. A fair coin is flipped until there are three heads. Use simulation to estimate the expected number of flips.

2. A fair coin is flipped until there are three consecutive heads. Use simulation to estimate the expected number of flips.

3. An unfair coin with $\Pr[H] = .3$ is flipped repeatedly until both heads and tails land up. Use simulation to estimate the expected number of flips.

4. An unfair coin with $\Pr[H] = .6$ is flipped repeatedly until there are either two consecutive heads or two consecutive tails. Use simulation to estimate the expected number of flips.

5. A basketball player makes free throws with probability .4 except after making two consecutive free throw shots. In that case, the probability of her making the next shot is .6. Use simulation to estimate the expected number of points she will score from free throws if she attempts 15 free throws in a game.

6. Suppose you have an unfair coin with $\Pr[H] = .7$ and a box containing three red and five green balls. A trial of an experiment consists of flipping the coin and noting which side lands up. If it is heads, then one ball is selected from the box and its color is noted; if the coin lands with tails up, then two balls are selected from the box and their colors are noted. In each case, the balls are replaced in the box. The experiment is repeated 20 times. Use a simulation model to provide an estimate of the expected number of green balls selected.

7. Suppose you have an unfair coin with $\Pr[H] = .8$ and two boxes each containing colored balls. Box A contains two red balls and three green balls, and box B contains two red balls and five green balls. A trial of an experiment consists of flipping the coin and noting which side lands up. If it is heads, then one ball is selected from box A and its color is noted; if the coin lands with tails up, then two balls are selected from box B and their colors are noted. In each case, the balls are replaced in the same box. The

experiment is repeated 20 times. Use a simulation model to provide an estimate of the expected number of green balls selected.

8. Use the techniques of Chapter 3 to determine the state vector after ten transitions for a Markov chain with transition matrix $\mathbf{P}$ of Example 4.11 and the initial state vector $\mathbf{x}_0 = [.1 \quad .6 \quad .1 \quad .2]$ of Example 4.12.

9. Use the techniques of Chapter 3 to answer the questions in Example 4.9. That is, determine the expected number of rolls of the die before the game ends, and determine the probability that Jessica wins.

10. Use the techniques of Chapter 3 to determine the 25-step transition probabilities for the Markov chain with transition matrix as in Example 4.11. Compare these values to those obtained by using a simulation with 1000 runs.

11. Use simulation to estimate the ten-step transition probabilities for a Markov chain with the following transition matrix.

$$\begin{bmatrix} .6 & .3 & 0 & .1 \\ .5 & 0 & .5 & 0 \\ 0 & .2 & .2 & .6 \\ .2 & 0 & .8 & 0 \end{bmatrix}$$

12. Use simulation to estimate the long-range occupancy probabilities for a Markov chain with the following transition matrix.

$$\begin{bmatrix} 0 & .4 & .4 & .2 \\ .2 & .2 & 0 & .6 \\ .5 & .5 & 0 & 0 \\ .3 & .5 & .2 & 0 \end{bmatrix}$$

13. An absorbing Markov chain has the following transition matrix.

$$\begin{bmatrix} .4 & .3 & 0 & .1 & 0 & .2 \\ 0 & 1 & 0 & 0 & 0 & 0 \\ .3 & 0 & .5 & 0 & .2 & 0 \\ 0 & .2 & .2 & .3 & 0 & .3 \\ 0 & 0 & 0 & 0 & 1 & 0 \\ .2 & 0 & .1 & .1 & .6 & 0 \end{bmatrix}$$

(a) If the system begins in state 3, use simulation to estimate the number of transitions before it first reaches an absorbing state.
(b) Same question as in part (a) if the system begins in state 4.
(c) If the system begins in state 3, use simulation to estimate the probability that it will be absorbed in state 2.

14. Use the techniques of Chapter 3 to determine the ten-step transition matrix for the Markov chain in Exercise 11. What is the maximum difference between the entries obtained in the simulation and the theoretical values?

15. Use the techniques of Chapter 3 to compare the theoretical long-range occupancy probabilities of the matrix in Exercise 12 with those obtained by the simulation in that exercise.

16. Use the techniques of Chapter 3 to answer the questions posed in Exercise 13.

4.4 Generating Values of Continuous Random Variables

The task of generating random numbers whose distribution function is continuous arises in many simulations, and it is important to have techniques that apply in this situation. For example, the length of a telephone call or the length of time it takes for your car to be serviced at a quick-oil-change facility may be assumed to be a random variable whose values can be any number in an interval. The technique we introduced for discrete random variables—summarized in equation (4.1)—can be applied, at least theoretically, to the case of continuous distribution functions.

We begin by summarizing the notation used in this case. Let X be a random variable, and let f_X and F_X denote its density and distribution functions, respectively. Then, for numbers a and b,

$$\Pr[a \le X \le b] = \int_a^b f_X(s)\,ds$$

and

$$F_X(x) = \int_{-\infty}^x f_X(s)\,ds$$

We have already introduced the concept of a random variable U uniformly distributed on $[0, 1)$. We now make that concept more precise.

Definition 4.1 A random variable U *uniformly distributed* on $[0, 1)$ has density and distribution functions defined as

$$f_U(x) = \begin{cases} 1, & 0 \le x \le 1 \\ 0, & \text{otherwise} \end{cases} \quad \text{and} \quad F_U(x) = \begin{cases} 0, & -\infty < x < 0 \\ x, & 0 \le x < 1 \\ 1, & x \ge 1 \end{cases}$$

The rationale for this terminology will become clear in a moment when we describe how to generate a sequence of random numbers with a given continuous distribution.

To verify that the function F_U is the distribution function for the random variable U, we note that $f_U(x)$ is 0 for $x < 0$, and consequently,

$$F_U(x) = \int_{-\infty}^x f_U(s)\,ds = \int_{-\infty}^x 0\,ds = 0 \quad \text{for } -\infty < x < 0$$

Next, $f_U(x) = 1$ for $0 \le x < 1$, and consequently, for $0 \le x < 1$, we have

$$F_U(x) = \int_{-\infty}^x f_U(s)\,ds = \int_{-\infty}^0 0\,ds + \int_0^x 1\,ds = 0 + x = x$$

Finally, $f_U(x)$ is 0 for $x \ge 1$, and consequently,

$$F_U(x) = \int_{-\infty}^x f_U(s)\,ds = \int_{-\infty}^0 0\,ds + \int_0^1 1\,ds + \int_1^x 0\,ds = 1$$

and the verification is complete.

A useful method for generating a sequence of random numbers $\{x_i\}$ that is statistically indistinguishable from a sequence of values of a given random variable X is known as the *inverse transform* method. We describe and apply the method in the special case that the distribution function F_X has an inverse (thus the name of the method), but it can be extended to more general distribution functions. For a random variable X, the range of F_X includes the interval $(0, 1)$, and if the sequence $\{u_i\}$ is a sequence of values of the random variable U that is uniformly distributed on $[0, 1)$, then the sequence $\{x_i\}$ with

$$(4.2) \qquad\qquad x_i = F_X^{-1}(u_i), \quad i = 1, 2, \dots$$

has the desired property. The verification of this assertion is a straightforward probability argument and is omitted.

In this discussion, we frequently use the phrase "$\{x_i\}$ is distributed as X" as shorthand for the following more precise statement: "$\{x_i\}$ is a sequence of numbers that is statistically indistinguishable from a sequence of values of the random variable X."

Remark The extension of the method to situations where F_X does not have an inverse is straightforward, but it involves the concept of an infimum (inf) of a set of numbers. Suppose that F_X is a distribution function defined on $(-\infty, +\infty)$ and set $\tilde{F}_X(u) = \inf(w: F_X(w) \geq u)$. Then $\{x_i\}$ can be generated as above with $\tilde{F}_X(u)$ playing the same role as $F_X^{-1}(u)$.

The application of the technique given by (4.2) provides a rationale for the terminology we introduced earlier for a continuous uniform distribution. Indeed, if $\{u_i\}$ is a sequence of random numbers uniformly distributed on $[0, 1)$, and if $\{x_i\}$ is a sequence of numbers defined by equation (4.2) (with $X = U$ so that $F_X^{-1} = F_U^{-1}$), then $x_i = u_i$ for each i.

It is frequently useful to have values of a random variable distributed on an interval $[a, b)$ that have the same statistical properties as the values of U on $[0, 1)$. Such values are said to be uniformly distributed on $[a, b)$, and this can also be accomplished by using equation (4.2).

EXAMPLE 4.14 Values of a random variable X uniformly distributed on an interval $[a, b)$ can be generated using equation (4.2) and the density and distribution functions

$$f_X(x) = \begin{cases} \dfrac{1}{(b-a)}, & a \leq x < b \\ 0, & \text{otherwise} \end{cases} \quad \text{and} \quad F_X(x) = \begin{cases} 0, & -\infty < x < a \\ \dfrac{(x-a)}{(b-a)}, & a \leq x < b \\ 1 & x \geq b \end{cases}$$

Using the inverse of $F_X(x)$, we see that the sequence of numbers $\{x_i\}$, with $x_i = a + (b-a)u_i$ and $\{u_i\}$ uniformly distributed on $[0, 1)$, is uniformly distributed on $[a, b)$. ∎

EXAMPLE 4.15 Let X be a random variable with density function

$$f(x) = \begin{cases} 2x, & 0 \leq x < 1 \\ 0, & \text{otherwise} \end{cases}$$

Then

$$F_X(x) = \begin{cases} 0, & -\infty < x < 0 \\ x^2, & 0 \le x < 1 \\ 1, & x \ge 1 \end{cases}$$

Applying the method, using expression (4.2), and recognizing again that F_X has an inverse in this case, we conclude that if $\{u_i\}$ is uniformly distributed on $[0, 1)$, then the sequence $\{x_i\}$ with $x_i = \{u_i\}^{1/2}$ is distributed as X.   ■

An important class of random variables—random variables that arise frequently in applications and are useful as approximations of more general random variables—are those with density functions that are piecewise constant and positive on an interval. Let X be a random variable with density function

(4.3)
$$f_X(x) = \begin{cases} c_j, & x_{j-1} \le x < x_j, \quad j = 1, 2, \dots, n \\ 0, & \text{otherwise} \end{cases}$$

where $c_j > 0$, $a = x_0 < x_1 < \cdots < x_n = b$. For example, see Figure 4.15 where $n = 4$.

We develop a formula for the distribution function F_X in this situation. For notational simplicity, in the remainder of this discussion we use F rather than F_X to denote the distribution function of X. Set $p_0 = 0$ and

$$p_j = \int_{x_{j-1}}^{x_j} f_X(s)\, ds = c_j(x_j - x_{j-1}), \quad j = 1, 2, \dots, n$$

and $F_0 = 0$, $F_j = \sum_{k=1}^{j} p_k$, $j = 1, 2, \dots, n$. Then we have

(4.4)
$$F(x) = \sum_{k=0}^{j-1} p_k + \int_{x_{j-1}}^{x} c_j\, ds = F_{j-1} + c_j(x - x_{j-1})$$

where $j = \max\{k : x_{k-1} \le x\}$.

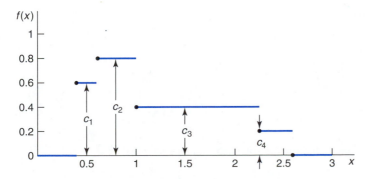

Figure 4.15

EXAMPLE 4.16 Let X be a random variable with density function as shown in Figure 4.15. In particular, suppose that

$$f_X(x) = \begin{cases} 0.6, & 0.4 \le x < .6 \\ 0.8, & 0.6 \le x < 1.0 \\ 0.4, & 1.0 \le x < 2.2 \\ 0.2, & 2.2 \le x < 2.6 \\ 0, & \text{otherwise} \end{cases}$$

Then

$$F_X(x) = \begin{cases} 0, & x < 0.4 \\ 0.6(x - 0.4), & 0.4 \le x < 0.6 \\ 0.12 + 0.8(x - 0.6), & 0.6 \le x < 1.0 \\ 0.44 + 0.4(x - 1.0), & 1.0 \le x < 2.2 \\ 0.92 + 0.2(x - 2.2), & 2.2 \le x < 2.6 \\ 1.0, & 2.6 \le x \end{cases}$$
■

■ Summary of the Method

Let X be a random variable with a piecewise constant density function as in (4.3). To generate a set of random numbers $\{x\}$ distributed according to X, set $F(x) = u$ and use expression (4.4). Solving (4.4) for x yields

$$x = x_{j-1} + (u - F_{j-1})/c_j, \quad F_{j-1} \le u < F_j$$

An algorithm that implements this method is as follows (see Figure 4.16):

1. Generate u, $0 \le u < 1$.
2. Determine j using the relation $\sum_{k=0}^{j-1} p_k \le u < \sum_{k=0}^{j} p_k$.
3. For the value of j determined in step 2, set $x = x_{j-1} + (u - \sum_{k=1}^{j-1} p_k)/c_j$.

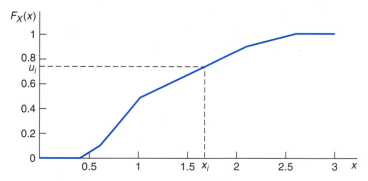

Figure 4.16

EXAMPLE 4.17 Let X be a random variable with density function

$$f_X(x) = \begin{cases} 0.50, & 0 \le x < 1 \\ 0.25, & 1 \le x < 3 \\ 0, & \text{otherwise} \end{cases}$$

Using the notation (and method) described above, we have $x_0 = 0$, $x_1 = 1$, and $x_2 = 3$. We also have $c_1 = 0.5$ and $c_2 = 0.25$, so $p_0 = 0$, $p_1 = \int_0^1 (c_1)\,ds = .5$, and $p_2 = \int_1^3 (c_2)\,ds = .5$. This gives $F_0 = 0$, $F_1 = 0 + .5$, and $F_2 = 1$, and, finally,

$$x = 0 + \frac{(u - 0)}{c_1} = 2u \quad \text{for } 0 \le u < 0.5$$

and

$$x = 1 + \frac{(u - .5)}{c_2} = 1 + 4(u - 0.5) \quad \text{for } 0.5 = u < 1$$

Then, if $\{u_i\}$ is uniformly distributed on $[0, 1)$, the set $\{x_i\}$ where

$$x_i = \begin{cases} 2u_i, & 0 \le u_i < 0.50 \\ 1 + 4(u_i - 0.50), & 0.50 \le u_i < 1 \end{cases}$$

is distributed according to X. ∎

Next, we illustrate the application of the method using the inverse of the distribution function in a situation where the density function is not piecewise constant. The general approach is similar to that used in Example 4.17, and the critical step is finding the inverse of the distribution function.

EXAMPLE 4.18 Let X be a random variable with density function

$$f_X(x) = \begin{cases} 0.2, & 1 \le x < 2 \\ 0.2 + 0.1(x - 2), & 2 \le x < 4 \\ 0.4 - 0.4(x - 4), & 4 \le x < 5 \\ 0, & \text{otherwise} \end{cases}$$

Then the distribution function of X is

$$F_X(x) = \begin{cases} 0, & x < 1 \\ 0.2(x - 1), & 1 \le x < 2 \\ 0.05x^2, & 2 \le x < 4 \\ 1 - 0.2(5 - x)^2, & 4 \le x < 5 \\ 1, & 5 \le x \end{cases}$$

The graph of the distribution function has the same general features as Figure 4.16. Specifically, it is a nondecreasing function defined for all x, and each of the subintervals $[1, 2)$, $[2, 4)$, and $[4, 5)$ has a corresponding interval on the probability axis on which it has an inverse that can be easily determined. That is, we solve the equation $u = F_X(x)$ for x in terms of u for u in each of the subintervals $0 \le u < 0.2$, $0.2 \le u < 0.8$, $0.8 \le u < 1$. The details follow.

For $0 \le u < 0.2$, we have $u = 0.2(x - 1)$, which yields $x = 5u + 1$.

For $0.2 \le u < 0.8$, we have $u = 0.05x^2$, which yields $x = 2\sqrt{5u}$.

For $0.8 \le u < 1$, we have $u = 1 - 0.2(5 - x)^2$, which yields $x = 5 - \sqrt{5(1 - u)}$.

Using this inverse function, we generate a sequence of values $\{x_i\}$ distributed as X by using a sequence $\{u_i\}$ uniformly distributed on $[0, 1)$ and setting

$$x_i = \begin{cases} 5u_i + 1, & 0 \le u_i < 0.2 \\ 2\sqrt{5u_i}, & 0.2 \le u_i < 0.8 \\ 5 - \sqrt{5(1 - u_i)}, & 0.8 \le u_i < 1 \end{cases}$$

∎

In many applications there is an empirical density (or distribution function), and the first step in the simulation process is to find an approximation to the empirical function for which this technique is applicable. A useful approach is to use piecewise linear functions as approximations, for in this case, finding the associated inverse function reduces to solving a quadratic equation.

EXAMPLE 4.19 Management of the Sundimes Coffee Shop has collected data on arrival times for early-morning customers, and the density function of arrival times—based on the data—is shown in Figure 4.17(a) for the period from 7:00 to 8:30 a.m. In Figure 4.17, times are given in minutes measured from 7:00 a.m. The goal is to formulate a simulation model for the number of customers to be served during this period. As a first step, the empirical arrival times are approximated by a random variable X with density function shown in Figure 4.17(b) and defined as follows:

$$f_X(x) = \begin{cases} 0, & x < 0 \\ (0.25/225)x, & 0 \le x < 15 \\ 0.25/15, & 15 \le x < 45 \\ (0.25/675)(90 - x), & 45 \le x < 90 \\ 0, & 90 \le x \end{cases}$$

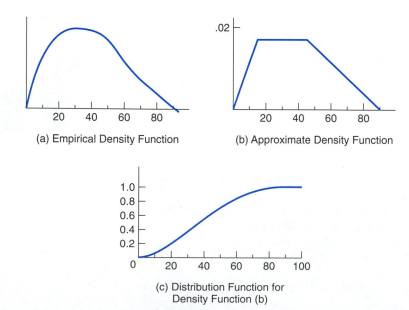

(a) Empirical Density Function

(b) Approximate Density Function

(c) Distribution Function for Density Function (b)

Figure 4.17

The distribution function of X is shown in Figure 4.17(c) and is defined by

$$(4.5) \qquad F_X(x) = \begin{cases} 0, & x < 0 \\ (0.25/450)x^2, & 0 \leq x < 15 \\ 0.125 + (.5/30)(x - 15), & 15 \leq x < 45 \\ 1 - (0.25/1350)(90 - x)^2, & 45 \leq x < 90 \\ 1, & 90 \leq x \end{cases}$$

In this situation, the distribution function has an inverse on each of the subintervals given in Equation (4.5). Each of these subintervals is associated with a subinterval of the vertical axis—the axis showing the cumulative probability. The endpoints of the intervals on the x-axis are 0, 15, 45, and 90, and these points correspond to the points 0, .125, .625, and 1 on the probability axis. The equation $u = F_X(x)$ can be solved for x in terms of u on each of the subintervals, and we have

$$(4.6) \qquad x = \begin{cases} \sqrt{(1800u)}, & 0 \leq u < .125 \\ 15 + 60(u - .125), & .125 \leq u < .625 \\ 90 - 30\sqrt{6(1 - u)}, & .625 \leq u < 1 \end{cases}$$

Consequently, given a sequence $\{u_i\}$ uniformly distributed on $[0, 1)$, then the sequence $\{x_i\}$ defined by Equation (4.6) is distributed as X. ∎

The method of using the inverse of the distribution function F_X is not always easy to implement, even with relatively simple density functions. For example, suppose that X is a random variable with density function

$$f(x) = \begin{cases} 0.3 + 0.6x + 1.2x^2, & 0 \leq x < 1 \\ 0, & \text{otherwise} \end{cases}$$

A direct application of the method described above involves solving the cubic equation $0.3x + 0.3x^2 + 0.4x^3 = u$ for x in terms of u. This is not straightforward, and an alternative approach is useful in such situations:

Suppose we have a random variable X whose density function f_X can be written as a finite sum of functions f_j, each of which has a special characteristic. In particular, suppose $f_X(x) = \sum_{j=1}^{n} f_j(x)$, where $f_j \geq 0$ and $\varphi_j(x) = \int_{-\infty}^{x} f_j(s)\, ds$, has an inverse that can be computed using standard techniques (such as algebraic operations, square roots, logarithms, ...), $j = 1, 2, \ldots, n$. Next, define $p_j = \int_{-\infty}^{\infty} f_j(s)\, ds$ for $j = 1, 2, \ldots, n$, $F_0 = 0$, and $F_k = \sum_{j=1}^{k} p_j$, $k = 1, 2, \ldots, n$. Then $0 = F_0 < F_1 < F_2 < F_3 < \cdots < F_n = 1$, and we can generate a set of random numbers distributed according to X as follows:

Let $\{u\}$ be uniformly distributed on $[0, 1)$. Then the set of numbers $\{x\}$, where $x = \varphi_j^{-1}(u - F_{j-1})$ for $F_{j-1} \leq u < F_j$, is distributed according to X.

EXAMPLE 4.20 Let X be a random variable with density function

$$f_X(x) = \begin{cases} 0.3 + 0.6x + 1.2x^2, & 0 \leq x < 1 \\ 0, & \text{otherwise} \end{cases}$$

and set $f_1(x) = 0.3$, $f_2(x) = 0.6x$, and $f_3(x) = 1.2x^2$ for $0 \le x < 1$. Using the process described just above, we have

$$p_1 = .3, \quad p_2 = .3, \quad \text{and } p_3 = .4$$

and

$$\varphi_1(x) = .3x, \quad \varphi_2(x) = .3x^2, \quad \text{and } \varphi_3(x) = .4x^3, \quad 0 \le x < 1$$

Consequently, if $\{u\}$ is distributed uniformly, then the set $\{x\}$ with

$$x = \begin{cases} u/.3, & u \le .3 \\ [(u/.3) - 1]^{.5}, & .3 \le u < .6 \\ [2.5u - 1.5]^{1/3}, & .6 \le u < 1 \end{cases}$$

is distributed according to X. ∎

  There are several continuously distributed random variables that arise naturally in applications. For example, the exponential distribution was used in Section 2.7, and the normal distribution was mentioned earlier in this section. The normal distribution is of special importance because it arises so frequently in applications and because of the central limit theorem. The latter result asserts (as a special case) that if $X_1, X_2, \ldots, X_n$, denote independent random variables with finite mean μ and positive variance σ^2, and if $X(n) = \sum_{k=1}^{n} X_k$, then the random variable $\sqrt{n}(X(n) - \mu)/\sigma$ has a limiting normal distribution (as n becomes large) with mean 0 and variance 1. This means that for sufficiently large n, the random variable $\sqrt{n}(X(n) - \mu)/\sigma$ is distributed as the unit normal distribution (mean 0 and variance 1).

  Because they arise so frequently in applications, there are a large number of special methods that can be used to generate values of the widely used continuous random variables. Usually there is no method that should be preferred in all cases, and the choice will usually be influenced by the specific circumstances. As an illustration, we provide an example of an algorithm to generate values of the standard normal random variable—mean 0 and variance 1: Let $U_1, \ldots, U_{12}$ be independent random variables uniformly distributed on $[0, 1)$. Then $X = \sum_{i=1}^{12} U_i - 6$ has mean 0 and variance 1 and is approximately normally distributed.

Exercises 4.4

1. Generate ten random numbers using a uniform distribution on $[0, 1)$, and use these ten numbers to create ten random numbers from the distribution given by the following density function:

$$f_X(x) = \begin{cases} 0.25, & -3 \le x < 1 \\ 0, & \text{elsewhere} \end{cases}$$

2. Describe a method of generating random numbers from the distribution given by the density function

$$f_X(x) = \begin{cases} (x + 1)/2, & -1 \le x \le 1 \\ 0, & \text{elsewhere} \end{cases}$$

3. A continuous random variable X has the following density function.

$$f_X(x) = \begin{cases} 0.5x, & 0 \le x < 1 \\ 0.5, & 1 \le x < 2 \\ 0.25, & 2 \le x < 3 \\ 0, & \text{otherwise} \end{cases}$$

(a) Find the distribution function for X and graph it.
(b) Suppose you are given five values of a random variable U, uniformly distributed on [0, 1): 0.4683, 0.6498, 0.2054, 0.4131, 0.8245. Use these values to generate five values distributed as X.

4. A random variable X has the density function

$$f_X(x) = \begin{cases} x - 1 & 1 \le x < 2 \\ 3 - x & 2 \le x < 3 \\ 0 & \text{otherwise} \end{cases}$$

(a) Find the distribution function F_X for the random variable X.
(b) If $\{u_i\}$ is uniformly distributed on [0, 1), find a formula for values $\{x_i\}$ distributed according to the random variable X.
(c) Implement the results of part (b), and generate 50 values randomly distributed with distribution function X.

5. A random variable X has the density function

$$f_X(x) = \begin{cases} 0.4 & 0 \le x < 1 \\ 0.24(2 - x) + 0.16 & 1 \le x < 2 \\ 0.16 & 2 \le x < 4 \\ 0 & \text{otherwise} \end{cases}$$

(a) Find the distribution function F_X for the random variable X.
(b) Graph the density and distribution functions of X.
(c) If $\{u_i\}$ is uniformly distributed on [0, 1), find a formula for values $\{x_i\}$ distributed according to the random variable X.
(d) Implement the results of part (c), and generate 50 values randomly distributed with distribution function X.

6. A random variable X has the density function

$$f_X(x) = \begin{cases} (x + 1)/2 & -1 \le x \le 0 \\ 1/2 + (3/8)x^{1/2} & 0 < x \le 1 \\ 0 & \text{otherwise} \end{cases}$$

(a) Find the distribution function F_X for the random variable X.
(b) Graph the density and distribution functions of X.
(c) If $\{u_i\}$ is uniformly distributed on [0, 1), find a formula for values $\{x_i\}$ distributed according to the random variable X.
(d) Implement the results of part (c), and generate 50 values randomly distributed with distribution function X.

7. A random variable X has the density function

$$f_X(x) = \begin{cases} 0.05x & 0 < x \le 2 \\ 0.1 + 0.4(x-2) & 2 < x \le 3 \\ 0.5 & 3 < x \le 4 \\ 0.5 - 1.25(x-4) & 4 < x \le 4.4 \\ 0 & \text{otherwise} \end{cases}$$

(a) Find the distribution function F_X for the random variable X.
(b) Graph the density and distribution functions of X.
(c) If $\{u_i\}$ is uniformly distributed on $[0, 1)$, find a formula for values $\{x_i\}$ distributed according to the random variable X.

8. The manager of a convenience store collects data on late-afternoon customer arrivals. She finds that there is a uniform distribution of customers from 4:20 to 7:40 p.m. (a period of length 200 minutes) and an additional group that arrives between 4:50 and 7:25 p.m. with peak arrivals between 5:50 and 6:05 p.m. With a continuous random variable X defined as the arrival time of a customer who arrives between 4:20 and 7:40, she approximates the density function of X as follows (times measured in minutes beginning at 4:20 p.m.):

$$f(x) = \begin{cases} 0, & x < 0 \\ 0.002, & 0 \le x < 50 \\ 0.002 + 0.0002(x-50), & 50 \le x < 90 \\ 0.01, & 90 \le x < 105 \\ 0.01 - 0.0001(x-105), & 105 \le x < 185 \\ 0.002, & 185 \le x < 200 \\ 0, & 200 \le x \end{cases}$$

(a) Find a formula for the distribution function of X.
(b) If $\{u_i\}$ is uniformly distributed on $[0, 1)$, find a formula for values $\{x_i\}$ distributed according to the random variable X.

9. With the random variable X defined as in Exercise 8:

(a) Graph the distribution function of X.
(b) Implement the results of Exercise 8(b), and find a sequence $\{x_i\}$ of 100 values distributed according to X.
(c) Group the values $\{x_i\}$ determined in part (b) into groups of values in the following intervals (times measured in minutes from 4:20 p.m.):

$$[0, 25), \ [25, 50), \ [50, 70), \ [70, 90), \ [90, 105), \ [105, 125),$$
$$[125, 145), \ [145, 165), \ [165, 185), \ \text{and} \ [185, 200)$$

and plot the results as a histogram. Adjust the vertical scale of your histogram to denote the fraction of the total number, and compare the adjusted histogram with the graph of f_X.

10. Change the random variable in Example 4.20 so that

$$f_X(x) = \begin{cases} 0.2 + 0.6x + 2x^3, & 0 \le x \le 1 \\ 0, & \text{otherwise} \end{cases}$$

Using the method described in that example, generate 100 values of x distributed as X.

11. Consider the random variable X with the density function

$$f_X(x) = \begin{cases} e^{-1} + e^{x-1}, & 0 \le x \le 1 \\ 0, & \text{otherwise} \end{cases}$$

Generate 100 values of x distributed as X.

4.5   Applications and Validation of Simulation Modeling

Simulation models are usually constructed for the same reason as other types of models: They provide a tool to help us understand more about a situation than we could otherwise. In this section we illustrate the use of simulation models in three extended examples, two in an economic/business environment and one in a medical setting.

■ Estimating Customer Flow in a Retail Store

Marvin's Music and Video occupies prime space in the Metroburg Mall, and the store is open ten hours each day: 11:00 a.m. to 9:00 p.m. Large numbers of potential customers pass in front of the store, and Marvin knows that the likelihood of those potential customers becoming actual customers is influenced by the level of activity in his store: It increases (or declines) as the number of the people in his store increases (or declines). His marketing consultant believes that Marvin's customers fall into two groups: core customers—those who come to the mall to shop at Marvin's—and casual customers—mall visitors who pass in front of his store and are attracted into the store by activity or advertising.

We begin by considering the core customers. The pattern of traffic varies throughout the day, with busy periods around lunch and in the evening and a lull in the afternoon. On the basis of survey data, we suppose that the number of core customers who arrive during each half-hour is described by three random variables, each of which applies to several of the half-hour periods, with the following means and standard deviations:

(4.7)
$$\begin{cases} \mu = 10 \text{ and } \sigma = 2.3 & \text{from 11:00 a.m. to 12:00 p.m.} \\ \mu = 15 \text{ and } \sigma = 2.8 & \text{from 12:00 p.m. to \ \ 1:00 p.m.} \\ \mu = \ \ 5 \text{ and } \sigma = 1.6 & \text{from \ \ 1:00 p.m. to \ \ 5:00 p.m.} \\ \mu = 15 \text{ and } \sigma = 2.8 & \text{from \ \ 5:00 p.m. to \ \ 8:00 p.m.} \\ \mu = 10 \text{ and } \sigma = 2.3 & \text{from \ \ 8:00 p.m. to \ \ 9:00 p.m.} \end{cases}$$

For simplicity, we assume that the random variables generating these data are independent Bernoulli trials with parameters n and p. With this assumption and the relations $\mu = np$ and $\sigma = \sqrt{np(1-p)}$, we can determine (for each time interval) approximate values for n and p when $\mu > \sigma^2$. For example, in the first half-hour we can use $n = 21$ and $p = .48$.

Similarly, we assume that the numbers of casual customers who arrive each half-hour are distributed as binomial random variables, where n depends on the general level of mall traffic and p depends on the total number of customers (core and casual) who are in the store in the preceding half-hour. In particular, suppose that the number of casual customers in half-hour k ($k = 1, 2, \ldots, 20$) is a binomial variable where n is

(4.8)
$$\begin{cases} 100 \text{ per half-hour} & \text{from } 11{:}00 \text{ a.m. to } 12{:}00 \text{ p.m.} \\ 150 \text{ per half-hour} & \text{from } 12{:}00 \text{ p.m. to } 1{:}00 \text{ p.m.} \\ 50 \text{ per half-hour} & \text{from } 1{:}00 \text{ p.m. to } 5{:}00 \text{ p.m.} \\ 150 \text{ per half-hour} & \text{from } 5{:}00 \text{ p.m. to } 8{:}00 \text{ p.m.} \\ 100 \text{ per half-hour} & \text{from } 8{:}00 \text{ p.m. to } 9{:}00 \text{ p.m.} \end{cases}$$

and

(4.9) $p = \begin{cases} .01 \text{ during the half-hour } 11{:}00 \text{ a.m. to } 11{:}30 \text{ a.m.} \\ .002(\text{total number of customers in store in half hour } (k-1)) + .01, \\ \text{for time intervals } k \text{ with } k > 1 \end{cases}$

We note explicitly that for each half-hour in the period 11:30 a.m. through 9:00 p.m., the probability p depends on the *total* number of customers in the store during the preceding half-hour as shown in Equation (4.9).

The task is to use a simulation model to predict the expected number of customers in one 10-hour day, and to predict how that number varies as Marvin implements strategies to increase the number of customers.

A simulation of the customer traffic in the store during one day proceeds as follows: The day is divided into 20 successive half-hour intervals, the first beginning at 11:00 a.m. and the last concluding at 9:00 p.m. During the first half-hour (11:00 to 11:30 a.m.), the arrivals of core customers are generated using binomial trials with the data in display (4.7) above, and the arrivals of casual customers are generated using binomial trials with n equal to 100 per half-hour and p equal to .01. The number of core customers, the number of casual customers, and the total number of customers are recorded. Beginning at 11:30, the arrivals of core customers are again generated using binomial trials with the data in (4.7), but casual customers are generated using the values of n given in (4.8) and the values of p computed using (4.9). Note that the number of casual customers in the period 11:30 a.m. to 12:00 p.m. is determined using a value of p that depends on the total number of customers in the period 11:00 to 11:30 a.m. This dependence continues throughout the remaining time periods.

Table 4.18 shows the results from a simulation consisting of 1000 runs of one day each. In that table we number the half-hour segments from 11:00 a.m. to 11:30 a.m. through 8:30 p.m. to 9:00 p.m. as time segments 1 through 20, respectively, and we show the mean numbers of core and casual customers for that segment over the entire 1000 days. Also, for each day we record the total core customers, the total casual customers, and the total combined customers. The mean number of total customers for this simulation is 280.6, and the standard deviation is 17.1. This information provides a response to one of the tasks—namely, to predict the expected number of customers.

In simulations of this sort, it is frequently important to know how the results change if the parameters change by small amounts. The task of providing this information is known as a *sensitivity analysis*. For example, suppose the probability of attracting casual customers increases

from .002(total number of customers in half hour $(k-1)/2) + .01$ for $k \geq 2$,
to .0025(total number of customers in half hour $(k-1)/2) + .01$ for $k \geq 2$

Using Table 4.18 we see that for most time segments the mean total number of customers in the store is between 6 and 23. Consequently the probability defined in the first line above

Table 4.18

Time Segment	Mean Number of Core Customers	Mean Number of Casual Customers	Mean Number of Total Customers
1	10.33	0.98	11.31
2	10.22	3.29	13.50
3	14.80	5.63	20.43
4	15.03	7.81	22.85
5	4.94	2.83	7.77
6	4.87	1.25	6.12
7	4.83	1.08	5.91
8	4.92	1.13	6.05
9	4.96	1.08	6.04
10	4.92	1.14	6.05
11	4.87	1.10	5.97
12	4.99	1.09	6.08
13	14.74	3.32	18.06
14	14.90	6.82	21.72
15	14.94	7.90	22.84
16	14.77	8.38	23.14
17	14.84	8.62	23.46
18	14.88	8.40	23.28
19	10.08	5.67	15.75
20	10.15	4.12	14.28
totals	198.96	81.63	280.59

Standard deviation of total customers $= 17.10$

is between .022 and .056. Similarly, the probability defined in the second line is between .025 and .068, and the increase from the first to the second is between 13% and 32%. The simulation results are that the mean number of casual customers increases to 103.3 and the total number of customers increases from about 280 to about 302.3. From Marvin's point of view, the value of attracting the additional customers is likely to depend on the expected net profit gain (increased revenue less increased cost).

The cost of attracting existing mall customers into Marvin's store is likely to be a cost that Marvin incurs himself. On the other hand, the cost of attracting new customers to the mall is likely to be shared by all the stores in the mall. How does the mean number of customers in Marvin's store change as a result of a change in the level of mall traffic? Suppose that the general mall traffic increases by 10%. That is, suppose that the number of casual customers in half-hour k is a binomial random variable where n is

110 per half-hour   from 11:00 a.m. to 12:00 p.m.
165 per half-hour   from 12:00 p.m. to 1:00 p.m.
 55 per half-hour   from 1:00 p.m. to 5:00 p.m.
165 per half-hour   from 5:00 p.m. to 8:00 p.m.
110 per half-hour   from 8:00 p.m. to 9:00 p.m.

and p is as in Equation (4.9). Then, the summary results (from data in a table similar to Table 4.18) are

Mean number of core customers	= 199.1
Mean number of casual customers	= 92.1
Mean number of total customers	= 291.2
Standard deviation of total customers	= 18.1

The number of casual customers in this case is about 10% higher than in the case summarized in Table 4.18. The total number of customers has increased by about 3.8%. This information responds to another of the original questions.

In the absence of more detailed data on the numbers of core customers, the assumption that the numbers of core customers are distributed as binomial trials seems reasonable. However, the alternative that the numbers are normally distributed is also a reasonable assumption. This option is the topic of Exercise 3 at the end of this section.

One of the basic assumptions of the model is that the probability of a person in the mall who is not one of Marvin's core customers becoming a casual customer depends on the number of people in the store in the preceding time segment. This assumption makes simulation a natural way to model the situation. If the probabilities are constant or depend only on the time segment, then analytical techniques can be used, although the details may be somewhat complex. Some of the latter situations are the topics of exercises.

■ Meeting Demands in a Fitness Center

Individuals who become members of a fitness center do so for a variety of reasons, but one reason is often the availability of specialized facilities or equipment. Consequently, management of the center is sensitive to the need for meeting member demand for equipment. At the same time, providing specific equipment requires funds to purchase and maintain it and space to house it. Indeed, because of financial and space constraints, making a decision to purchase a specific piece of equipment may mean that other equipment cannot be purchased.

We consider a situation where management of a fitness center is evaluating whether it has enough stationary bicycles to meet the demand. There are currently five bicycles, and the demand in the early morning appears to exceed the capacity. Management believes members are frustated by the need to wait for bicycles in the period from 6:00 a.m. to 7:30 a.m., and the goal of the study is to determine whether it is economically desirable to add more bicycles. Data have been collected on equipment utilization between 6:00 a.m. and 7:30 a.m., and the data include arrival times and excercise times. The data represent observations of a stochastic process (actually several stochastic processes), and to simplify the modeling process, it is helpful to make assumptions about the underlying distributions for these random variables.

Suppose that data are collected and reviewed with a goal of making reasonably simple assumptions about arrival rates. On the basis of the review, we assume that the rate of arrivals in the period 6:00 to 7:30 a.m. varies, with a peak arrival rate during the 30-minute period from 6:30 to 7:00 and lower rates before 6:30 and after 7:00. More specifically, we assume that arrivals have an exponential distribution with an average of 30 arrivals in the period from 6:00 to 6:30, an average of 45 arrivals in the period from 6:30 to 7:00, and an average of 20 arrivals in the period from 7:00 to 7:30.

Suppose that most of the members use the bicycles as part of an exercise program that includes other activities, and we assume that the period of excercise time spent on the bicycles is of fixed duration for all members: 8 minutes. Finally, we assume that if no bicycle is available when a member arrives, the member will wait for one to become available if there are three or fewer members waiting, but if there are four members already waiting, then the new arrival will leave.

With these assumptions, we can simulate a single day as follows: We consider each minute of the 90-minute period, one minute after another, and we determine whether one or more of the members currently using a bike completes a session during that minute. If so, each of the bicycles released becomes available for a new user. If there are members waiting, as many as possible are assigned bikes; those in line the longest are assigned first. Then, members arriving during that minute are considered, and if more than one member arrives in a minute, they are considered one after another. For each arrival, we check whether there is a bicycle not being used by another person. If so, the member begins to use the bicycle. If there is no unused bicycle, then the member joins the queue of people waiting for a bicycle if there are three or fewer people already waiting. If there are four people waiting, then the new arrival leaves. Note that the protocol for this simulation is for those who finish using a bicycle during a particular minute to do so before newly arrived members look for a bicycle. This is, of course, only one of several such assumptions that could be made.

With the original task in mind, it is appropriate to record the total number of members who use the equipment each day, the total number who do not use it because there are four others waiting when they arrive, and the average waiting time for members who use the equipment. We assume that no member tries to use the equipment more than once a day. We will refer to members who leave without using a bicycle as *disappointed*.

We repeat the simulation for 1000 runs, and we have the following results:

$$
\begin{array}{ll}
\text{Average number who try to use a bicycle} & = 94.60 \\
\text{Average number who are disappointed} & = 37.15 \\
\text{Average percent who are disappointed} & = 38.71\% \\
\text{Average waiting time} & = 4.1 \text{ minutes}
\end{array}
$$

The average percentage is computed by computing the percent disappointed each day and then averaging these percentages over the 30 days. This is not (necessarily) the same as the ratio of the average number who are disappointed to the average number who tried to use a bicycle.

For the particular goal of this discussion, it is the average percentage of disappointed members that is of interest. In other situations, what should be noted might be the average number of disappointed members or the number (or percentage) of members who succeed in riding a bicycle.

The questions that the managers of the center ask are "What are the economic implications of these results, and how do these results change if more equipment is added?" To respond to these questions, we look at the economics of the situation. Suppose the yearly membership dues for the facility are $360, or $30 per month. Also, suppose that members who use the stationary bikes do so ten times a month and that half of those who are disappointed more than three times in a month will cancel their membership. Finally, assume that the yearly costs for purchase and maintenance of each stationary bicycle is $1200, or $100 per month.

With these assumptions, we can investigate the economics of the current situation. Recall that the member's decision criterion results in half of the disappointed members dropping their membership if bicycles are unavailable more than 30% of the time—that is, unavailable more that three times a month. Let s be the probability that a member will be disappointed when trying to use a bicycle on a random visit. Then, viewing multiple attempts to ride a bicycle as independent repetitions of a binomial trial with disappointment probability s, we find that the probability of four or more disappointments in a month (ten visits to the fitness center) is

$$(4.10) \qquad 1 - (1-s)^{10} - 10s(1-s)^9 - 45s^2(1-s)^8 - 120s^3(1-s)^7$$

Half of the members with four or more disappointments will cancel, and the probability of this event is .5 times the expression (4.10). The economic cost to the center is $30 per month for each membership cancelled. Using the simulation results for the current situation, we estimate s as .387, so the probability that a random member who attempts to use the bicycles will cancel is .2923. Also, we have the average number of members who attend each day—approximately 94.5 in the simulation leading to the data in Table 4.19. Thus, in this situation, the cost of cancellations is about $830 per month. This substantially exceeds the cost of installing another bicycle, and the next step is to use simulations to determine how many additional bicycles should be installed to maximize net income—that is, dues from continuing members less new costs.

We repeat the simulation with additional stationary bicycles, using a total of 6 through 12 bicycles. We assume that the arrival schedules do not depend on the number of bicycles, an assumption that could be changed if there were data on which to base other assumptions about arrivals. Because the arrivals are given by a stochastic process, the average number of members who try to use bicycles varies slightly from one simulation to another. For the eight simulations reported here, that number was between 93.94 and 95.14. Using data derived from the simulations, we have the results summarized in Table 4.19. The cost and revenue data in the table are monthly figures.

Judging on the basis of the results shown in Table 4.19, management would achieve higher net revenue with more bicycles. The highest net revenue would be achieved for eight bicycles, although seven, eight, or nine bicycles would all give improvements of about 20% or more above the current level.

Table 4.19

Number of Bicycles	Percent of Disappointed Members	Probability of Cancellation	Expected Cost of Cancellation	Cost of New Bicycles	Expected Net Revenue
5	38.7	.292	$830	0	$2006
6	30.1	.177	504	$100	2231
7	22.1	.0804	228	200	2407
8	15.9	.0301	86	300	2449
9	11.2	.0095	27	400	2408
10	7.0	.0018	5	500	2330
11	4.6	.0004	1	600	2234
12	2.9	.000	0	700	2135

As with all models, the details of conclusions may depend on the values of the parameters that occur in the model. The dependence of the conclusions on parameter values in the fitness center model is pursued in the exercises.

■ Modeling the Spread of a Communicable Disease

Diseases contracted by humans can be classified in a variety of ways, and one of the most important is in terms of the transmission mechanism. Some diseases are acquired from a source other than an infected individual; an example is food poisoning acquired by consuming contaminated food. Some are transmitted indirectly through a vector; for instance, malaria is transmitted by mosquitos. And some are transmitted by bacteria or viruses directly from an infected individual. In this discussion we model the transmission of a disease that is passed from an individual who has the disease and is capable of passing on the infection directly to an individual who is susceptible—that is, capable of acquiring the disease. We begin the discussion by introducing more precise terminology and notation.

We suppose that we have a population of fixed size that can be partitioned into three groups of individuals: individuals who have the disease, individuals who are susceptible to catching the disease, and individuals who are immune to the disease. We suppose that no one enters or leaves the population (no births, deaths, or migration in or out) during the time interval of interest. Although our model can be modified to incorporate these features, they add complexity to the discussion, and our goals can be met by considering the simpler situation. Depending on the nature of the disease, the group of those who have the disease may consist of two disjoint subgroups: those who are infected and are capable of spreading the disease and those who are infected but are not capable of spreading it. In several common diseases, immediately after an individual is infected there is a period of *latency* during which the individual cannot transmit the disease to anyone else, followed by a period of infectiousness during which transmission is possible. The period of infectiousness continues until the individual recovers and becomes immune. For some diseases, individuals may become susceptible again after recovering from one infection, but our model does not incorporate this feature. Our model would be appropriate for many infections caused by viral parasites, where individuals who recover have long-term immunity to reinfection. In addition, from a clinical perspective, after an individual contracts the disease, there may be a period during which there are no symptoms, followed by a period during which the individual does display symptoms. This period when symptoms are displayed normally overlaps, but is not identical to, the period of infectiousness. Again depending on the disease, during the period when symptoms are present, the activities of a symptomatic individual may be restricted either voluntarily, as when the symptoms make some activities unpleasant or impossible, or through a quarantine.

For the model discussed here, we suppose that the length of the latent period and the length of the period of infectiousness are random variables whose probability density functions are known. Finally, we suppose that, given the date of infection and the latent period, the date of onset of symptoms and the length of the symptomatic period are random variables whose probability density functions are known. For simplicity, in all cases we suppose that these random variables are binomial random variables with known means and variances. An alternative assumption would be that these random variables are normal with known means and variances.

We suppose that the social dynamics of our population are such that—aside from individuals with symptoms, whose activities may be restricted—all meetings between two members of the population are equally likely. In practice, of course, the situation would probably be much more complicated than this. For example, in the case of a disease that is propagated through a group of children of elementary school age, the total population may consist of several (possibly overlaping) subpopulations: the students who take the same bus to school, the students who are in the same class, the students who eat lunch together, and so on. In a situation such as this, meetings of (at least) two types should be considered: meetings involving students in the same group and meetings involving students in different groups. However, the main idea can be illustrated in the simpler situation of a population consisting of a single group, and we shall take this approach. Accordingly, we assume that our population consists of a number of individuals and that all members of the population who are not subject to quarantine behave the same as far as contacts with others are concerned. Also, we assume there is no immigration or emigration. In the more complex situation where some individuals move into and out of the population being modeled, it is important to keep track of the characteristics of each such individual.

A basic concept in a model of direct transmission is that of *contact*. In our discussion, a contact is a meeting involving two members of the population such that for some individuals, one infectious and one susceptible, disease transmission results. For example, a brief meeting between two individuals may not result in transmission, whereas in some circumstances, two individuals having lunch together will result in transmission. Specifically, for the present model we suppose that during each time step, each infectious individual meets with others at random, and the number of contacts is given by a binomial random variable with $n = 5$ and $p = .2$; there is an average of one contact per time step. For each contact, an individual is selected at random from among those not quarantined. If the individual selected is susceptible, then the susceptible individual becomes infected with probability i_p, the infection probability of the model. We suppose that the same probability i_p applies to every contact involving an infectious individual. Thus the transmission component of the model depends on the number of contacts per unit time and on the probability that a contact results in transmission. In turn, these parameters depend on the social structure of the population and on the characteristics of the disease.

We construct a simulation model in which the numbers of susceptible, infectious, and immune individuals are tracked through time. Here, as in many similar situations, it is customary to follow the disease until there are no longer any infectious individuals. In our models, we suppose that a relatively small number of infected individuals enter the population at a specific time, which we take as time 0. We assume that the specific latent period, onset of symptoms, and immunity for each of these infected individuals is determined by the random variables described above. As time continues, our simulation tracks the number of individuals in each of the following subgroups:

- susceptible (those who are capable of acquiring the disease but have not yet done so)
- infected (those who have the disease, whether or not they are capable of spreading it)
- infectious (those who have the disease and are capable of spreading it)
- symptomatic (those who have the disease and show symptoms)
- immune (those who have had the disease but are no longer infected)

For our model, we suppose that initially there are no immune individuals, and that individuals become immune only by contracting the disease and recovering from it. For the model

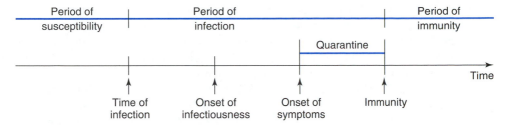

Figure 4.18

discussed here, we have the following:

- The time between infection and infectiousness (also referred to as the incubation period) is a binomial random variable with probability .5 and number of trials (time steps) equal to 5. Consequently the incubation period ranges from 0 time steps to 5 time steps with a mean of 2.5.
- The time between infectiousness and the onset of symptoms is a binomial random variable with probability .5 and number of trials 5.
- The time between the onset of symptoms and immunity is a binomial random variable with probability .3 and number of trials 10.
- If an individual is quarantined, the quarantine becomes effective with the onset of symptoms. Individuals who complete their period of quarantine return immediately to the population and are eligible for contacts.

The relationship between these times is shown in Figure 4.18.

One of our goals in this discussion is to use a simulation model to assess the effects of a quarantine on the spread of the disease. Here we use the term *quarantine* to denote behavior, voluntary or involuntary, that removes infected individuals from the population. There may be voluntary changes in behavior of individuals who have the disease; infected individuals may be unable or unwilling to maintain their usual activities. Also, infected individuals may voluntarily respond to recommendations from public health agencies to limit their contact with others. There may also be involuntary changes in behavior resulting from government public health decisions. Events that occurred in the spring of 2003 during an outbreak of SARS (severe acute respiratory syndrome) illustrate a range of societal responses to an infectious disease perceived to be dangerous. Here we study the effects of withdrawing a fraction of the infected individuals as soon as they develop symptoms. The quarantine continues until the individual recovers and becomes immune. Because the onset of symptoms lags the beginning of the infectious period, this approach will not halt the spread of the disease, but it does retard that spread. The extent of the reduction in the total number eventually infected depends on the effectiveness of the quarantine. In our simulation, we select symptomatic individuals at random to be quarantined. The probability that an infected individual is selected for quarantine does not depend on any of the other parameters associated with that individual.

For our simulation, we consider a population consisting of 1000 individuals, and at time 0 we begin monitoring the population after a small number of them have been infected. We suppose that each of these "initial infectives" acquired the disease sometime in the five time steps preceding the time labeled 0 and that the incidence of infection is uniformly spread over that period. The data reported in Table 4.20 are for three situations: populations of initial infectives consisting of 5, 10, and 20 individuals, with all others susceptible. Each run of the simulation begins at time 0 with a population consisting of the specified number m of initial infectives and $1000 - m$ susceptible individuals. During each time step, the contacts of infectious individuals are determined, and the results of the contacts—the newly infected individuals, if any—are recorded. The size of each subpopulation (susceptible, infected, and immune) is recorded at each time step. The simulation run ends when the number of infected individuals is 0. The simulation described here was run 100 times for each set of parameter values.

First, we consider the impact of various quarantine policies on the spread of the disease, and we use effectiveness probabilities from 0 through 100 in steps of 10. An effectiveness probability of 0 means, of course, that there is no quarantine at all; we refer to the data for no quarantine as *baseline* data. An effectiveness of 10 means that 10% of individuals showing symptoms are quarantined. At each time step, those quarantined are selected at random from among those who first show symptoms during that time step, and the selection of those in quarantine is the first act during each time step. As soon as the quarantine ends, the individual returns to the population open to contacts. Recall that individuals who have been infected cannot be reinfected, and thus they are not susceptible. Data for $m = 5, 10, 20$ and $i_p = .2$ are shown in Table 4.20. In that table, we show the average number of immune individuals when the simulation ends (the total number of individuals who were infected during the period before the disease died out) and the average duration of the disease (the time before the number of infected individuals is first equal to 0). We also show these numbers as a percent of the average number (or duration) in the baseline data—that is, when there is no quarantine at all.

Second, we consider the variation in the average number of infected individuals with the initial number of infectives for low values of quarantine effectiveness. The values $m = 5$, 10, and 20 amount to 0.5%, 1%, and 2% of the population size, and the corresponding values of the average number of infected individuals when the disease dies out are 92.65, 166.59, and 229.28. That is, increasing the number of individuals who are initially infected does not increase the average number of infected individuals proportionately. This result is consistent with intuition, because higher numbers of initial infectives mean that contacts between two infected individuals are more likely, and such contacts do not result in any new infectives.

Finally, we consider how the total number infected varies with quarantine effectiveness (QE). Data in Table 4.20 show that the largest changes in the total number infected occur as the quarantine effectiveness increases from 0 through 40. Indeed, the total number of infected drops by about 75% as the quarantine effectiveness increases from 0 to 40 for $m = 5$ and $m = 10$, and the drop is about 67% for $m = 20$. It may be more appropriate to look at the members of the population that are infected after time 0. We refer to these individuals as *newly infected*. If an exponential decline is fit to the data on newly infected individuals, then the best least-squares fits have rates of decline of 0.7211, 0.7161, and 0.7456 for $m = 5$, 10, and 20, respectively (Exercise 7). A somewhat better fit to these data is given by a logistic curve. The rationale and the details are pursued in Exercises 8 and 9.

The results of different runs of the simulation show high variability for small numbers of initial infectives ($m = 5$, for example). Indeed, with a small number of infectives there is a larger probability that the number of infected individuals never increases much above 5

Table 4.20

m	5		10		20	
	Average Total Infected		Average Total Infected		Average Total Infected	
Quarantine Effectiveness	Number	% of Baseline	Number	% of Baseline	Number	% of Baseline
Baseline	92.65	100.0	166.6	100.0	229.3	100.0
10	67.78	73.2	114.2	68.5	178.3	77.8
20	46.45	50.1	72.8	43.7	124.9	54.5
30	31.04	33.5	59.8	35.9	98.0	42.7
40	21.48	23.2	42.1	25.3	75.6	33.0
50	17.79	19.2	31.0	18.6	60.4	26.3
60	15.13	16.3	25.7	15.4	54.0	23.5
70	13.12	14.2	23.5	14.1	45.6	19.9
80	10.63	11.5	20.2	12.1	37.1	16.2
90	9.99	10.8	17.6	10.6	35.2	15.4
100	7.91	8.5	14.8	8.9	30.9	13.5
	Average Duration		Average Duration		Average Duration	
Quarantine Effectiveness	Number	% of Baseline	Number	% of Baseline	Number	% of Baseline
Baseline	105.6	100.0	111.4	100.0	85.1	100.0
10	65.5	62.0	77.1	69.2	70.5	82.8
20	49.9	47.3	59.9	53.8	58.6	68.9
30	30.4	28.8	41.0	36.8	42.2	49.6
40	23.4	22.2	28.6	25.7	31.9	37.5
50	18.4	17.4	21.6	19.4	24.7	29.0
60	16.1	15.2	14.1	12.7	19.4	22.8
70	9.5	9.0	10.4	9.3	14.7	17.3
80	7.4	7.0	9.4	8.4	10.4	12.2
90	6.0	5.7	7.0	6.3	9.4	11.0
100	4.0	3.8	4.7	4.2	6.8	8.0

and the disease dies out in a few time steps. However, at the same time, the disease may persist even with small numbers of initial infectives, and there may be runs with long disease durations and relatively large total numbers of infected. For example, among the 100 runs with $i_p = 0.2$, $m = 5$, and quarantine effectiveness $= 0$ (no quarantines), there were 38 runs with total numbers infected of 20 or less and 62 runs with total number infected of 60 or less. In the same simulation, there were 10 runs with total numbers infected of at least 250 and 25 runs with duration of disease at least 100 time steps. For these parameters, the standard deviation in the total number infected is 105.62, which exceeds the mean of 92.65, and the standard deviation of the duration is 54.44, which is about 80% of the mean duration of 68.33.

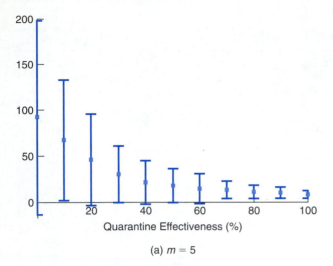

Quarantine Effectiveness (%)

(a) $m = 5$

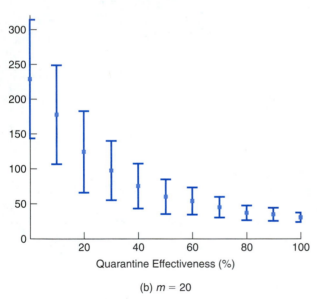

Quarantine Effectiveness (%)

(b) $m = 20$

Figure 4.19

The results of the simulations (average and $+/-$ 1-standard-deviation error bars for total numbers infected) are shown in Figure 4.19 for quarantine effectiveness values (in percent) from 0 through 100 in steps of 10. Figure 4.19(a) displays results for $m = 5$, indicative of the case of small numbers of initial infectives, and Figure 4.19(b) displays results for $m = 20$, indicative of the case of relative large numbers of initial infectives. The high variability in the results for small numbers of initial infectives is illustrated by the relation between the average and the standard deviation for the case $m = 5$. When there are 20 initial infectives, there is somewhat less variability in the outcomes of the simulation runs, and the standard deviations are significantly smaller than the means for all values of quarantine effectiveness. Figure 4.19 also illustrates the point made above about the relation between quarantine effectiveness and size of the epidemic as measured by total number infected. That is, one might conclude that most of the effects of a quarantine occur for values of quarantine effectiveness of 40 or less and that these effects are more pronounced for small numbers of initial infectives.

For many diseases, some individuals will behave in ways that amount effectively to a "self-quarantine," and with no other intervention this might result in quarantine effectiveness of 10 or 20—possibly more if the symptoms are pronounced. On the other hand, an effectiveness above 70 would probably require stringent reporting and enforcement. Either option has its costs—hence the interest in these models.

■ Verification and Validation of Simulation Models and Interpretation of Output

Verification and validation are activities that are best integrated with the model-building process whether the resulting model is an analytical or a simulation model, and the total process will be more efficient and effective when this is done. Verification is concerned with building a model that is useful for the intended purposes. It involves

- Assessing the face validity of the model. Are the assumptions consistent with observations and empirical data? For instance, if a model for population growth is one in which the growth is determined completely by the reproductive capability of the members of the population, then do the data show that there are no constraints on the growth that arise from limitations on crucial resources? Another example is a situation where the patient is taking two medications for different purposes. In each case, the dosage is adjusted for the severity of the condition and the characteristics of the individual (age and weight, for example). It is often assumed that taking one medication does not influence the effectiveness of the other, and it is important to know whether this is a legitimate assumption. If not, what adjustment to the model should be made?
- Verifying that the assumptions have been correctly translated into the mathematical structure. For instance, in constructing simulation models, it is commonly assumed that random variables are distributed as one of the standard distributions. Are the features of the assumed distribution consistent with the real model of the situation?
- Verifying that the analysis is correct—that is, that the methods of solution are the right methods and that they are carried out correctly. If the model is a complex one, this may be a substantial task. It may be useful to begin this verification by considering a slightly simpler problem to confirm that the solution methods are accomplishing the desired goal. For instance, if a complex network model is being analyzed, it may be worthwhile to verify a solution algorithm in a simple network.
- Verifying that the results of the analysis or simulation are correctly interpreted in terms of the original real-world situation. For instance, the results of a sensitivity analysis may be valid only if the actual solution of the perturbed situation is the same as—or very close to—the solution of the original problem. If this does not hold, then the results of the analysis may have a different meaning.
- Reviewing the methods employed to estimate parameters used in the model and confirming that the values used in the model are consistent with information about the original situation.

In addition, the verification of simulation models involves some steps that are specific to such models. A particularly important step is confirming the logic of the basic simulation core. The fundamental idea behind a simulation model is to use a computer to mimic the behavior of a system. To do this, it is essential to identify correctly the structure of the system being modeled and then to mimic this structure correctly in a computer program. It is helpful to use such basic programming principles as beginning with a simplified version of the system, writing complex programs in modules, inserting output (print or display) statements judiciously to confirm that the program is doing what is intended, initially running the program with simplifying assumptions or parameter values where the results are known or can easily be confirmed, and so forth. For instance, earlier in the chapter we discussed several simulation models for situations where analytical models could also be constructed. The agreement of the predictions resulting from the study of both models gives us confidence that the simulation core is doing what is intended.

Output analysis is the study of data produced in a simulation. A simulation is usually conducted to gain information about some quantity associated with the system being simulated. How many tokens are selected before all four letters are acquired? How many people visit a fitness center? How many people are infected by a disease? A single run of the simulation produces a specific set of data—perhaps a single number, perhaps a set of numbers that provide information on different aspects or behaviors of the system being studied. A second independent run produces another set of data giving additional information on the same set of system attributes. Typically, we continue with additional independent runs until we have N data sets. The task of assessing the meaning of these data sets is in general a complex one, and there is a substantial literature on the subject. We discuss only a single simple setting, and we describe a useful criterion for determining how many runs should be carried out to achieve certain goals. Suppose that the same quantity is measured on each simulation run and that a single value, x_i, is collected on the ith run, $1 \leq i \leq N$. These values are values of a random variable X, and in many applications the distribution function, mean μ, and variance σ^2 are unknown. The task is to use these values, the x_i, to estimate the latter quantities. We suppose that the runs are independent, a legitimate assumption if different sequences of random numbers are used and the initial conditions are identical. Then the quantity

$$X(N) = \frac{1}{N} \sum_{i=1}^{N} x_i$$

is known as the *sample mean,* and it provides a point estimator for the (unknown) *population mean μ* based on the values $\{x_i\}$. It is a consequence of the central limit theorem that the quantity

(4.11) $$z = (X(N) - \mu)/(\sigma/\sqrt{N})$$

is approximately distributed as the unit normal random variable (mean 0 and variance 1). In applications, the population standard deviation σ is frequently unknown, and we need an estimate for this quantity as well. We define the *sample variance* as

$$S^2 = \frac{1}{N-1} \sum_{i=1}^{N} [x_i - X(N)]^2$$

In the situation considered here (N repetitions of an experiment with a stationary distribution), we use the sample variance as an (unbiased) estimator of the population variance σ^2. Once we replace the population standard deviation σ in Equation (4.11) by the estimate S, the resulting distribution is approximately Student's t distribution. However, for large values of N, say $N \geq 30$, the discrepancy with the normal is small, and in our examples and exercises, N will generally be much larger than 30. Consequently, we can assume that the distribution (4.11) is approximately normal, and the usual techniques of interval estimation can be applied. For example, an approximate 95% confidence interval for the mean μ is provided by the interval

(4.12) $$(X(N) - 1.96S/\sqrt{N}, \ X(N) + 1.96S\sqrt{N})$$

This result is frequently used in the following way. Suppose the goal is to determine the number of runs for which the approximate 95% confidence interval has a specified—presumably small—length h. Then a value of N such that $2 \times 1.96S/\sqrt{N} < h$ will be satisfactory. The term $1.96S/\sqrt{N}$ in expression (4.12) arises from the approximation provided by the central limit theorem and the fact that a random value of the unit normal random variable falls in the interval $(-1.96, 1.96)$ with probability .95.

We illustrate these ideas by returning to the absorbing Markov chain in Example 4.13. In that example the system began in state 1, and we set X equal to the number of transitions before reaching an absorbing state. For the simulation output reported in Section 4.3, we have $N = 1{,}000{,}000$, and we found the sample mean to be 8.1873 and the sample variance S^2 to be 49.5775. Using the result above, we have

$$X(N) - 1.96S/\sqrt{N} = 8.1735 \quad \text{and} \quad X(N) + 1.96S/\sqrt{N} = 8.2011$$

and consequently, the approximate 95% confidence interval for the mean number of transitions to absorption is the interval (8.1735, 8.2011). As we noted in Section 4.3, the number of transitions to absorption obtained by theoretical methods is 8.1819.

Exercises 4.5

1. Develop a computational model (for example, a spreadsheet) for Marvin's Music and Video, and compare the results of that model with the results reported in the discussion of the simulation model. Base the computational model on the assumption developed in this section, and use the average number of customers in the store in time period k to predict the number in period $k + 1$.

2. Suppose Marvin wants to rely solely on an increase in general mall traffic to increase his total number of customers by 10%. Use the results of the discussion in this section to estimate the percent increase required.

3. Construct a simulation model for Marvin's Music and Video with the same general approach used in this section, but replace the assumption that the numbers of core customers are distributed as a binomial random variable with the assumption that the numbers of these customers are normally distributed with means and standard deviations in various time blocks as given in the discussion.

Table 4.21

Number of Bicycles	Percent of Disappointed Members
4	48.9
5	38.6
6	29.5
7	21.9
8	15.4
9	10.9
10	7.2
11	4.5
12	2.8

4. Repeat the analysis of the fitness center model when the assumption that half of the members who are disappointed more than three times a month will cancel their memberships is replaced by the assumption that

   (a) 30% of such members will cancel.
   (b) 70% of such members will cancel.

5. Repeat the analysis of the fitness center model when the assumption that half of the members who are disappointed more than three times a month will cancel their memberships is replaced by the assumption that

   (a) half of the members who are disappointed more than four times a month will cancel their membership.
   (b) half of the members who are disappointed more than two times a month will cancel their membership.

6. The data reported in Table 4.19 depend, of course, on the assumption about arrivals. If that assumption is altered, then it is expected that the data will change. The data in Table 4.21 are derived from a simulation in which arrivals are assumed to be distributed with an exponential distribution of interarrival times with an average of

   - 30 arrivals in the period from 6:00 to 6:30.
   - 45 arrivals in the period from 6:30 to 7:30.
   - 20 arrivals in the period from 7:00 to 7:30.

   Assume that the average number of arrivals in each simulation is about 95.6 and that other assumptions in the simulation model discussed in this section continue. Complete an analysis similar to that in the text, using the data in Table 4.21.

7. Verify the conclusion regarding exponential fits to the data in Table 4.20 in the epidemic model discussed above. In particular, show that if an exponential decline is fit to the data on newly infected individuals shown in Table 4.20, then the best least-squares fits have rates of decline of 0.7211, 0.7161, and 0.7456 for $m = 5$, 10, and 20, respectively.

8. Consider the model for disease transmission.

   (a) Develop a rationale for using a logistic function as a fit for data on total newly infected individuals.

Table 4.22

Value of r	Probability
0	.2
0.01	.5
0.02	.2
0.05	.1

(b) Find the best logistic fit for the data on total newly infected individuals in Table 4.20.

9. Consider a simulation of a population model similar to that discussed in Section 4.1, and suppose that the random variable describing the population growth has the density function shown in Table 4.22 (compare with Table 4.1). A simulation of the growth of this population through 20 time steps based on 10,000 runs has mean 132.13 and standard deviation 62.63. A simulation based on 100,000 runs has mean 131.98 and standard deviation 62.53. Find the 95% confidence interval for the population size based on each simulation. Also, find the expected value $E[r]$ of r and the population size based on a deterministic model using a growth rate equal to $E[r]$.

CHAPTER 5

Linear Programming Models

5.0   Introduction

Linear programming models, as well as more general optimization models, are among the most widely used models for business and government decision making. These models are appropriate for problems arising in a variety of settings, some of which fit naturally into the structure of these models, and for other settings that can be converted into the appropriate structure. The utility of these models rests largely on two facts:

- The basic assumptions are valid, or nearly valid, for many situations, and the models serve as useful approximations even when the assumptions are not quite valid.
- There are very useful and powerful computational techniques for solving the resulting mathematical problems.

Indeed, the way the ideas and methods originated is an interesting example of the interplay of theory, application, and computation.

Problems that can be expressed in mathematical form as linear programming problems were studied in the 1930s, and an example of one of them is introduced in Section 5.1—the diet problem. However, it was the large and complex problems in areas such as scheduling, resource allocation, and transportation, which arose in World War II, that convinced decision makers of the usefulness of linear programming in solving "real-world" problems. In the postwar industrial development of the mid-1940s, the techniques that had proved so valuable during the war were applied to comparable situations arising in the civilian economy. At first, however, the usefulness of the methods was constrained by the lack of effective and efficient ways to solve the resulting mathematical problems. This shortcoming was partially resolved in the late 1940s with the invention of the simplex algorithm by George Dantzig. The solution of problems the size of those that regularly occur in applications became truly feasible in the 1950s with the introduction and development of computers. The remarkable coincidence— in a period of little over a decade—of the recognition of the common mathematical structure of many naturally occurring decision problems, the discovery of an algorithm to solve the resulting mathematical problems, and the invention of technology that made the algorithm applicable to the large-scale problems that actually arise in practice, is one of the great events in modern applied mathematics.

In this chapter we describe some situations in which linear programming models have been useful, we introduce the basic characteristics of these models, and we explore the accompanying mathematical concepts. In the first section we introduce several situations wherein mathematical programming problems arise; in Section 5.2 we introduce the mathematical structure of such problems and some of the fundamental mathematical results; in Sections 5.3 and 5.4 we provide examples of ways in which the basic concepts can be extended to incorporate other aspects of the situations that give rise to the original models. In Section 5.5 we introduce and study a topic closely related to linear programming: networks and flows. Here we also study the job assignment problem, which can be attacked using either linear programming or networks and flows.

It is important to note that we do not discuss the many computational methods that have been developed for linear programming problems. We expect the reader to have access to such algorithms via a package such as Maple, Mathematica, or the spreadsheet Excel with Solver, and we assume that such a package will be used for homework exercises.

5.1 Formulation of Linear Programming Problems

Among the most important characteristics of mathematical programming models is their capacity to incorporate features of real-world situations and their versatility. We illustrate these characteristics with several examples. We use these examples as an opportunity to discuss the model-building activity, and we emphasize the connection between features of the original situation and the mathematical expressions of the model. After examining several examples, we formulate a general class of models that includes these examples.

■ A Diet Problem

The situation we consider first is a straightforward version of one that arises in many circumstances wherein the goal is to provide food to people in a way that meets nutritional requirements and financial objectives. The specific situation is as follows: Suppose that there are m nutritional requirements—for instance, amounts of various vitamins, minerals, and total calories—that must be fulfilled with foods selected from a set of n foods—say apples, beef, and carrots—in such a way that the cost is as small as possible.

Our first assumption is that the amount of a nutrient in a unit of food is known for each combination of nutrient and food, and that the amount does not vary from one unit of food to another. In fact, the amounts of nutrients in foods do vary somewhat, depending on the exact type of food and on the growing conditions, harvesting, and storage of each unit of food. That is, the amount of vitamin D in a kilogram of oranges may depend on the variety of orange, where it was grown, the conditions under which it was grown, and how long ago it was harvested. However, in comparison with other aspects of the situation, and recognizing that we can distinguish among subvarieties of foods if we choose, it is reasonable to assume that the amount of nutrient i in 1 unit of food j is a known constant, and we denote it by $a_{ij}, i = 1, 2, \ldots, m; j = 1, 2, \ldots, n$.

Our second assumption is that the amount of nutrient i in x units of food j is xa_{ij}. That is, 2 units of food j contain $2a_{ij}$ units of nutrient i, and .5 unit of food j contains $.5a_{ij}$ unit of nutrient i. Note that this assumption includes an assumption about the ability of a human to process nutrients. It means that a person gains twice the amount of vitamin D from 2 units of oranges as from 1 unit. This assumption is likely to be valid over only a limited range of values of x. For example, a person is unlikely to extract 10 times as much vitamin D from 1 kg of oranges consumed in a single day as from 100 g of oranges. Also, we assume that the amount of nutrient i obtained from x units of food j and y units of food k is $xa_{ij} + ya_{ik}$. Again, for specific foods and nutrients, this assumption may be valid for a range of values x and y but invalid for other values of x and y.

Suppose that for each nutrient i there is a Recommended Daily Allowance b_i, $i = 1$, $2, \ldots, m$. In practice, the Recommended Daily Allowance may depend on the size, age, activity-level, and possibly other attributes of the person involved. For simplicity, we do not incorporate this feature into our model, although we could do so with an increase in complexity.

Finally, suppose that by a *diet* we mean a vector $\mathbf{x}$ with n coordinates, the jth coordinate giving the number of units of the jth food included, $j = 1, 2, \ldots, n$. It follows from the meaning of $\mathbf{x}$ that each coordinate of $\mathbf{x}$ must be nonnegative. With this definition and the assumptions above, the amount of nutrient i included in the diet $\mathbf{x} = [x_1 \quad x_2 \quad \cdots \quad x_n]$ is

$$a_{i1}x_1 + a_{i2}x_2 + \cdots + a_{in}x_n = \sum_{j=1}^{n} a_{ij}x_j$$

and the constraint that the diet must meet the Recommended Daily Allowance condition is

$$\sum_{j=1}^{n} a_{ij}x_j \geq b_i, \quad \text{for } i = 1, 2, \ldots, m$$

It is convenient to summarize these inequalities using matrix notation. We define an $m \times n$ matrix $\mathbf{A}$ by setting the ij-entry in $\mathbf{A}$ equal to a_{ij}, for $i = 1, 2, \ldots, m$ and $j = 1$, $2, \ldots, n$, and we define a (column) m-vector $\mathbf{b}$ by setting the ith coordinate of $\mathbf{b}$ equal to b_i for $i = 1, 2, \ldots, m$. Using this notation and the convention that an inequality sign between vectors or matrices denotes coordinate-wise inequality, we find that the matrix-vector version of the inequalities is

(5.1) $\mathbf{A}\mathbf{x} \geq \mathbf{b}, \quad \mathbf{x} \geq \mathbf{0}$

where we list the nonnegativity condition on the coordinates as a separate constraint.

Remark   Vectors, such as $\mathbf{x}$ in the discussion above, can be represented as either a one-row matrix or a one-column matrix. In (5.1), the vector $\mathbf{x}$ needs to be a one-column matrix so that the product $\mathbf{A}\mathbf{x}$ will be defined. Our convention throughout this chapter is that if a vector needs to be a one-row matrix, then it is a one-row matrix, and if it needs to be a one-column matrix, then it is a one-column matrix.

The final component of the model has to do with the economics of the situation. Suppose that we know the cost c_j of 1 unit of food j, $j = 1, 2, \ldots, n$. Also, suppose that the cost of x units of food j is xc_j; that is, there is no discount for quantity. Finally, suppose that the cost of x units of food j and y units of food k is $xc_j + yc_k$; that is, there are no discounts for combining purchases. We define a cost vector $\mathbf{c}$ to be an n-vector whose jth coordinate is c_j, $j = 1, 2, \ldots, n$. With these assumptions, the cost of a diet $\mathbf{x}$ is

(5.2) $x_1c_1 + x_2c_2 + \cdots + x_nc_n = \mathbf{x} \cdot \mathbf{c}$

Now we can use this notation and terminology to formulate a mathematical problem corresponding to the situation described above. Given an $m \times n$ matrix $\mathbf{A}$, a vector $\mathbf{b}$ in $\mathbb{R}^m$, and a vector $\mathbf{c}$ in $\mathbb{R}^n$, find a vector $\mathbf{x}$ that satisfies (5.1) and makes $\mathbf{x} \cdot \mathbf{c}$ as small as possible. In the case of the diet problem introduced above, the entries in the matrix $\mathbf{A}$ are the amounts of nutrients in the various foods, the entries in the vector $\mathbf{b}$ are the Recommended Daily Allowances of the nutrients, and the entries in the vector $\mathbf{c}$ are the costs of the foods. This problem is an example of a "standard minimum problem." We will formally define such problems in the next section.

■ A Resource Allocation Problem

Many situations in business can be modeled as problems in which resources are allocated to production processes. Obvious applications are in manufacturing—for example, in oil refineries and automobile manufacturing—but situations with similar characteristics arise in many other areas. Suppose that we have m resources and n production processes, and that when the jth production process is operated at unit level—that is, when 1 unit of the jth good is produced—it requires a_{ij} units of the ith resource, $i = 1, 2, \ldots, m$. For example, 20 pounds of steel may be needed to make 1 engine of a certain size.

A production schedule $\mathbf{x} = [x_1 \quad x_2 \quad \cdots \quad x_n]$ is a vector with n coordinates, one coordinate corresponding to each production process, and the jth coordinate is the level or intensity at which the jth process is operated. That is, x_j is the number of units of good j to be produced in the period of time covered by the schedule (for example, in one day, one week, or even one year). By the definition of x_j, it follows that $x_j \geq 0$ for each j, or, equivalently, $\mathbf{x} \geq \mathbf{0}$.

We suppose that the production of x_j units of good j requires $a_{ij}x_j$ units of the ith resource, $i = 1, 2, \ldots, m$. That is, there are no savings in resources from volume production. Also, we assume that if we produce x_j units of good j and x_k units of good k, then $a_{ij}x_j + a_{ik}x_k$ units of resource i will be required, $i = 1, 2, \ldots, m$. That is, no resources can be saved through joint production. These are clearly assumptions that may be valid for some production processes and not for others. They are, however, essential assumptions for the model to be a *linear* one. Suppose that the resources are available in limited supply and that for each resource i there is an amount b_i available, $i = 1, 2, \ldots, m$. Then, if the amounts of resources used do not exceed the amounts available, the production schedule $\mathbf{x}$ must satisfy

$$(5.3) \qquad a_{i1}x_1 + a_{i2}x_2 + \cdots + a_{in}x_n \leq b_i, \quad i = 1, 2, \ldots, m$$

As in the preceding example, we introduce an $m \times n$ matrix $\mathbf{A}$ whose ij-entry is a_{ij} for $i = 1, 2, \ldots, m$, and $j = 1, 2, \ldots, n$, and a vector $\mathbf{b}$ whose ith coordinate is b_i, $i = 1, 2, \ldots, m$. Then the expressions (5.3) can be expressed in matrix form as

$$(5.4) \qquad \qquad \mathbf{Ax} \leq \mathbf{b}$$

Also, assume that the profit associated with 1 unit of good j is c_j for $j = 1, 2, \ldots, n$, and define a profit vector $\mathbf{c}$ by $\mathbf{c} = [c_1 \quad c_2 \quad \cdots \quad c_n]$. Then the total profit associated with the production schedule $\mathbf{x}$ is $x_1c_1 + x_2c_2 + \cdots + x_nc_n = \mathbf{x} \cdot \mathbf{c}$.

The resource allocation problem is to find a production schedule $\mathbf{x}$ that satisfies (5.4), $\mathbf{x} \geq \mathbf{0}$, and for which $\mathbf{x} \cdot \mathbf{c}$ is as large as possible. This problem is an example of a "standard maximum problem."

Table 5.1

Variety	Amount (in grams) of				Net Profit ($)
	Peanuts	Raisins	Seeds	Dried Fruit	
A	80	20	50	0	$0.14
B	60	20	40	30	$0.12
C	30	0	80	40	$0.08
Amount Available (kg)	1000	600	500	300	

EXAMPLE 5.1   To illustrate these ideas, we consider a simple example, one that has few resources and products but has the basic characteristics just described. A firm produces three varieties of a snack food (trail mix) used by bikers and hikers. The trail mix is prepared from various combinations of peanuts, raisins, seeds, and dried fruit. Each package of trail mix produced weighs 150 grams, and the compositions of the three varieties are shown in Table 5.1. Also, suppose that on a particular day, the amounts (in kilograms) of ingredients available to the firm are as shown in Table 5.1. Finally, suppose that on this day, the net profit (price received minus cost of production) for each of the varieties is as shown in Table 5.1.

We assume that the task of the manager is to determine how many packages of each variety of trail mix to produce to yield the maximum net profit. To formulate the problem in mathematical form, we begin by defining the variables to be the quantities the manager is to determine: the number of packages of each variety of trail mix to produce. We set

$$x_1 = \text{number of packages of variety A to be produced}$$
$$x_2 = \text{number of packages of variety B to be produced}$$
$$x_3 = \text{number of packages of variety C to be produced}$$

The variables x_1, x_2, and x_3 are often called the *decision variables,* because they are set by the manager's decision about how much to make of each product. It follows from the definition that each of the variables must be nonnegative:

(5.5) $x_1 \geq 0, \quad x_2 \geq 0, \quad x_3 \geq 0$

Also, with this definition, the fact that the amount of each ingredient that can be used is constrained by the amount available leads to the following inequalities:

(5.6)
$$\begin{cases} 80x_1 + 60x_2 + 30x_3 \leq 1{,}000{,}000 & \text{constraint on amount of peanuts} \\ 20x_1 + 20x_2 \phantom{{}+ 30x_3} \leq 600{,}000 & \text{constraint on amount of raisins} \\ 50x_1 + 40x_2 + 80x_3 \leq 500{,}000 & \text{constraint on amount of seeds} \\ 30x_2 + 40x_3 \leq 300{,}000 & \text{constraint on amount of dried fruit} \end{cases}$$

where the amounts are measured in grams. Finally, the manager is to select a production schedule that makes the net profit as large as possible. That is, the task is to maximize

(5.7) $0.14x_1 + 0.12x_2 + 0.08x_3$

subject to the constraints (5.5) and (5.6).

Using the notation of equation (5.4), we have

$$(5.8) \quad \mathbf{x} = \begin{bmatrix} x_1 \\ x_2 \\ x_3 \end{bmatrix}, \quad \mathbf{A} = \begin{bmatrix} 80 & 60 & 30 \\ 20 & 20 & 0 \\ 50 & 40 & 80 \\ 0 & 30 & 40 \end{bmatrix}, \quad \mathbf{b} = \begin{bmatrix} 1{,}000{,}000 \\ 600{,}000 \\ 500{,}000 \\ 300{,}000 \end{bmatrix}, \quad \mathbf{c} = \begin{bmatrix} 0.14 & 0.12 & 0.08 \end{bmatrix}$$

where the entries in $\mathbf{A}$ and $\mathbf{b}$ are in grams, and the units of $\mathbf{c}$ are dollars per package. Note that the vector $\mathbf{x}$ with $x_1 = 10{,}000$, $x_2 = 0$, and $x_3 = 0$ satisfies (5.5) and (5.6), and so does the vector $\mathbf{x}$ with $x_1 = 2000$, $x_2 = 10{,}000$, and $x_3 = 0$. For $\mathbf{x} = [10{,}000 \quad 0 \quad 0]$, $\mathbf{x} \cdot \mathbf{c} = \$1400$, and for $\mathbf{x} = [2000 \quad 10{,}000 \quad 0]$, $\mathbf{x} \cdot \mathbf{c} = \$1480$. In fact the latter $\mathbf{x}$ is a solution to the problem, but we have no way of verifying that at this point in the discussion. ■

■ A Transportation Problem

A special case of a transportation problem was discussed in Section 2.4. Here we pose a more general problem and express it in a form similar to equation (5.4).

Suppose we have a product that is produced at m factories or sources and is to be transported to n stores for sale. Suppose that there is an amount s_i of the product available at factory i, $i = 1, \ldots, m$, and that amount d_j is required at store j, $j = 1, \ldots, n$. Also suppose that the cost of moving 1 unit of the product from factory i to store j is c_{ij}, $i = 1, \ldots, m$, and $j = 1, \ldots, n$. Finally, suppose that the problem is to determine how many units of the product to ship from each factory to each store so that the total shipping cost is as small as possible.

The quantities to be determined, the amount to be shipped from factory i to store j, for $i = 1, \ldots, m$, and $j = 1, \ldots, n$ will be denoted by x_{ij}. A *shipping schedule* $\mathbf{x}$ is a vector with $m \times n$ coordinates, where the first n coordinates are $x_{11}, \ldots, x_{1n}$, the next n coordinates are $x_{21}, \ldots, x_{2n}$, and the last n coordinates are $x_{m1}, \ldots, x_{mn}$.

The total amount of product shipped from factory i is the sum of the amounts shipped from factory i to each of the stores:

$$\text{Total amount of product shipped from factory } i = x_{i1} + x_{i2} + \cdots + x_{in} = \sum_{j=1}^{n} x_{ij}$$

We assume that products are shipped from the factories to the stores, and consequently, each of the x_{ij} must be nonnegative. Also, the amount shipped from factory i cannot exceed the amount available at factory i:

$$(5.9) \qquad x_{i1} + x_{i2} + \cdots + x_{in} \leq s_i, \quad i = 1, \ldots, m$$

Finally, the amount required at each of the stores must be provided:

$$(5.10) \qquad x_{1j} + x_{2j} + \cdots + x_{mj} \geq d_j, \quad j = 1, \ldots, n$$

We assume that the cost of shipping k units of the product is k times the cost of shipping 1 unit and that the cost of multiple shipments is the sum of the costs of the individual shipments. With these assumptions, the cost of the shipping schedule $\mathbf{x}$ is

$$(5.11) \qquad \mathbf{x} \cdot \mathbf{c} = c_{11} x_{11} + c_{12} x_{12} + \cdots + c_{mn} x_{mn} = \sum_{i=1}^{m} \sum_{j=1}^{n} c_{ij} x_{ij}$$

where the cost vector is

$$\mathbf{c} = [c_{11} \quad c_{12} \quad \cdots \quad c_{1n} \quad c_{21} \quad \cdots \quad c_{2n} \quad \cdots \quad c_{m1} \quad c_{m2} \quad \cdots \quad c_{mn}]$$

This problem is now similar in form to the earlier examples. The goal is to find a nonnegative vector $\mathbf{x}$, a shipping schedule, that satisfies (5.9) and (5.10) and for which (5.11) is as small as possible. Whereas the resource allocation problem was a maximization problem, this is a minimization problem. Both are examples of a specific type of linear programming problem.

In typical linear programming problems—for example, the problems known as standard maximum problems—the decision variables are constrained to be nonnegative, but they are not constrained in any other way. However, in many very important problems, the variables have additional constraints. For example, they might be constrained to be integers, and in some cases, the integer must be either a 0 or a 1. In our next example we discuss an important problem wherein each decision variable must be either a 0 or a 1. It is an example of a class of problems called covering problems.

EXAMPLE 5.2   The SHORT-HOP airline is about to begin service in the north central United States. The airline decides it will begin by flying to the following nine cities: Dayton (D), Fort Wayne (F), Green Bay (GB), Grand Rapids (GR), Kenosha (K), Marquette (M), Peoria (P), South Bend (S), and Toledo (T). Company managers are aware that it is very difficult to fly directly between all pairs of these cities, so they would like to mimic the national airlines and fly all flights into and out of a small number of hub cities. They make the following assumptions:

1. All cities should be within 200 miles of a hub city.
2. The number of hub cities should be as small as possible.

The problem for management is to decide on the appropriate number of hubs and also to decide where they should be located.

Although this problem may not appear to be a linear optimization problem, in fact it is very similar in mathematical structure to the other problems we have considered in this section. To show this we first need some data. In Table 5.2 we list the distances (straight-line distances) between all pairs of these cities. Next, using Table 5.2, we are able to formulate a linear problem that, when solved, will tell us how many hubs we need and which cities should be the hubs. First, as decision variables, we define $x_1, x_2, \ldots, x_9$ by the rule $x_i = 1$ if a city i is a hub, and $x_i = 0$ if a city i is not a hub. Here the cities are numbered as in

Table 5.2

	D	F	GB	GR	K	M	P	S	T
D	0	102	381	232	272	494	294	170	134
F	102	0	279	133	175	395	235	72	91
GB	381	279	0	159	134	143	273	215	298
GR	232	133	159	0	115	262	255	94	139
K	272	175	134	115	0	275	156	103	229
M	494	395	143	262	275	0	416	341	387
P	294	235	273	255	156	416	0	186	320
S	170	72	215	94	103	341	186	0	139
T	134	91	298	139	229	387	320	139	0

Table 5.3

CITY	1	2	3	4	5	6	7	8	9
1	1	1	0	0	0	0	0	1	1
2	1	1	0	1	1	0	0	1	1
3	0	0	1	1	1	1	0	0	0
4	0	1	1	1	1	0	0	1	1
5	0	1	1	1	1	0	1	1	0
6	0	0	1	0	0	1	0	0	0
7	0	0	0	0	1	0	1	1	0
8	1	1	0	1	1	0	1	1	1
9	1	1	0	1	0	0	0	1	1

Table 5.2, alphabetically, so city $1 = $ Dayton and city $9 = $ Toledo. For example, the vector [0 0 0 1 0 1 0 0 1] represents hubs at Grand Rapids, Marquette, and Toledo.

For each choice of the variables x_i, $i = 1, \ldots, 9$, the sum, $\sum_{i=1}^{9} x_i$, is the total number of hubs. Hence our objective is to minimize the objective function

$$f(\mathbf{x}) = \sum_{i=1}^{9} x_i, \quad \text{where } \mathbf{x} = \begin{bmatrix} x_1 \\ \vdots \\ x_9 \end{bmatrix}$$

For constraints on the variables x_i (the vector $\mathbf{x}$), we recall that every one of our nine cities is to be within 200 miles of a hub. We need to express this as a constraint on $\mathbf{x}$, and as a first step, we need to know, for each city, which other cities are within 200 miles. This information is obtained from Table 5.2 and shown in Table 5.3, where an entry of 1 in row i and column j means that city i and city j are within 200 miles of each other. An entry of 0 means they are not within 200 miles of each other.

The entries in Table 5.3 form a matrix, $\mathbf{A} = [a_{ij}]$. We note from Table 5.3 that it will be impossible to use only one hub, because no row has all entries equal to 1. Next we consider the product $\mathbf{Ax}$ and note that the ith row of $\mathbf{Ax}$ represents the number of hubs within 200 miles of city i. This follows because the ith row is the sum $\sum_{j=1}^{9} a_{ij}x_j$, each product $a_{ij}x_j$ is either 0 or 1, and the product is 1 only if both $a_{ij} = 1$ and $x_j = 1$. We now have a simple way to express our constraint that every city be within 200 miles of a hub:

$$\mathbf{Ax} \geq \begin{bmatrix} 1 \\ 1 \\ \vdots \\ 1 \end{bmatrix}$$

If we let $\mathbf{U}$ denote the column vector with nine coordinates and with all entries of 1, then our mathematical problem is the following:

Find a vector $\mathbf{x}$, with all coordinates 0 or 1 such that $\mathbf{Ax} \geq \mathbf{U}$ and $\mathbf{x} \cdot \mathbf{U}$ is a minimum.

The problem we have just formulated is an example of a linear programming problem that includes the additional constraint that the decision variables must have values of 0 or 1. This is a relatively small problem, with only nine variables, and it can be easily solved by hand or by using a program such as Excel or Maple (Exercise 9). However, in general, large problems that are constrained to have only integer or binary variables can be very difficult to solve quickly. One reference that discusses methods for solving these problems is Wayne Winston's *Operations Research* (PWS-Kent Publishing, 1991). ■

Exercises 5.1

In each of the following situations, formulate (but do not solve) a linear programming problem whose solution provides the desired information.

1. A national consulting firm plans to standardize its consulting activities by creating four types of consulting groups, each group consisting of consultants specializing in financial issues, technological issues, or operational issues. The composition of the groups is as follows:

   • A group of type A consists of 6 consultants with these specialties: 3 financial, 1 technological, 2 operational.
   • A group of type B consists of 5 consultants with these specialties: 2 financial, 2 technological, 1 operational.
   • A group of type C consists of 3 consultants with these specialties: 1 financial, 1 technological, 1 operational.
   • A group of type D consists of 7 consultants with these specialties: 2 financial, 3 technological, 2 operational.

   The firm has 50 consultants with financial expertise, 40 with technological expertise, and 30 with operational expertise. Finally, the firm estimates that each week, a group of type A will generate net profit of $5500, a group of type B will generate $4000, a group of type C will generate $3000, and a group of type D will generate $6000.

   Management is interested in determining how many groups of each type should be created to yield the maximum net profit.

2. Between 8 a.m. and noon on Monday, a telephone company has a demand for 500 calls from Boston to Phoenix and for 400 calls from Washington, DC, to Phoenix. The calls can be routed through either Indianapolis or Memphis and then through either Denver or Houston. The capacities between these cities are shown in Table 5.4.

   Find the maximum number of telephone calls that can be routed from Boston to Phoenix and from Washington to Phoenix using this capacity.

   Remark In this situation we consider only calls that originate in Boston and Washington and terminate in Phoenix, not calls that originate or terminate in any other cities.

Table 5.4

Boston to Indianapolis	300	Boston to Memphis	200
Washington to Indianapolis	250	Washington to Memphis	250
Indianapolis to Denver	180	Indianapolis to Houston	220
Memphis to Denver	160	Memphis to Houston	260
Denver to Phoenix	500	Houston to Phoenix	400

3. An accounting firm has expanded its client base, and new accountants must be added. The current workforce consists of 100 experienced accountants, and the plan is to add a total of 50 newly hired accountants, for a total workforce of 150, over a period of five weeks. Make the following assumptions:

   - New accountants begin work on Monday each week and on no other days.
   - A total of 150 accountants, experienced and trained new accountants, are to be available for work at the beginning of week 6.
   - New accountants must be trained before they can provide client service, and it takes one experienced accountant two weeks to train one new employee. During this time, neither the experienced accountants nor the new employees provide any service on client projects.
   - An experienced accountant or a trained new accountant can handle one project each week.
   - Experienced accountants are paid $800 per week, and new employees are paid $600 per week for both training and project work.
   - The contracted business for the next five weeks is $d_1, d_2, \ldots, d_5$ projects per week, and this demand must be met on a weekly basis.

   Find a hiring schedule that meets the company's goals with minimum salary costs.

4. A cargo barge has two compartments for storing cargo: forward and rear. These compartments have the limitations on volume and weight shown in Table 5.5. To enable the barge to safely pass through some shallow areas, 40% of the weight must be in the forward compartment and 60% in the rear. The barge owner has 3 cargoes available for the next trip, and the data on these cargoes are shown in the lower section of Table 5.5. Any portion of each cargo may be shipped. The shipper needs to know how much of each cargo to ship and how to load the cargo in the two compartments in order to maximize profit for this barge for the trip.

   How would the problem change if the balancing condition were replaced by the requirement that not less than 40% and not more than 45% of the weight be in the forward compartment?

Table 5.5

Barge Data		
	Capacity	
Compartment	Weight (tons)	Space (cu ft)
Forward	8	5000
Rear	12	7000

Cargo Data			
Cargo	Weight (tons)	Volume (cu ft/ton)	Profit ($/ton)
1	14	500	100
2	22	700	130
3	9	400	90

Table 5.6

Time of Day (24-hour clock)	Period	Minimum Number of Security Personnel Required during Period
2–6	1	2
6–10	2	5
10–14	3	8
14–18	4	10
18–22	5	4
22–2	6	3

5. A catering company requires a known number of cloth napkins for meals to be served during the week. The goal is to meet the demand at minimum cost. Suppose the napkins required are 100, 80, 120, 150, and 130 for Monday through Friday, respectively. New napkins can be purchased for $0.12 each, and used napkins can be laundered. Napkins can be laundered for next-day service at $0.06 each, for use the second day at $0.04 each, and for use the third day at $0.03 each. At the end of the week, all used napkins can be sold for $0.01 each.

6. An airport security firm has the responsibility of assigning security personnel to meet the contractual commitments shown in Table 5.6.

   Period 1 immediately follows period 6 to maintain continuous service.

   Each security person works eight continuous hours. The assignment task is to determine how many people should start work at the beginning of each period. Formulate a linear optimization problem whose solution gives the minimum number of security people to start work at the beginning of each period so that the staffing requirements are met.

7. A medical laboratory has a "clean room" that requires high-efficiency filters in its air-handling system. Each time the room is used, a new filter is installed. Used filters can be cleaned. The laboratory has a schedule of activities for the next five weeks, and from that schedule the filter requirements can be deduced. Filters are ordered for one week at a time, and used filters are collected and processed (either sent out for cleaning or held for sale) at the end of each week. The goal is to meet the requirements for filters at minimum cost.

   Data for the situation are as follows: Filter requirements for the next five weeks are 15, 12, 18, 22, and 20 respectively. New filters can be purchased for $35 each, and used filters can be cleaned and reconditioned, the costs depending on how quickly the cleaning is done. Filters can be cleaned for the next week for $18 each, for use the second week at $12 each, and for the third week at $9 each. At the end of the five-week period, used filters can be sold for $4 each.

8. The transportation problem was formulated with the assumption that the total supply is at least as large as the total demand. Show how the model can be modified to replace this assumption with the (apparently more restrictive) assumption that supply equals demand. That is, show that if you can solve the transportation problem with supply equal to demand, then you can use that solution to solve a transportation problem with supply at least as large as demand.

9. Solve the SHORT-HOP airline problem, Example 5.2. How many hubs are needed, and where should they be located? Also solve the problem if the distance to a hub is at most 150 miles instead of 200 miles.

10. Modify the SHORT-HOP airline problem by adding a tenth city, Duluth (DU). Create the appropriate table of distances, and find the number of hubs needed and where they should be located. Use distance to a hub of at most 275 miles.

5.2   Linear Programming Problems and Duality

Each of the problems formulated in Section 5.1 is of the following form: maximize (or minimize) a linear function subject to nonnegativity conditions and inequality constraints on the variables. It will be helpful to identify one such problem formally and use it as a basis for discussion.

> Given an $m \times n$ matrix $\mathbf{A}$, an m-vector $\mathbf{b}$, and an n-vector $\mathbf{c}$, the *Standard Maximum Problem,* determined by $\mathbf{A}$, $\mathbf{b}$, and $\mathbf{c}$, denoted for convenience by SMP[$\mathbf{A}$, $\mathbf{b}$, $\mathbf{c}$], is the problem

(5.12)
$$\text{maximize} \quad \mathbf{x} \cdot \mathbf{c}$$
$$\text{subject to} \quad \mathbf{x} \geq \mathbf{0} \text{ and } \mathbf{A}\mathbf{x} \leq \mathbf{b}$$

> The *feasible set* for this standard maximum problem is $\{\mathbf{x} \in \mathbb{R}^n \mid \mathbf{x} \geq \mathbf{0} \text{ and } \mathbf{A}\mathbf{x} \leq \mathbf{b}\}$. The function $\mathbf{x} \cdot \mathbf{c}$ is called the *objective function* for the problem.

We will use this problem, the standard maximum problem, as a basic problem, and later on we will see how to convert other problems into this form. Other discussions may adopt other forms of linear programming problems as basic, and in such cases the results may have slightly different forms. However, the primary results have the same substance.

One of the striking features of linear programming is the fact that it is possible to gain information about one problem—a standard maximum problem, for example—by studying it together with another problem, a problem referred to as a *dual problem.*

> The dual of the standard maximum problem (5.12) is the problem

(5.13)
$$\text{minimize} \quad \mathbf{y} \cdot \mathbf{b}$$
$$\text{subject to} \quad \mathbf{y} \geq \mathbf{0} \text{ and } \mathbf{A}^T\mathbf{y} \geq \mathbf{c}$$

In (5.13), $\mathbf{A}^T$ denotes the transpose of the matrix $\mathbf{A}$, and if $\mathbf{A}$ is an $m \times n$ matrix, then $\mathbf{A}^T$ is an $n \times m$ matrix. The vector $\mathbf{y}$ is an m-vector.

The dual of a standard maximum problem is an example of a standard minimum problem. The definition of a standard minimum problem is as follows:

> Given a $p \times q$ matrix $\mathbf{D}$, a p-vector $\mathbf{e}$, and a q-vector $\mathbf{f}$, the standard minimum problem, determined by $\mathbf{D}$, $\mathbf{e}$, and $\mathbf{f}$, denoted for convenience by smp[$\mathbf{D}$, $\mathbf{e}$, $\mathbf{f}$], is the problem

$$\text{minimize} \quad \mathbf{y} \cdot \mathbf{f}$$
$$\text{subject to} \quad \mathbf{y} \geq \mathbf{0} \text{ and } \mathbf{D}\mathbf{y} \geq \mathbf{e}$$

> The *feasible set* for the smp[$\mathbf{D}$, $\mathbf{e}$, $\mathbf{f}$] is $\{y \in \mathbb{R}^q \mid \mathbf{y} \geq \mathbf{0} \text{ and } \mathbf{D}\mathbf{y} \geq \mathbf{e}\}$. The function $\mathbf{y} \cdot \mathbf{f}$ is called the *objective function* for this problem.

Using this definition of a standard minimum problem and our definition of a dual minimum problem, we see that the dual of the SMP[**A**, **b**, **c**] is the smp[$\mathbf{A}^T$, **c**, **b**].

EXAMPLE 5.3 In Example 5.1 (Section 5.1) we formulated a standard maximum problem for a resource allocation situation involving trail mix. The original problem was given in equations (5.5), (5.6), and (5.7) and in matrix form in (5.8). The dual problem is

$$\text{minimize} \quad 1{,}000{,}000y_1 + 600{,}000y_2 + 500{,}000y_3 + 300{,}000y_4$$

$$\text{subject to} \quad \begin{cases} y_1 \geq 0, \ y_2 \geq 0, \ y_3 \geq 0, \ y_4 \geq 0 \\ 80y_1 + 20y_2 + 50y_3 \geq 0.14 \\ 60y_1 + 20y_2 + 40y_3 + 30y_4 \geq 0.12 \\ 30y_1 + 80y_3 + 40y_4 \geq 0.08 \end{cases}$$

The vector **y** with $y_1 = 0.0027$, $y_2 = 0$, $y_3 = 0$, and $y_4 = 0$ is feasible for the dual problem, as is the vector **y** with $y_1 = 0$, $y_2 = 0$, $y_3 = 0.003$, and $y_4 = 0$. There is no feasible vector **y** with only the y_2 coordinate positive, and there is no feasible vector **y** with only the y_4 coordinate positive (Exercise 1). ■

Not every linear programming problem has a solution; some do and some do not. One of our goals will be to find criteria for determining which problems have solutions. First, we give an example of a simple problem that has no solution.

EXAMPLE 5.4 Consider the following standard maximum problem:

$$\text{maximize} \quad x_1 + x_2 + x_3$$

$$\text{subject to} \quad \begin{cases} x_1 \geq 0, \ x_2 \geq 0, \ x_3 \geq 0 \\ x_1 - 2x_2 \leq -2 \\ -2x_1 + x_2 \leq -2 \\ 10x_1 + 10x_2 - x_3 \leq 1 \end{cases}$$

It can be verified easily that the vector $\mathbf{x} = [2 \quad 2 \quad 40]$ is in the feasible set for this problem. In fact, for all $t \geq 2$, the vector $\mathbf{x} = [t \quad t \quad 20t]$ is feasible, the objective function $x_1 + x_2 + x_3$ has the value $22t$, and this can be made arbitrarily large by taking t large. Therefore, the problem has no solution; that is, there is no maximum for the objective function over the feasible set. ■

For the problem of Example 5.4, the dual problem is

$$\text{minimize} \quad -2y_1 - 2y_2 + y_3$$

$$\text{subject to} \quad \begin{cases} y_1 \geq 0, \ y_2 \geq 0, \ y_3 \geq 0 \\ y_1 - 2y_2 + 10y_3 \geq 1 \\ -2y_1 + y_2 + 10y_3 \geq 1 \\ -y_3 \geq 1 \end{cases}$$

However, the last of the constraints, $-y_3 \geq 1$, is inconsistent with the nonnegativity condition $y_3 \geq 0$, and consequently, the feasible set for the dual problem is empty. In fact, in this case, the feasible set of the dual must be empty, because if both the feasible set for a standard maximum problem and the feasible set for its dual are not empty, then both problems have solutions. This result, which is known as the Fundamental Duality Theorem, is an important theoretical result in the study of linear programming problems.

THEOREM 5.1 (Fundamental Duality Theorem)   Given a standard maximum problem, SMP[$\mathbf{A}, \mathbf{b}, \mathbf{c}$], if the feasible sets for both the standard problem and its dual are not empty, then both problems have solutions. In addition:

1. For any $\mathbf{x}$ feasible for SMP[$\mathbf{A}, \mathbf{b}, \mathbf{c}$] and any $\mathbf{y}$ feasible for the dual, $\mathbf{x} \cdot \mathbf{c} \le \mathbf{y} \cdot \mathbf{b}$.
2. If $\mathbf{x}^*$ is feasible for SMP[$\mathbf{A}, \mathbf{b}, \mathbf{c}$] and $\mathbf{y}^*$ is feasible for the dual, and if $\mathbf{x}^* \cdot \mathbf{c} = \mathbf{y}^* \cdot \mathbf{b}$, then $\mathbf{x}^*$ and $\mathbf{y}^*$ are optimal for the two problems, respectively.
3. If $\mathbf{x}^*$ and $\mathbf{y}^*$ are optimal for the SMP[$\mathbf{A}, \mathbf{b}, \mathbf{c}$] and its dual, then $\mathbf{x}^* \cdot \mathbf{c} = \mathbf{y}^* \cdot \mathbf{b}$.   ∎

A proof of this complete result would divert us from our primary objective and is omitted, but we will prove parts (1) and (2) below. First, however, it is important to note the geometry of the result, and we give a simple example in which both of the feasible sets can be graphed in two dimensions.

EXAMPLE 5.5   (Geometric aspects of linear optimization: A simple example)   Let

$$\mathbf{A} = \begin{bmatrix} 1 & 2 \\ 3 & 4 \end{bmatrix}, \quad \mathbf{b} = \begin{bmatrix} 5 \\ 12 \end{bmatrix}, \quad \mathbf{c} = \begin{bmatrix} 5 \\ 8 \end{bmatrix}$$

Consider the problem of maximizing the quantity $L(\mathbf{x}) = \mathbf{x} \cdot \mathbf{c} = 5x_1 + 8x_2$, where

$$\mathbf{x} = \begin{bmatrix} x_1 \\ x_2 \end{bmatrix}$$

is subject to the constraints $\mathbf{x} \ge 0$ and $\mathbf{A}\mathbf{x} \le \mathbf{b}$. These constraints are given by the four inequalities

(5.14)
$$\begin{aligned} x_1 + 2x_2 &\le 5 \\ 3x_1 + 4x_2 &\le 12 \\ x_1 &\ge 0 \\ x_2 &\ge 0 \end{aligned}$$

We want to examine the set of solutions of (5.14) from a geometrical point of view. This set of solutions is the set of feasible vectors for the linear optimization problem posed above. We begin by considering each of the inequalities in (5.14) to be the determining equation for a set of points in the x_1-x_2 plane. Specifically, we associate with each inequality the set of points in $\mathbb{R}^2$ that satisfy that inequality. For example, the first inequality, $x_1 + 2x_2 \le 5$, determines the set $F_1 = \{\mathbf{x} \mid x_1 + 2x_2 \le 5\}$. If we graph the four sets obtained in this manner, we obtain the graphs shown in Figure 5.1. Each set is a half-plane.

A vector $\mathbf{x}$ is feasible if it satisfies all four inequalities of (5.14). In other words, it must be in each of the sets F_1, F_2, F_3, and F_4. Hence the set of feasible vectors F is given by $F = F_1 \cap F_2 \cap F_3 \cap F_4$. This set is shown in Figure 5.2.

The set F is bounded by four lines, and these lines determine four corner points of F: $(0, 0)$, $(0, \frac{5}{2})$, $(2, \frac{3}{2})$, and $(4, 0)$. If we evaluate the function $L(\mathbf{x}) = 5x_1 + 8x_2$ at these corner points, we obtain $L(0, 0) = 0$, $L(0, \frac{5}{2}) = 20$, $L(2, \frac{3}{2}) = 22$, and $L(4, 0) = 20$. Thus, at least among the corner points of F, the maximum of L is 22. Since there are infinitely

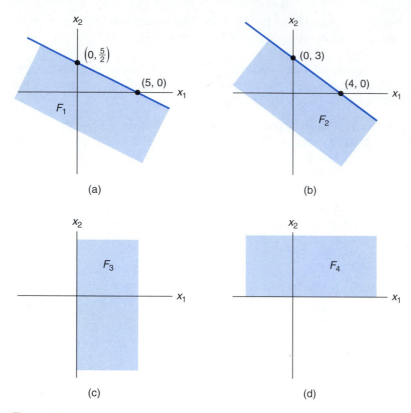

Figure 5.1

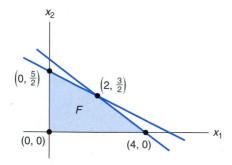

Figure 5.2

many points of F that are not corner points, it is not obvious that the maximum of L over the entire set F is also 22. We shall prove this result in a more general setting later in the section. We prove it in this special case by considering the dual problem.

The dual problem in this case is the problem of minimizing $\mathbf{y} \cdot \mathbf{b}$, where

$$\mathbf{y} = \begin{bmatrix} y_1 \\ y_2 \end{bmatrix}$$

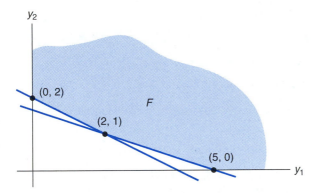

Figure 5.3

is subject to the constraints $\mathbf{A}^T\mathbf{y} \geq \mathbf{c}$ and $\mathbf{y} \geq \mathbf{0}$. These constraints are given by the following inequalities:

(5.15)
$$\begin{aligned} y_1 + 3y_2 &\geq 5 \\ 2y_1 + 4y_2 &\geq 8 \\ y_1 &\geq 0 \\ y_2 &\geq 0 \end{aligned}$$

As in the original problem, each inequality determines a set of points in the plane. We take the intersection of these four sets to obtain the set of feasible vectors for the dual problem. This set is shown in Figure 5.3. The corner points of the feasible set for the dual problem are the points $(0, 2)$, $(2, 1)$, and $(5, 0)$. If we evaluate the function $L'(\mathbf{y}) = \mathbf{y} \cdot \mathbf{b} = 5y_1 + 12y_2$ at these corner points, we obtain the values $L'(0, 2) = 24$, $L'(2, 1) = 22$, and $L'(5, 0) = 25$. Therefore, on the corner points, the function L' has a minimum of 22 at the point $(2, 1)$. However, it now follows that this is actually a minimum of L' for all feasible points in F' because of our information about the original problem. Indeed, we know that $L'(2, 1) = 22 = L(2, \frac{3}{2})$, and according to the Fundamental Duality Theorem, $L'(\mathbf{y}) \geq L(\mathbf{x})$ for all feasible vectors $\mathbf{x}$ and $\mathbf{y}$. Therefore, the vectors

$$\mathbf{x} = \begin{bmatrix} 2 \\ \frac{3}{2} \end{bmatrix} \quad \text{and} \quad \mathbf{y} = \begin{bmatrix} 2 \\ 1 \end{bmatrix}$$

are optimal vectors for the original and dual problems, respectively. Thus we have shown that, at least in this special case, it is possible to find the optimal vectors by checking only a finite set of special vectors. Although we will not prove all of the Fundamental Theorem, we will try to show that the situation of this example is not unusual and that, in fact, similar statements can be made about more general problems where the vectors are not in $\mathbb{R}^2$. ∎

Let $[\mathbf{A}, \mathbf{b}, \mathbf{c}]$ be a standard maximum problem, and let $F = \{\mathbf{x} \in \mathbb{R}^n \mid \mathbf{x} \geq \mathbf{0}$ and $\mathbf{A}\mathbf{x} \leq \mathbf{b}\}$ be the feasible set for that problem. If $\mathbf{x} \in F$, then $\mathbf{A}\mathbf{x} \leq \mathbf{b}$, and consequently there is a vector $\mathbf{w}$, $\mathbf{w} \geq \mathbf{0}$, such that $\mathbf{A}\mathbf{x} + \mathbf{w} = \mathbf{b}$. The vector $\mathbf{w}$ is called the *slack vector* for the problem.

Likewise, the set $F' = \{\mathbf{y} \in \mathbb{R}^m \mid \mathbf{y} \geq \mathbf{0}$ and $\mathbf{A}^T\mathbf{y} \geq \mathbf{c}\}$ is the feasible set for the dual problem. If $\mathbf{y} \in F'$, then, as above, there is a vector $\mathbf{v}$, $\mathbf{v} \geq \mathbf{0}$, such that $\mathbf{A}^T\mathbf{y} - \mathbf{v} = \mathbf{c}$. The vector $\mathbf{v}$ is called the *surplus vector* for the dual problem.

Consider the relations

(5.16) $$\mathbf{Ax} + \mathbf{w} = \mathbf{b} \quad \text{and} \quad \mathbf{A}^T\mathbf{y} - \mathbf{v} = \mathbf{c}$$

If we take the dot product of the first equation with any vector $\mathbf{y} \in \mathbb{R}^m$, then we have

$$\mathbf{y} \cdot (\mathbf{Ax} + \mathbf{w}) = \mathbf{y} \cdot \mathbf{b} \quad \text{or} \quad \mathbf{y} \cdot \mathbf{Ax} + \mathbf{y} \cdot \mathbf{w} = \mathbf{y} \cdot \mathbf{b}$$

Similarly, if we take the dot product of the second equation in (5.16) with any vector $\mathbf{x} \in \mathbb{R}^n$, then we have

$$\mathbf{x} \cdot (\mathbf{A}^T\mathbf{y} - \mathbf{v}) = \mathbf{x} \cdot \mathbf{c} \quad \text{or} \quad \mathbf{x} \cdot \mathbf{A}^T\mathbf{y} - \mathbf{x} \cdot \mathbf{v} = \mathbf{x} \cdot \mathbf{c}$$

Next, a straightforward expansion of $\mathbf{y} \cdot \mathbf{Ax}$ and $\mathbf{x} \cdot \mathbf{A}^T\mathbf{y}$—that is, expressing each of the matrix products as a sum of products of the entries—shows that $\mathbf{y} \cdot \mathbf{Ax} = \mathbf{x} \cdot \mathbf{A}^T\mathbf{y}$ (Exercise 2). Then, subtracting $\mathbf{x} \cdot \mathbf{A}^T\mathbf{y} - \mathbf{x} \cdot \mathbf{v} = \mathbf{x} \cdot \mathbf{c}$ from $\mathbf{y} \cdot \mathbf{Ax} + \mathbf{y} \cdot \mathbf{w} = \mathbf{y} \cdot \mathbf{b}$, we have

(5.17) $$\mathbf{y} \cdot \mathbf{w} + \mathbf{x} \cdot \mathbf{v} = \mathbf{y} \cdot \mathbf{b} - \mathbf{x} \cdot \mathbf{c}$$

Expression (5.17) is known as Tucker's duality relation, and it holds for any vectors $\mathbf{x} \in F$, $\mathbf{y} \in F'$ and any associated slack and surplus vectors $\mathbf{w}$ and $\mathbf{v}$, respectively.

In the case that $\mathbf{x} \in F$ and $\mathbf{y} \in F'$, the vectors $\mathbf{x}$, $\mathbf{y}$, $\mathbf{w}$, and $\mathbf{v}$ are all nonnegative, and it follows that $\mathbf{y} \cdot \mathbf{w} = y_1 w_1 + y_2 w_2 + \cdots + y_m w_m \geq 0$ and $\mathbf{x} \cdot \mathbf{v} = x_1 v_1 + x_2 v_2 + \cdots + x_n v_n \geq 0$. From (5.17), it follows that

(5.18) $$\mathbf{y} \cdot \mathbf{b} - \mathbf{x} \cdot \mathbf{c} \geq 0 \quad \text{or} \quad \mathbf{y} \cdot \mathbf{b} \geq \mathbf{x} \cdot \mathbf{c}$$

To summarize, if $\mathbf{x} \in F$ and $\mathbf{y} \in F'$, then the value of the objective function for the standard maximum problem evaluated at $\mathbf{x}$ is no larger than the value of the objective function of the dual problem evaluated at $\mathbf{y}$. This proves the first part of Theorem 5.1.

There is a consequence of the inequality (5.18) that is quite useful, and in fact it is also part of the Fundamental Theorem, part (2). Because it is so important, we prove (2): Let $\mathbf{x} \in F$, $\mathbf{y} \in F'$, and suppose $\mathbf{x} \cdot \mathbf{c} = \mathbf{y} \cdot \mathbf{b}$. Next, let $\mathbf{x}'$ be any other vector in F. Then, for the vector $\mathbf{y}$ of the Fundamental Theorem, by (5.18) we have $\mathbf{y} \cdot \mathbf{b} \geq \mathbf{x}' \cdot \mathbf{c}$. By the hypothesis of (2), $\mathbf{y} \cdot \mathbf{b} = \mathbf{x} \cdot \mathbf{c}$, and consequently $\mathbf{x} \cdot \mathbf{c} \geq \mathbf{x}' \cdot \mathbf{c}$. This means that $\mathbf{x}$ is a solution of the standard maximum problem. The proof that $\mathbf{y}$ is a solution of the dual problem is similar.

■ Complementary Slackness

One of the consequences of the Fundamental Theorem is that there is a relationship between constraints of a standard maximum problem that are satisfied as strict inequalities and dual variables that have the value zero. The relationship follows from the equality (5.17). If $\mathbf{x}^*$ and $\mathbf{y}^*$ are optimal for a standard maximum problem and its dual, respectively, then using the Fundamental Theorem we know $\mathbf{x}^* \cdot \mathbf{c} = \mathbf{y}^* \cdot \mathbf{b}$. Next, when we apply (5.17) to $\mathbf{x}^*$ and $\mathbf{y}^*$, it follows that

$$\mathbf{y}^* \cdot \mathbf{w} + \mathbf{x}^* \cdot \mathbf{v} = \mathbf{y}^* \cdot \mathbf{b} - \mathbf{x}^* \cdot \mathbf{c} = 0$$

where $\mathbf{w}$ and $\mathbf{v}$ are the slack vectors for the standard problem and its dual, respectively. Since each of the vectors $\mathbf{x}^*$, $\mathbf{w}$, $\mathbf{y}^*$, and $\mathbf{v}$ have nonnegative coordinates, it follows that $\mathbf{y}^* \cdot \mathbf{w} = 0$ and $\mathbf{x}^* \cdot \mathbf{v} = 0$. The latter equalities require that for each positive coordinate of $\mathbf{y}^*$, the corresponding coordinate of $\mathbf{w}$ must be 0 (recall that $\mathbf{y}^*$ and $\mathbf{w}$ are both vectors in $\mathbb{R}^m$)

and for each positive coordinate of $\mathbf{x}^*$, the corresponding coordinate of $\mathbf{v}$ must be 0 (recall that $\mathbf{x}^*$ and $\mathbf{v}$ are both vectors in $\mathbb{R}^n$). This relationship between positive coordinates in the vectors $\mathbf{x}^*$ and $\mathbf{y}^*$ and zero coordinates in $\mathbf{w}$ and $\mathbf{v}$ is known as *complementary slackness*. Another way to express this is to say that for every inequality in the original problem that is satisfied as a strict inequality for an optimal vector $\mathbf{x}^*$, the corresponding dual variable has the value zero. Likewise, for every inequality in the dual problem that is satisfied as a strict inequality for an optimal vector $\mathbf{y}^*$, the corresponding variable in the original problem has the value zero.

■ General Linear Programming Problems

We have restricted our discussion to standard maximum problems, but much of our work is applicable to a wider array of problems. For instance, suppose that we have a problem similar to a standard maximum problem but with an equality constraint rather than all inequality constraints. That is, one of the constraints in the feasible set is of the form

$$x_1 a_{i1} + x_2 a_{i2} + \cdots + x_n a_{in} = \sum_{j=1}^{n} x_j a_{ij} = b_i$$

This equality constraint can be written as a pair of inequality constraints:

$$x_1 a_{i1} + x_2 a_{i2} + \cdots + x_n a_{in} \le b_i \quad \text{and} \quad x_1 a_{i1} + x_2 a_{i2} + \cdots + x_n a_{in} \ge b_i$$

or

$$x_1 a_{i1} + x_2 a_{i2} + \cdots + x_n a_{in} \le b_i \quad \text{and} \quad -x_1 a_{i1} - x_2 a_{i2} - \cdots - x_n a_{in} \le -b_i$$

That is, a single equality constraint is replaced by two inequality constraints. If we make this replacement for every equality constraint, then we have a set of constraints in which all are inequalities, as required for a standard maximum problem, but with a larger coefficient matrix $\mathbf{A}$. Replacing each equality constraint by two inequalities adds one row to the matrix $\mathbf{A}$ for each such replacement.

EXAMPLE 5.6 A linear optimization problem is as follows:

$$\text{maximize} \quad 2x_1 + x_2$$

$$\text{subject to} \quad \begin{cases} x_1 \ge 0, \ x_2 \ge 0 \\ x_1 + 2x_2 \le 6 \\ 5x_1 + 2x_2 \le 20 \\ 2x_1 - x_2 = 2 \end{cases}$$

This can be written as a standard maximum problem by writing the third constraint, an equality constraint, as a pair of inequality constraints. The equality constraint $2x_1 - x_2 = 2$ is equivalent to the pair of constraints

$$2x_1 - x_2 \le 2 \quad \text{and} \quad 2x_1 - x_2 \ge 2$$

or

$$2x_1 - x_2 \le 2 \quad \text{and} \quad -2x_1 + x_2 \le -2$$

When we replace the equality constraint by the last pair of inequalities, the problem is written in the form of a standard maximum problem. The matrix $\mathbf{A}$ and the vector $\mathbf{b}$ for the latter problem are

$$A = \begin{bmatrix} 1 & 2 \\ 5 & 2 \\ 2 & -1 \\ -2 & 1 \end{bmatrix} \qquad b = \begin{bmatrix} 6 \\ 20 \\ 2 \\ -2 \end{bmatrix}$$ ∎

To continue the discussion, we construct the dual of the standard maximum problem of Example 5.6. Because there are four inequality constraints, there are four dual variables, and the dual problem is

minimize $6y_1 + 20y_2 + 2y_3 - 2y_4$

subject to $\begin{cases} y_1 \geq 0, \; y_2 \geq 0, \; y_3 \geq 0, \; y_4 \geq 0 \\ y_1 + 5y_2 + 2y_3 - 2y_4 \geq 2 \\ 2y_1 + 2y_2 - y_3 + y_4 \geq 1 \end{cases}$

We note that the objective function and the inequality constraints can be written as

objective function $6y_1 + 20y_2 + 2(y_3 - y_4)$

inequality constraints $\begin{cases} y_1 + 5y_2 + 2(y_3 - y_4) \geq 2 \\ 2y_1 + 2y_2 - (y_3 - y_4) \geq 1 \end{cases}$

That is, the variables y_3 and y_4 do not occur independently, but always as the difference $(y_3 - y_4)$. Therefore, we can replace $y_3 - y_4$ by another variable, call it y_5. Since y_3 and y_4 are both nonnegative, the difference $y_3 - y_4$ can be any real number. That is, the variable y_5 is unconstrained. Summarizing, the dual of the standard maximum problem of Example 5.6 is the problem

minimize $6y_1 + 20y_2 + 2y_5$

subject to $\begin{cases} y_1 \geq 0, \; y_2 \geq 0, \; y_5 \text{ unconstrained} \\ y_1 + 5y_2 + 2y_5 \geq 2 \\ 5x_1 + 2x_2 \leq 20 \\ 2y_1 + 2y_2 - y_5 \geq 1 \end{cases}$

The results of this specific example are valid in general, and we formulate the result as a statement about a type of linear programming problem we call a *General Linear Programming Problem*. For our definition of General Problems, it will be useful to have an additional bit of notation. Let A be an $m \times n$ matrix. We denote the rows of A by $a_1, a_2, \ldots, a_m$, and we denote the columns of A by $a^1, a^2, \ldots, a^n$. Note that each a_i is an n-vector and each a^j is an m-vector.

Given an $m \times n$ matrix A, an m-vector b, and an n-vector c, a subset P of the integers $\{1, 2, \ldots, n\}$, and a subset Q of the integers $\{1, 2, \ldots, m\}$, the associated *General Maximum Problem* is

maximize $x \cdot c$

subject to $\begin{cases} x_i \geq 0, & \text{for } i \in P \\ x \cdot a_j \leq b_j, & \text{for } j \in Q \\ x \cdot a_j = b_j, & \text{for } j \in Q' \end{cases}$

where Q' is the set of integers in $\{1, 2, \ldots, m\}$ that are not in Q.

Using the ideas and techniques of Example 5.6 and the discussion following that example, we can transform this into a standard maximum problem. That standard maximum problem has a dual as defined at the beginning of this section. Using the same techniques, we can write the dual problem as follows:

$$\text{minimize} \quad \mathbf{y} \cdot \mathbf{b}$$

$$\text{subject to} \quad \begin{cases} y_i \geq 0, & \text{for } i \in Q \\ \mathbf{y} \cdot \mathbf{a}^j \geq c_j, & \text{for } j \in P \\ \mathbf{y} \cdot \mathbf{a}^j = c_j, & \text{for } j \in P' \end{cases}$$

where P' is the set of integers in $\{1, 2, \ldots, n\}$ that are not in P.

A convenient way to remember this is that each unconstrained variable in the original problem is associated with an equality constraint in the dual, and each equality constraint in the original problem is related to an unconstrained variable in the dual.

EXAMPLE 5.7 Formulate the dual of the following general maximum problem:

$$\text{maximize} \quad 2x_1 + 6x_2 - 4x_3$$

$$\text{subject to} \quad \begin{cases} x_1 \geq 0 \quad (x_2 \text{ and } x_3 \text{ are unconstrained}) \\ 5x_1 + 5x_2 + x_3 \leq 10 \\ x_1 - x_2 + 5x_3 = 5 \end{cases}$$

Since there are two inequality/equality constraints, there are two dual variables. The first constraint is an inequality, so the first dual variable must be nonnegative. The second constraint is an equality constraint, so the second dual variable is unconstrained. Since the first variable in the original problem is nonnegative, the first inequality/equality constraint in the dual problem is an inequality constraint. Since the second and third variables in the original problem are unconstrained, the second and third inequality/equality constraints in the dual problem are equality constraints. The dual problem is

$$\text{minimize} \quad 10y_1 + 5y_2$$

$$\text{subject to} \quad \begin{cases} y_1 \geq 0, \ y_2 \text{ unconstrained} \\ 5y_1 + y_2 \geq 2 \\ 5y_1 - y_2 = 6 \\ y_1 + 5y_2 = -4 \end{cases} \qquad ∎$$

Exercises 5.2

1. In Example 5.4, show that there is no feasible vector $\mathbf{y}$ with only the y_2 coordinate positive and that there is no feasible vector $\mathbf{y}$ with only the y_4 coordinate positive.

2. Let $\mathbf{A}$ be an $m \times n$ matrix, $\mathbf{x} \in \mathbb{R}^n$, and $\mathbf{y} \in \mathbb{R}^m$. Expand $\mathbf{y} \cdot (\mathbf{A}\mathbf{x})$ and $\mathbf{x} \cdot (\mathbf{A}^T \mathbf{y})$ and show that $\mathbf{y} \cdot (\mathbf{A}\mathbf{x}) = \mathbf{x} \cdot (\mathbf{A}^T \mathbf{y})$.

3. A standard maximum problem is as follows:

$$\text{maximize} \quad x + 3y$$

$$\text{subject to} \quad \begin{cases} x \geq 0, \ y \geq 0 \\ 2y - 5x \leq 5 \\ 3x - 5y \leq 12 \end{cases}$$

(a) Show that the feasible set for the linear programming problem is not empty.

(b) Show *directly* that the linear programming problem has no solution. That is, show that there are feasible vectors for which the objective function takes arbitrarily large values.

(c) Show that the feasible set for the dual problem is empty.

4. A SMP is as follows:

$$\text{maximize} \quad 4x - y$$

$$\text{subject to} \quad \begin{cases} x \geq 0, \ y \geq 0 \\ 3x - y \leq 6 \\ 5x - y \geq 5 \end{cases}$$

(a) Show that the feasible set for this problem is not empty.

(b) Show *directly* that the problem has no solution. That is, show that there are feasible vectors for which the objective function takes arbitrarily large values.

(c) Show that the feasible set for the dual problem is empty.

5. A SMP is as follows:

$$\text{maximize} \quad x - y$$

$$\text{subject to} \quad \begin{cases} x \geq 0, \ y \geq 0 \\ 3x - y \leq 6 \\ 5x - y \geq 5 \end{cases}$$

(a) Show that the feasible set F for this problem is not empty. Find the corner points of F, and evaluate the objective function at the corner points.

(b) Find the feasible set F' for the dual problem, find the corner points of F', and solve the dual problem by evaluating the objective function at the corner points.

(c) Show that there is a corner point (x, y) of the set F, and a corner (u, v) of the set F', such that the objective function of the original problem evaluated at (x, y) has the same value as the objective function for the dual evaluated at (u, v). Use Theorem 5.1 to conclude that (x, y) is a solution of the original problem and that (u, v) is a solution of the dual problem.

6. A standard minimum problem is as follows:

$$\text{minimize} \quad 18x_1 - 15x_2 + 20x_3 + 9x_4$$

$$\text{subject to} \quad \begin{cases} x_1 \geq 0, \ x_2 \geq 0, \ x_3 \geq 0, \ x_4 \geq 0 \\ 2x_1 - 10x_2 + 4x_3 + 3x_4 \geq 36 \\ 3x_1 - 5x_2 + 5x_3 - 6x_4 \geq 24 \end{cases}$$

(a) Solve this problem using Maple or an equivalent package—for example, using "solver" in Excel.

(b) Solve the dual problem graphically and verify that $\mathbf{x}^* \cdot \mathbf{c} = \mathbf{y}^* \cdot \mathbf{b}$, where $\mathbf{x}^*$ is a solution of the original problem and $\mathbf{y}^*$ is a solution of the dual.

7. Find the dual of the following general maximum problem:

$$\text{maximize} \quad x_1 + 5x_2 - 2x_3 + 2x_4 - 3x_5$$

$$\text{subject to} \quad \begin{cases} x_2 \geq 0, \ x_3 \geq 0, \ x_5 \geq 0 \\ 2x_1 + x_2 - 5x_4 \leq 20 \\ 2x_1 - 3x_3 + 2x_4 - x_5 \geq -5 \\ 3x_2 + 5x_3 + 2x_5 = 10 \end{cases}$$

8. State and solve the problem that is the dual of the following general maximum problem.

$$\text{maximize} \quad u_1 + u_2 + v_1 + v_2$$

$$\text{subject to} \quad \begin{cases} u_1 + v_1 = 1 \\ u_1 + v_2 = 2 \\ u_2 + v_1 = 3 \\ u_2 + v_2 = 5 \end{cases}$$

9. Write the following general maximum problem as a standard maximum problem, and find the matrices and vectors $\mathbf{A}$, $\mathbf{b}$, and $\mathbf{c}$.

$$\text{maximize} \quad 12x_1 + 3x_2 + -2x_3 + x_4$$

$$\text{subject to} \quad \begin{cases} x_1 \geq 0, \ x_4 \geq 0 \\ 10x_1 + 5x_2 + x_3 \leq 120 \\ 2x_2 + 5x_3 + x_4 \leq 20 \\ 12x_1 - 2x_2 + 5x_4 = 80 \end{cases}$$

10. Find the dual of the general maximum problem posed in Exercise 9.

11. Find the dual of the following general maximum problem:

$$\text{maximize} \quad 2x_1 - x_2 + 5x_3 - x_4 + 8x_5$$

$$\text{subject to} \quad \begin{cases} x_2 \geq 0, \ x_4 \geq 0, \ x_5 \geq 0 \\ 2x_2 + 5x_4 - 4x_5 = 20 \\ x_1 + 2x_2 - 3x_4 \leq 10 \\ 2x_1 + 3x_3 - 4x_4 - x_5 \geq -8 \end{cases}$$

12. A standard minimum problem is as follows:

$$\text{minimize} \quad 25x_1 - x_2 + 18x_3 - x_4 + 40x_5$$

$$\text{subject to} \quad \begin{cases} x_1 \geq 0, \ x_2 \geq 0, \ x_3 \geq 0, \ x_4 \geq 0, \ x_5 \geq 0 \\ x_1 - x_2 + x_3 + 4x_5 \geq 1 \\ 5x_1 - 2x_3 - x_4 + x_5 \geq 1 \end{cases}$$

(a) Solve this problem using Maple or an equivalent package.
(b) Solve the dual problem graphically and verify that $\mathbf{x}^* \cdot \mathbf{c} = \mathbf{y}^* \cdot \mathbf{b}$.
(c) State and verify the complementary slackness conditions in this example.

13. A SMP is as follows:

$$\text{maximize} \quad x - y$$

$$\text{subject to} \quad \begin{cases} x \geq 0, \ y \geq 0 \\ 2y - 5x \leq 5 \\ 3x - 4y \leq 12 \end{cases}$$

(a) Show that the feasible set for the linear programming problem is not empty.
(b) Show *directly* that the linear programming problem has no solution. That is, show that there are feasible vectors for which the objective function takes arbitrarily large values.
(c) Show that the feasible set for the dual problem is empty.

5.3 Duality, Sensitivity, and Uncertainty

Considering a linear programming problem and the associated dual simultaneously is extremely helpful in many situations, including both theoretical discussions and specific applications. In this section we consider examples that illustrate the utility of dual problems. We begin by developing an interpretation of the dual problem in two instances, both examples that we have discussed earlier.

EXAMPLE 5.8 (An interpretation of the dual of the diet problem)   The diet problem introduced in Section 5.1 is the problem of finding a least-cost diet that meets certain nutritional constraints. Specifically, in Section 5.1, we considered n foods and m nutrients and the requirements that the diet provide the recommended daily allowance of each of the m nutrients. We made the following assumptions:

1. The nutritional content of each food was known. And we denoted by a_{ij} the number of units of nutrient i in 1 unit of food j.
2. b_i units of nutrient i, $i = 1, 2, \ldots, m$, are required.
3. The cost of 1 unit of food j is c_j, $j = 1, 2, \ldots, n$.

The task is to find a diet $\mathbf{x} = [x_1 \quad x_2 \quad \cdots \quad x_n]$, where x_i is the number of units of food i consumed, $i = 1, 2, \ldots, n$, that satisfies the nutritional requirements and for which the cost is as small as possible. In Section 5.1 this problem was formulated as follows:

$$\text{minimize} \quad \mathbf{x} \cdot \mathbf{c} = x_1 c_1 + x_2 c_2 + \cdots + x_n c_n$$
$$\text{subject to} \quad \begin{cases} x_i \geq 0, & \text{for } i = 1, 2, \ldots n \\ \displaystyle\sum_{j=1}^{n} x_j a_{ij} \geq b_i, & \text{for } i = 1, 2, \ldots, m \end{cases}$$

The dual problem is

$$\text{maximize} \quad y_1 b_1 + y_2 b_2 + \cdots + y_m b_m$$
$$\text{subject to} \quad \begin{cases} y_i \geq 0, & \text{for } i = 1, 2, \ldots m \\ \displaystyle\sum_{j=1}^{m} y_j a_{ji} \leq c_i, & \text{for } i = 1, 2, \ldots, n \end{cases}$$

The meanings of the quantities a_{ji}, b_i, and c_i are as defined in the original problem. It remains to assign a meaning to the dual variables (y_j, $j = 1, \ldots, m$) and to interpret the objective function and the inequality constraints of the dual problem. To accomplish this, we begin with the objective function. We know that for solutions $\mathbf{x}$ and $\mathbf{y}$ of the original problem and the dual, respectively, we have $\mathbf{x} \cdot \mathbf{c} = \mathbf{y} \cdot \mathbf{b}$. Since $\mathbf{x} \cdot \mathbf{c}$ is a cost, it follows that $\mathbf{y} \cdot \mathbf{b} = y_1 b_1 + y_2 b_2 + \cdots + y_m b_m$ must also be a cost, measured in units of dollars, for example. Also, b_i is a number of units of nutrient i. And consequently, y_i must be measured in units of cost per unit of nutrient i. Continuing, each of the inequality constraints for the dual problem has the form $\sum_{j=1}^{m} y_j a_{ji} \geq c_i$, and since the right-hand side is a cost per unit of food i, the left-hand side must have the same units. Also, since a_{ji} is the amount of nutrient j in food i, this shows again that y_i is measured in units of cost per unit of nutrient i.

The next step is to use this information about the variable y_i to create an interpretation of the dual problem. This task does not have a single outcome, and one interpretation may be more useful than another. Here, we adopt the following view of the situation. Suppose the decision maker who is faced with the task of finding the least-cost diet is offered an alternative, the alternative of buying nutrient pills and providing the nutrients through pills rather than through food. We view the situation from the perspective of the person who is trying to sell the nutrient pills. The seller's task is to price the pills to make a sale and to maximize the value of the sale. To make the sale, the required amounts of nutrients (the b_i's) must be sold, and the value of the sale is therefore $y_1 b_1 + y_2 b_2 + \cdots + y_m b_m$. If the person preparing the diet, who is interested in minimizing the cost of the diet, is to purchase pills to provide all of the nutrients, then the pills must provide the nutrients at a cost no greater than the cost of providing the nutrients through food. This condition is $\sum_{j=1}^{m} y_j a_{ji} \geq c_i$. Consequently, the choice of the variables y_i in the dual problem is the same as the task of maximizing income through the sale of nutrient pills. Hence this setting provides an interpretation of the dual of the diet problem.   ■

EXAMPLE 5.9   (An interpretation of the dual of the transportation problem)   The context for this example is the version of the transportation problem discussed in Section 2.4. In that situation, there are two factories where mobile homes are assembled and three distribution centers where they are sold. We assume that the demand at each center is known: demand d_j at distribution center j, $j = 1, 2, 3$. To meet the demand, s_1 homes are assembled at factory 1 and s_2 homes are assembled at factory 2. The cost of moving one mobile home from factory i to distribution center j is c_{ij}, $i = 1, 2$, and $j = 1, 2, 3$.

The goal is to determine how many homes to move from each factory to each distribution center so that the total transportation cost,

$$(5.19) \qquad c_{11}x_{11} + c_{12}x_{12} + \cdots + c_{23}x_{23} = \sum_{i=1}^{2} \sum_{j=1}^{3} c_{ij} x_{ij}$$

is as small as possible. The transportation schedule is a vector

$$\mathbf{x} = [x_{11} \quad x_{12} \quad x_{13} \quad x_{21} \quad x_{22} \quad x_{23}]$$

We require that the number of mobile homes shipped from each factory not exceed the number produced there, and we require that the number of mobile homes shipped to each distribution center at least equal the demand at that center:

$$x_{i1} + x_{i2} + x_{i3} \leq s_i, \qquad i = 1, 2 \qquad \text{and}$$
$$x_{1j} + x_{2j} \geq d_j, \qquad j = 1, 2, 3$$

The problem is a minimization problem, so the standard form must have all inequality constraints written in the form $\mathbf{Ax} \geq \mathbf{b}$. When written in this form, the supply and demand constraints are

$$\left. \begin{aligned} -x_{11} - x_{12} - x_{13} \geq -s_1 \\ -x_{21} - x_{22} - x_{23} \geq -s_2 \end{aligned} \right\} \qquad \text{and} \qquad \left. \begin{aligned} x_{11} + x_{21} \geq d_1 \\ x_{12} + x_{22} \geq d_2 \\ x_{13} + x_{23} \geq d_3 \end{aligned} \right\}$$

and the matrix $\mathbf{A}$ and vector $\mathbf{b}$ are

$$(5.20) \quad \mathbf{A} = \begin{bmatrix} -1 & -1 & -1 & 0 & 0 & 0 \\ 0 & 0 & 0 & -1 & -1 & -1 \\ 1 & 0 & 0 & 1 & 0 & 0 \\ 0 & 1 & 0 & 0 & 1 & 0 \\ 0 & 0 & 1 & 0 & 0 & 1 \end{bmatrix} \quad \mathbf{b} = \begin{bmatrix} -s_1 \\ -s_2 \\ d_1 \\ d_2 \\ d_3 \end{bmatrix}$$

The objective function (5.19) is one of the form $\mathbf{c} \cdot \mathbf{x}$, where the vector $\mathbf{c}$ is

$$[c_{11} \quad c_{12} \quad c_{13} \quad c_{21} \quad c_{22} \quad c_{23}]$$

The vector $\mathbf{y}$ of the dual problem has five coordinates, and we denote them by $y_1, \ldots, y_5$. The dual problem is to maximize $\mathbf{y} \cdot \mathbf{b}$ subject to $\mathbf{y} \geq 0$, $\mathbf{y}\mathbf{A}^T \leq \mathbf{c}$. Using $\mathbf{A}$, $\mathbf{b}$, and $\mathbf{c}$ from the original problem, we find that the dual problem is

$$\text{maximize} \quad -s_1 y_1 - s_2 y_2 + d_1 y_3 + d_2 y_4 + d_3 y_5$$

$$\text{subject to} \quad \begin{cases} y_1 \geq 0, \ y_2 \geq 0, \ y_3 \geq 0, \ y_4 \geq 0, \ y_5 \geq 0 \\ -y_1 + y_3 \leq c_{11} \\ -y_1 + y_4 \leq c_{12} \\ -y_1 + y_5 \leq c_{13} \\ -y_2 + y_3 \leq c_{21} \\ -y_2 + y_4 \leq c_{22} \\ -y_2 + y_5 \leq c_{23} \end{cases}$$

The inequality constraints can also be written as

$$(5.21) \quad \begin{aligned} y_3 &\leq c_{11} + y_1 \\ y_4 &\leq c_{12} + y_1 \\ y_5 &\leq c_{13} + y_1 \\ y_3 &\leq c_{21} + y_2 \\ y_4 &\leq c_{22} + y_2 \\ y_5 &\leq c_{23} + y_2 \end{aligned}$$

Now, since $\mathbf{c} \cdot \mathbf{x}$ is a cost, $\mathbf{y} \cdot \mathbf{b}$ must also be a cost (recall that for $\mathbf{x}$, a solution of the original problem, and $\mathbf{y}$, a solution of the dual problem, $\mathbf{c} \cdot \mathbf{x} = \mathbf{y} \cdot \mathbf{b}$). Also, the coordinates of $\mathbf{b}$ are numbers of mobile homes, so the coordinates of $\mathbf{y}$ must be dollars per mobile home. Since each c_{ij} is a cost per mobile home—the cost of moving one home from factory i to distribution center j—the entries in the inequalities (5.21) have consistent units.

We develop an interpretation of the dual problem by considering the following situation. Suppose that one option for the manager in the original setting is to sell mobile homes at the factories and then purchase equivalent mobile homes for delivery at the distribution centers. Suppose that the manager can sell a home at factory 1 for a price of y_1 and a home at factory 2 for a price of y_2. Also, suppose that the manager can purchase a mobile home for delivery at distribution center 1 for a cost of y_3, a mobile home for delivery at distribution center 2 for a cost of y_4, and a mobile home for delivery at distribution center 3 for a cost of y_5. With these meanings of the variables y_i, we consider the expression $y_3 \leq c_{11} + y_1$. The left-hand side is the price the manager must pay for a mobile home delivered at distribution center 1, and the right-hand side is the sum of the price the manager will receive for a home at factory 1 and the cost of moving a home from factory 1 to distribution center 1. Thus the expression $y_3 \leq c_{11} + y_1$ represents the constraint that in order for the manager to use this option, the price paid for a home at distribution center 1 cannot be more than the sum of the price received for a mobile home at factory 1 and the cost of moving a home from factory 1 to distribution center 1. Otherwise, the manager will not sell and buy a home but instead will transport it from the factory to the distribution center. The interpretation of the other inequalities in (5.21) is similar. Finally, the expression $-s_1 y_1 - s_2 y_2 + d_1 y_3 + d_2 y_4 + d_3 y_5$ represents the gain to a reseller resulting from the purchase of the homes at the factories and the sale of equivalent homes at the distribution centers. If we assume that the reseller is interested in maximizing the profit from this purchase-and-resale activity, then the problem of finding prices y_i that yield maximum profit, subject to the condition that the manager will use the option of selling/purchasing, is precisely the dual problem formulated above. ∎

■ Changes in A, b, and c

In many situations the parameters in a linear programming model (the entries in the matrix **A** and the vectors **b** and **c**) are estimated from actual data. In such situations it is important to know how variations in the parameters affect the solution of the problem. There are many ways in which such questions can be asked, and some versions have complicated answers. However, some versions can be studied using relatively elementary ideas, the concepts introduced in this chapter. Our approach will again illustrate the interplay between the original problem and its dual.

As our first example, we consider a situation in which the problem is one of allocating resources among production activities to achieve maximum profit. The linear programming problem is a standard maximum problem:

$$\text{maximize } \mathbf{x} \cdot \mathbf{c} \quad \text{subject to } \mathbf{x} \geq 0 \text{ and } \mathbf{Ax} \leq \mathbf{b}$$

where the entries in the matrix **A** are production parameters, the entries in the vector **b** are amounts of resources available, and the entries in the vector **c** are profit per unit of product produced.

EXAMPLE 5.10 Consider the problem described in Example 5.1 (Section 5.1): the problem of allocating resources to producing packages of various types of trail mix. Specifically, the task is to allocate resources—peanuts, raisins, seeds, and dried fruit—to the production of trail mix of types A, B, and C to achieve maximum profit. The resulting linear programming problem is given in Section 5.1 [expressions (5.5), (5.6), and (5.7)] as

$$\text{maximize} \quad 0.14x_1 + 0.12x_2 + 0.08x_3,$$

$$\text{subject to} \quad \begin{cases} x_1 \geq 0, \ x_2 \geq 0, \ \text{and } x_3 \geq 0 & \\ 80x_1 + 60x_2 + 30x_3 \leq 1,000,000 & \text{(peanuts)} \\ 20x_1 + 20x_2 \leq 600,000 & \text{(raisins)} \\ 50x_1 + 40x_2 + 80x_3 \leq 500,000 & \text{(seeds)} \\ 30x_2 + 40x_3 \leq 300,000 & \text{(fruit)} \end{cases}$$

where

$$x_1 = \text{number of packages of variety A to be produced}$$
$$x_2 = \text{number of packages of variety B to be produced}$$
$$x_3 = \text{number of packages of variety C to be produced}$$

The solution of the problem (Exercise 1) is

$$\mathbf{x}^* = [x_1 \quad x_2 \quad x_3] = [2000 \quad 10{,}000 \quad 0]$$

and the optimal value of the objective function—the maximum profit in dollars—is 1480. Note that for the solution $\mathbf{x}^*$, the first and second of the inequalities (5.20) are satisfied as strict inequalities, and consequently, by complementary slackness, the first and second dual variables in the solution $\mathbf{y}^*$ of the dual problem have the value 0.

The dual of the trail mix problem, a minimization problem, is

minimize $\quad 1{,}000{,}000y_1 + 600{,}000y_2 + 500{,}000y_3 + 300{,}000y_4$

subject to $\quad \begin{cases} y_1 \geq 0, \ y_2 \geq 0, \ y_3 \geq 0, \ \text{and } y_4 \geq 0 \\ 80y_1 + 20y_2 + 50y_3 \geq 0.14 \\ 60y_1 + 20y_2 + 40y_3 + 30y_4 \geq 0.12 \\ 30y_1 + 80y_3 + 40y_4 \geq 0.08 \end{cases}$

The solution of this problem (Exercise 2) is

$$\mathbf{y}^* = [y_1 \quad y_2 \quad y_3 \quad y_4] = [0 \quad 0 \quad 0.002800 \quad 0.0002667]$$

and the optimal value of the objective function is again 1480.

Now, suppose that the trail mix manufacturer has an opportunity to purchase additional seeds. Thus we consider a change in the vector $\mathbf{b}$. It may be that with the additional seeds, the manufacturer will be able to produce additional packages of trail mix and make a larger net profit. How much should the manufacturer be willing to pay for additional seeds?

We can answer this question by considering the dual problem. We know that the maximum value of the objective function for the original problem is equal to the minimum value of the objective function for the dual. In this case, both values equal 1480. That is,

$$(5.22) \quad 1480 = \mathbf{x}^* \cdot \mathbf{c} = \max \ \mathbf{x} \cdot \mathbf{c} = \min \ \mathbf{y} \cdot \mathbf{b} = \mathbf{y}^* \cdot \mathbf{b}$$
$$= 1{,}000{,}000(0) + 600{,}000(0) + 500{,}000(0.0028) + 300{,}000(0.0002667)$$

Suppose that the manufacturer has an opportunity to purchase 1 additional unit of seeds. That is, there is the option of having 500,001 grams of seeds available to make trail mix. Using the relation (5.22), we see that if the vector $\mathbf{y}^* = [0 \quad 0 \quad 0.0028 \quad 0.0002667]$ remains optimal for the dual problem, then the minimum value of the objective function for the dual problem increases to $500{,}001(0.0028) + 300{,}000(0.0002667) = 1480.0028$, and consequently, the maximum value of the objective function of the original problem is also 1480.0028. Therefore, if the manufacturer pays 0.0028 or less for an additional gram of seeds, then the maximum net profit will be at least as large as in the original situation. This value, $.0028 per gram, is called the *shadow price* of seeds. Likewise, $.0002667 per gram is the shadow price for dried fruit—it is the price the manufacturer can afford to pay for 1 additional gram of dried fruit and have at least as large a net profit as in the original situation.

The shadow prices for peanuts and raisins are zero. In this example, the optimal production schedule is such that some peanuts and some raisins are unused. Accordingly, the manufacturer cannot increase the maximum net profit by having additional peanuts and raisins available.

In each case the conclusion is based on the assumption that the solution of the dual problem does not change when the objective function is altered by a small amount. Although this is a valid assumption in some cases, it does not always hold. In this example, it holds for increases in the amount of seeds available up to 650 kg and for increases in the amount of dried fruit available up to 375 kg.

Remark   The discussion above has ignored an important point about the existence of solutions of linear programming problems. When we talk about the shadow price for one of the resources, we note that this is the price one should not exceed to buy 1 more unit of that resource in an attempt to increase production and thus increase profit. However, there is not always a feasible solution of the original problem that will increase the profit. In each case, one must solve the new problem (with the additional resources) and check whether there is a new solution that gives an increase in profits.   ■

EXAMPLE 5.11   Each month a manufacturing firm discharges about 5 million liters of waste water into Mulberry Creek. The State Department of Environmental Management (SDEM) has tested the discharge and found that it does not meet current standards. The SDEM findings require that the quality of the discharge be improved. Specifically, for each 1 million liters of waste water discharge, the firm must reduce the volume of contaminants as follows:

- Metals must be reduced by 100 grams.
- Volatile compounds, or volatiles, must be reduced by 1000 grams.
- Suspended solids must be reduced by 200 kilograms.

The firm conducts an analysis of the options available and identifies the following alternative responses to address the SDEM's concerns:

1. Raw materials of higher quality can be purchased.
2. The manufacturing process can be refined.
3. A chemical treatment process can be used on the waste water before it is discharged.
4. The filters currently being used to filter the waste water before it is discharged can be replaced with high-efficiency filters.

The treatment methods can be applied individually or in combinations. A consulting firm provides information on the impact of each of the alternatives on the composition of the waste water. For each alternative, the maximum feasible reduction in the various contaminants is given in Table 5.7.

The cost of achieving the maximum feasible reduction using each alternative is shown in the "Cost" column in Table 5.7.

Table 5.7   Data for Example 5.11.

Treatment Alternative	Reduction (grams per million liters)			Cost
	Metals	Volatiles	Suspended Solids	
Better raw materials	400	800	100,000	10,000
Refined processing	600	500	100,000	15,000
Chemical treatments	400	500	0	7,500
High-efficiency filters	100	0	500,000	10,000

Suppose that each of the alternatives can be implemented at any level less than the maximum feasible and that the cost depends linearly on the level of implementation. That is, if the alternative of refined processing is used at a level of one-half the maximum feasible, then the effect is to reduce the amount of metals by 300 grams per million liters, the amount of volatiles by 250 grams per million liters, and the amount of suspended solids by 50,000 grams per million liters of waste water, and the cost of this is $7500. The decision variables for the problem are the levels (*level* being defined as fraction of the maximum feasible) at which the alternative treatments are implemented. We make the following definitions:

$$\mathbf{x} = \begin{bmatrix} x_1 \\ x_2 \\ x_3 \\ x_4 \end{bmatrix}$$

x_1 — Level at which the alternative of purchasing higher-quality raw materials is implemented

x_2 — Level at which the alternative of using refined manufacturing processes is implemented

x_3 — Level at which the alternative of using chemical treatments is implemented

x_4 — Level at which the alternative of using high-efficiency filters is implemented

And we note that each of the decision variables must be nonnegative and no larger than 1.

We suppose that the goal of the manufacturing firm is to use one or a combination of the treatments to reduce the contaminants to an acceptable level, and to do so at the least possible cost. With the definition of $\mathbf{x}$ given above, the problem the manufacturing firm faces can be expressed in the form of a linear programming problem as follows:

$$\text{minimize} \quad 10{,}000x_1 + 15{,}000x_2 + 7{,}500x_3 + 10{,}000x_4$$

$$\text{subject to} \quad \begin{cases} x_1 \geq 0, \ x_2 \geq 0, \ x_3 \geq 0, \ x_4 \geq 0 \\ x_1 \leq 1, \ x_2 \leq 1, \ x_3 \leq 1, \ x_4 \leq 1 \\ 400x_1 + 600x_2 + 400x_3 + 100x_4 \geq 100 \\ 800x_1 + 500x_2 + 500x_3 \geq 1000 \\ 100{,}000x_1 + 100{,}000x_2 + 500{,}000x_4 \geq 200{,}000 \end{cases}$$

The solution of this problem (Exercise 5) is

$$\mathbf{x}^* = [x_1 \quad x_2 \quad x_3 \quad x_4] = [1 \quad 0 \quad 0.4 \quad 0.2]$$

and the optimal value of the objective function is 15,000.

We consider next the possibility of uncertainty in the amounts of contaminants to be removed. We suppose that the manufacturer is required to reduce the amount of volatiles by 1001 grams per million liters of waste water rather than by 1000 grams per million liters. What will be the incremental cost of complying with the more stringent requirement?

As in the preceding example, we can answer this question by considering the dual of the original problem. The dual of this problem is the maximization problem

$$\text{maximize} \quad -y_1 - y_2 - y_3 - y_4 + 100y_5 + 1000y_6 + 200{,}000y_7$$

$$\text{subject to} \quad \begin{cases} y_1 \geq 0, \ y_2 \geq 0, \ y_3 \geq 0, \ y_4 \geq 0, \ y_5 \geq 0, \ y_6 \geq 0, \ y_7 \geq 0 \\ -y_1 + 400y_5 + 800y_6 + 100{,}000y_7 \leq 10{,}000 \\ -y_2 + 600y_5 + 500y_6 + 100{,}000y_7 \leq 15{,}000 \\ -y_3 + 400y_5 + 500y_6 \leq 7500 \\ -y_4 + 100y_5 + 500{,}000y_7 \leq 10{,}000 \end{cases}$$

The solution of this problem (Exercise 6) is

$$\mathbf{y}^* = [y_1 \quad y_2 \quad y_3 \quad y_4 \quad y_5 \quad y_6 \quad y_7] = [4000 \quad 0 \quad 0 \quad 0 \quad 0 \quad 15 \quad 0.02]$$

and the optimal value of the objective function is again 15,000.

We know that the minimum value of the objective function for the original problem is equal to the maximum value of the objective function for the dual. In this case both values equal 15,000. That is,

$$15{,}000 = \min \mathbf{x} \cdot \mathbf{c} = \max \mathbf{y} \cdot \mathbf{b}$$
$$= -1(4000) - 1(0) - 1(0) - 1(0) + 100(0) + 1000(15) + 200{,}000(0.02).$$

Next, since the vector $\mathbf{y}^*$ remains optimal in the case where the required reduction of volatiles is 1001, we have

$$\max \mathbf{y} \cdot \mathbf{b} = -1(4000) - 1(0) - 1(0) - 1(0) + 100(0) + 1001(15) + 200{,}000(0.02)$$
$$= 15{,}015$$

Thus, since $\min \mathbf{x} \cdot \mathbf{c} = \max \mathbf{y} \cdot \mathbf{b}$, we conclude that the minimum cost increases from $15,000 to $15,015. That is, removing an additional 1 gram of volatiles per million liters of waste water costs the manufacturer $15.

Similarly, if the goal is to remove 999 grams of volatiles per million liters of waste water, then the cost will be $14,985, a savings of $15. Again, this assumes that the vector $\mathbf{y}^*$ remains optimal for the dual problem, and in fact, it does remain optimal. ∎

As the last example of this section, we consider a situation wherein the uncertainty in parameter values is made an explicit part of the setting. There are several ways in which uncertain parameter values can be handled, and there are standard ways of handling problems of certain types. Here, we consider one example and several ways of approaching the underlying problem.

EXAMPLE 5.12 A corporation makes two types of high-efficiency filters for use in "clean room" applications. One unit of the first type, type A, requires 2 kilograms (kg) of filtering material, 50 grams of chemicals, and two electronic modules. One unit of the second type, type B, requires 5 kg of filtering materials, 50 grams of chemicals, and one electronic module. Suppose that the corporation has 500 kg of filtering materials, 10 kg of chemicals, and 220 electronic modules and that the profit is $5 on each type A filter and $9 on each type B filter.

If we assume that there is sufficiently large demand for each type of filter, then the solution of the following linear programming problem tells the company what allocation of resources results in maximum profit:

$$\text{maximize} \quad 5x + 9y$$

$$\text{subject to} \quad \begin{cases} x \geq 0, \ y \geq 0 \\ 2x + 5y \leq 500 \\ 50x + 50y \leq 10{,}000 \\ 2x + y \leq 220 \end{cases}$$

where x and y are the numbers of filters of types A and B, respectively.

The solution of this problem is $x = 75$ and $y = 70$, and the maximum profit is $1005.

Now suppose that instead of assuming the demand is sufficiently large, we assume the demand is uncertain. In particular, assume that we have the following information on the probabilities of various demands.

- There are two equally likely scenarios for demand of type A filters. In one of them the maximum possible demand is 50, and in the other the maximum possible demand is 120.
- There are three possible scenarios for type B filters. The possible maximum demands are 60, 80, and 100, and the associated probabilities are .25, .50, and .25, respectively.

We illustrate two approaches to this version of the problem. One approach is to include the information on probabilities by adding the expected demands as new constraints. In this case the expected demand for type A filters is 85, and the expected demand for type B filters is 80 (Exercise 10). Therefore, we add the new constraints $x \leq 85$, $y \leq 80$ to the problem formulated above. If we solve the resulting problem, we find that the solution is again $x = 75$ and $y = 70$. However, in this case the profit is not $1005. Indeed, 50 filters of type A will be sold with a probability of 1, and an additional 25 will be sold with probability .5. Also, 60 filters of type B will be sold with probability 1, and an additional 10 will be sold with probability .75. Consequently, the expected profit is

$$50 \cdot 5 + 25 \cdot 5 \cdot (.5) + 60 \cdot 9 + 10 \cdot 9 \cdot (.75) = 920$$

That is, the uncertain demand has reduced the expected profit.

In our second approach to the problem, we exploit the idea introduced above in a systematic way and take greater advantage of the information on demands. We begin by introducing variables defined to be the numbers of type A filters in each range of demand (up to 50 and between 50 and 120) that are sold. In particular, we set

$$x = x_1 + x_2, \quad \text{where } 0 \leq x_1 \leq 50 \text{ and } 0 \leq x_2 \leq 70$$

We use the same idea for the number of type B filters sold, and we set

$$y = y_1 + y_2 + y_3, \quad \text{where } 0 \leq y_1 \leq 60, \ 0 \leq y_2 \leq 20, \ \text{and } 0 \leq y_3 \leq 20$$

With these definitions, the expected profit is

$$5 \cdot x_1 + 5 \cdot (.5) \cdot x_2 + 9 \cdot y_1 + 9 \cdot (.75) \cdot y_2 + 9 \cdot (.25) \cdot y_3$$

The linear programming problem is

maximize $\quad 5 \cdot x_1 + 5 \cdot (.5) \cdot x_2 + 9 \cdot y_1 + 9 \cdot (.75) \cdot y_2 + 9 \cdot (.25) \cdot y_3$

subject to
$$\begin{cases}
x_1 \geq 0, \ x_2 \geq 0, \ y_1 \geq 0, \ y_2 \geq 0, \ y_3 \geq 0 \\
2x_1 + 2x_2 + 5y_1 + 5y_2 + 5y_3 \leq 500 \\
50x_1 + 50x_2 + 50y_1 + 50y_2 + 50y_3 \leq 10{,}000 \\
2x_1 + 2x_2 + y_1 + y_2 + y_3 \leq 220 \\
0 \leq x_1 \leq 50 \\
0 \leq x_2 \leq 70 \\
0 \leq y_1 \leq 60 \\
0 \leq y_2 \leq 20 \\
0 \leq y_3 \leq 20
\end{cases}$$

The solution of this problem is $x_1 = 50$, $x_2 = 0$, $y_1 = 60$, $y_2 = 20$, $y_3 = 0$, and the maximum profit is $925. This approach, the solution of a linear programming problem wherein the objective function is the expected profit, takes into account full information about the probability distribution of the random demands. Note that for the profit to be maximum, x_1 should take the value 50 before x_2 takes any positive value. This will be the case because the coefficient of x_1 in the objective function is larger than the coefficient of x_2, and both x_1 and x_2 have the same coefficients in the constraints. For similar reasons, y_1 should take the value 60 before either y_2 or y_3 takes positive values, and y_2 should take the value 20 before y_3 takes positive values. ■

The concepts and techniques introduced in this section illustrate some of the ways in which the basic ideas of linear programming can be extended and refined. Many of these topics are especially important in applications, and there is a substantial body of literature wherein they are considered. A few additional examples are the topics of exercises.

Exercises 5.3

1. Use Maple or an equivalent software package to solve the maximum problem formulated as (5.20) in Example 5.9. Use the following values for the supplies, demands, and costs: $s_1 = 40$, $s_2 = 60$, $d_1 = 25$, $d_2 = 55$, $d_3 = 20$, $c_{11} = 5$, $c_{12} = 10$, $c_{13} = 8$, $c_{21} = 10$, $c_{22} = 8$, and $c_{23} = 7$.

2. Use Maple or an equivalent software package to solve the minimum problem that is the dual of the problem solved in Exercise 1.

3. A manufacturer produces products A, B, and C using resources R_1, R_2, R_3, and R_4. The amounts of the various resources needed to produce 1 unit of each of the products, the total amount of each resource available, and the net profit per unit of each product are shown in the accompanying table. The goal is to use the available resources to produce products that yield maximum profit. You may suppose that fractional units of products are acceptable.

   (a) Formulate and solve a linear programming problem that gives the desired information.

   (b) If the manufacturer can purchase additional amounts of resource R_4 to increase net profit, what is the largest value of the price p that should be paid for one unit of R_4?

Resource \ Product	A	B	C	Amount of Each Resource Available
R_1	20	30	500	2400
R_2	15	40	200	1600
R_3	6	6	120	750
R_4	20	10	300	900
Net Profit per Unit of Product	$12	$35	$205	

(c) With the value of p determined in part (b), what is the largest amount of R_4 that the manufacturer can buy and be confident that net profit will increase by $p \times$ (increased amount of R_4)?

Hint for part (c): Modify the constraint for resource R_4 by increasing the 900 to $900+w$, where w is a variable. Also, modify the objective function by including (as a negative number) the cost of the increased amount of R_4; that is, include a term $-qw$, where q is any number smaller that the price p found in part (b). Show that for all values of $q < p$, the solution of the revised problem is assumed for a value w^* of w, $w^* > 0$. Then $900 + w^*$ is the answer to part (c).

4. Suppose that the situation in Example 5.10 is modified so that the manufacturer has 450 kilograms of peanuts and 450 kilograms of seeds to use in making trail mix; all other conditions are as in Example 5.10.

   (a) Solve the modified problem and find the maximum profit.
   (b) How much should the trail mix manufacturer be willing to pay for an additional gram of peanuts in order to be able to produce more trail mix and generate a larger profit?
   (c) If the manufacturer can purchase an additional 54 kilograms of peanuts at 90 percent of the price determined in part (b), how much additional net profit (increased profit less increased cost) will be generated? What production will yield this increased profit?
   (d) Suppose the manufacturer has an opportunity to purchase 60 kilograms at 90 percent of the price determined in part (b). What can you say about this situation?

5. In the situation of Exercise 4, how much more should the manufacturer be willing to pay for additional seeds to be able to produce more trail mix and generate a larger profit?

6. Use Maple or an equivalent software package to solve the minimization problem formulated as the original problem in Example 5.11.

7. Use Maple or an equivalent software package to solve the maximization problem formulated as the dual problem in Example 5.11.

8. Suppose that the situation in Example 5.11 is modified so that the manufacturer must reduce metals by 800 grams, and all other conditions are as in Example 5.11. Solve this modified problem and find the minimum cost of achieving the goals. Find the incremental cost of reducing the volatiles by 1020 grams per million liters of waste water in this modified situation.

9. In the situation described in Example 5.11, estimate the largest values of constants a and b so that the vector [4000 0 0 0 0 15 0.02] remains optimal for the dual problem when the volatiles must be reduced by $1000-a$ grams and by $1000+b$ grams, respectively.

10. In the situation described in Example 5.12, suppose that demand is uncertain and that:
    • It is equally likely that demand for type A filters is 50 and 120.
    • The demands for type B filters are 60, 80, and 100 with probabilities .25, .50, and .25, respectively.

    Show that in this case, the expected demand for type A filters is 85 and the expected demand for type B filters is 80.

Table 5.8

	Type of Luggage			Amount of
Resource	Duffle	Garment	Gear	Resource Available
Fabric (sq meters)	3	2.5	2	850
Aluminum (kg)	1.5	2	.5	500
Labor (min)	8	10	6	3000
Net profit per piece ($)	14	12	9	

11. In the situation described in Example 5.12, suppose that demand is uncertain and that:
    - The demands for type A filters are 40, 60, and 100 with probabilities .2, .3, and .5, respectively.
    - The demands for type B filters are 45, 55, and 110 with probabilities .3, .3, and .4, respectively.

    Find the expected demands for each type of filter, and solve the linear programming problem with the added constraint that the solution must not exceed the expected demands.

12. In the situation of Exercise 11, solve the linear programming problem using the information on the probabilities of the various demands. That is, use the method of the last part of Example 5.12.

13. A luggage manufacturer makes three types of luggage—garment bags, duffle bags, and gear bags—using fabric, aluminum, and labor. The data on amounts of materials required for one piece of each type of luggage are shown in Table 5.8. This table also includes the amounts of the three resources available each day and the net profit per piece of luggage. How much of each type of luggage should be made in order to maximize net profit?

14. In the setting of Exercise 13, how much should the manufacturer be willing to pay for additional units of fabric, aluminum, and labor? That is, find the shadow prices for the problem.

5.4 An Example of Integer Programming: A Job Assignment Problem

There are problems in every branch of science that are essentially combinatorial in nature. Certain of such problems that arise in the management sciences, especially allocation and search questions, have relatively well-developed theories associated with them. Other questions, particularly in the life sciences, are almost completely open. Many of these questions can be phrased using everyday language and yet involve interesting and frequently very deep combinatorial problems. It is possible to give the flavor of the type of questions that arise without embarking on a systematic study, and we shall give here an example of an assignment problem that is typical. This example also illustrates a practical concern common in these problems: the need to find an effective means of handling large-scale situations. That is, the task of obtaining a useful algorithm is both important and nontrivial. Most problems actually arising in practice require the use of computers, and an increase in the speed of computation has a very real effect on the types of algorithms that are feasible. We shall return to the discussion of algorithms for assignment problems in Section 5.5.

■ The Problem

The well-known firm MacroHard finds itself in the position of having many available positions at its worldwide facilities. It decides to use its scientific personnel in a large-scale recruiting drive. These scientists will travel to interview applicants and then report to the main office concerning the qualifications of the individuals interviewed. The personnel department must decide when enough people have been recruited so that all jobs can be filled. These recruiting activities will take skilled scientists away from their regular work, so it is important that the recruiting be as limited as possible.

■ A Real Model

Here we shall identify some specific problems and look more closely at some special cases. The problem can be illustrated by means of a simple diagram. Let points $J_1, \ldots, J_n$ represent available jobs, not necessarily distinct, and let points $A_1, \ldots, A_m$ represent applicants. Let the fact that applicant A_i is qualified for job J_k be indicated by connecting A_i to J_k with a line. This will make it easy to evaluate and compare the qualifications of the applicants, both singly and as a group. Figure 5.4 is an example of such a diagram or graph, with $m = 3$ and $n = 4$. In a situation involving many jobs and many applicants, a diagram such as this becomes quite complicated, but it may still be of some use.

There are certain approximations that facilitate the study of this problem. First, it may be difficult to decide when an applicant is really qualified for a job. In the general case, it would probably be best to rate people for jobs in such a way that someone could be described as "well qualified" as opposed to "barely qualified." However, in this section, we shall simply assume that an applicant either is qualified for a job or is not. There are no intermediate stages. In Section 5.5 we shall discuss some generalizations.

As a second approximation, we assume that each applicant will remain available while the firm evaluates the applicants' credentials. In reality, of course, a number of applicants might accept other jobs or become unavailable for other reasons.

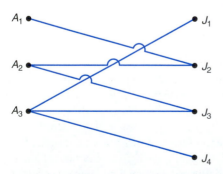

Figure 5.4

Subject to these approximations, an initial examination would seem to indicate that the problem can be easily handled by trial and error. One simply tries all the possible ways of assigning applicants to jobs, and then a check is made to see whether each job is filled by a qualified applicant under one of these assignments. In principle this is true. If a suitable assignment of applicants exists, then certainly it can be found by trial and error. However, the number of trials may be extremely large. The number of possible assignments increases very quickly with an increase in the numbers m and n, and it soon becomes virtually impossible to check all of them. For example, with $m = n$, there are $n!$ ways of assigning the applicants to the jobs. Even for n as small as 10, this gives more than 3 million possible assignments. Because both m and n are likely to exceed 100 in practice, even using a computer does not make the procedure practicable. A method more efficient than trial and error must be found.

■ Comments on the Model and Its History

There are two separate problems here. First, there is the problem of deciding when you have enough qualified applicants to fill the available jobs, or, equivalently, knowing how many jobs, of those available, can be filled with appropriate applicants. It may well be that some jobs cannot be filled with the current set of applicants. Second, there is the problem of actually assigning the applicants to the jobs. For large numbers m and n, this is a nontrivial problem, and some thought is necessary to decide how to proceed. This situation is a good example of the difference between proving that a problem has a solution (giving an existence proof that there are enough applicants) and actually constructing a solution (giving an algorithm for assigning the applicants to the jobs). In this section, we shall concentrate on giving necessary and sufficient conditions for a given set of applicants being capable of filling all the available jobs. Certain aspects of the algorithm will be considered in Section 5.5.

An obvious necessary condition is that there be as many applicants as there are jobs. But this clearly is not sufficient, because all applicants might be qualified for the same job and only for that one. It seems, therefore, that we need a condition that the applicants are as a group qualified for all jobs. However, even this is not sufficient. For example, consider the situation depicted in Figure 5.5.

This situation raises another point. Although applicant A_3 is (probably) qualified for the position as a janitor, it is unlikely that she or he would accept the position if it were offered. This motivates the following definition.

Definition 5.1   An applicant is *appropriate* for a job if qualified for the job and willing to accept it if it is offered. A set of applicants is *complete* if, for each subset of k jobs, there are at least k applicants who are appropriate for the jobs, in the sense that each applicant is appropriate for at least one of the jobs, $k = 1, 2, \ldots, n$.

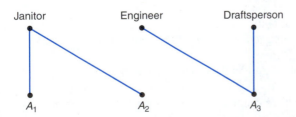

Janitor   Engineer   Draftsperson

A_1 A_2 A_3

Figure 5.5

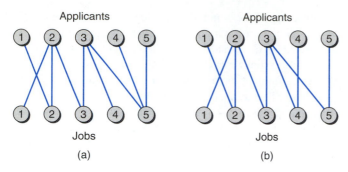

Figure 5.6

The definition of completeness may be difficult to check, and for a given set of applicants and jobs, small changes in the appropriateness of applicants for jobs may move a set of applicants from not complete to complete, or vice versa. In Figure 5.6, the applicants in part (a) are not complete for the jobs, but with one change in appropriateness for applicant 4, the set becomes complete in part (b).

We can now state the main result of this section as a theorem. As we noted earlier, this theorem does not provide a practical method of assigning applicants to jobs. We turn to that question in Section 5.5.

THEOREM 5.2 A necessary and sufficient condition for a job assignment problem to be solvable is that the set of applicants is complete.

Proof. Proving the "necessity" part of the theorem is quite simple. It is enough to point out that if we can fill all the jobs, then corresponding to each subset of k of the vacant jobs, we automatically have a set of k appropriate applicants. We just take the k applicants who were assigned to these jobs. Proving the "sufficiency" part is much harder. Our proof is given by an induction on n, the total number of jobs to be filled. We suppose that the set of applicants is complete.

$n = 1$. In this case there is only one job to be filled. Since the set of applicants is complete, there is an applicant who is appropriate for the job. We assign this applicant, and the problem is solved.

$n = 2$. There are two vacant jobs, J_1 and J_2. Since the set of applicants is complete, we know that there are at least two applicants, each of whom is appropriate for at least one of the two jobs J_1 and J_2. Also, there is at least one applicant appropriate for J_1 and at least one appropriate for J_2. We conclude (from the definition of completeness) that there are two applicants A_1 and A_2, for whom one of the following statements is true:

(a) Both A_1 and A_2 are appropriate for both jobs.
(b) One applicant is appropriate for both jobs, and the other is appropriate for only one job.
(c) One applicant is appropriate only for J_1, and the other is appropriate only for J_2.

For each of these cases, it is clear that we can make a job assignment in which both jobs are filled by appropriate applicants.

Now we assume that the theorem holds for $n = 1, 2, \ldots, k$ and prove that it holds for $n = k + 1$. There are two cases to consider.

Case 1   We suppose that the set of applicants is complete with "room to spare." Thus, given any ℓ jobs, $1 \le \ell \le k$, there are always *more* than ℓ applicants who are appropriate for these jobs. In this case, we begin by choosing a specific vacant job. By the completeness condition, we know that some applicant is appropriate for this job. Assign this applicant to the chosen job. Now consider the remaining k jobs. For each ℓ, $1 \le \ell \le k$, there must be at least ℓ applicants who find these jobs appropriate, because originally there were more than ℓ and only one applicant has been removed. By the induction hypothesis, we can fill these jobs with appropriate applicants.

Case 2   Suppose that the set of applicants is not complete with room to spare. Thus there is a subset B of jobs containing ℓ jobs, $1 \le \ell \le k$, and there are exactly ℓ applicants appropriate for these jobs; that is, no applicant outside of this set of ℓ applicants is appropriate for any job in B. Now $1 \le \ell \le k$ and the set of applicants for the jobs in B is complete (Exercise 3), and therefore, by the induction hypothesis, we can assign jobs in B. Consider the remaining $k + 1 - \ell$ jobs. Since $1 \le k + 1 - \ell \le k$, the proof of the assertion will be finished if we can show that the remaining applicants are complete for the remaining jobs. If not, there would exist a subset D of m jobs, $1 \le m \le k + 1 - \ell$, for which there does not exist a set of m applicants (not assigned to jobs in B) each of whom is appropriate for some job in D. But then there does not exist a set of at least $m + \ell$ applicants each of whom is appropriate for a job in $D \cup B$. This contradicts the completeness of the original set of applicants, and therefore no such set D can exist.

The proof is now complete by induction.   ∎

The proof of Theorem 5.2 is straightforward in that it involves no ideas outside those used in the definition of the problem. Also, it gives a flavor of the proofs of many "existence theorems" for assignment problems. However, the method of the proof is not useful in actually assigning applicants to jobs. In Section 5.5 we turn to that topic and give an alternative proof that *does* lead to a method for assigning applicants to jobs. In that discussion, we provide a more obvious connection between the assignment problem of this section and the linear programming problems of the earlier sections in this chapter, a topic that is also pursued in Exercises 7 and 8.

Standard maximum problems have the property that if x_1 and x_2 are two points in the feasible set, and if x_1 is "close to" x_2, then the value of the objective function at x_1 is "close to" the value of the objective function at x_2. This fact holds for many other linear programming problems, but it does not hold for integer linear programming problems, where "close" is measured using the usual distance in $\mathbb{R}^n$. In particular, if x is the solution of a standard maximum problem, and if y is the solution of the same problem with the additional constraint that the (new) feasible set consists only of the points in the feasible set with integer coordinates, then x and y need not be "close." In fact, if z is the point in the feasible set for the SMP with integer coordinates that is closest to x, then z need not be close to y, and in fact, the value of the objective function at z may be much smaller than the value of the objective function at y. We illustrate this point with a simple example.

EXAMPLE 5.13 Consider the SMP

$$\text{maximize} \quad 2x_1 + 78x_2$$

$$\text{subject to} \quad \begin{cases} x_1 \geq 0, \ x_2 \geq 0 \\ x_1 + 40x_2 \leq 80 \\ x_1 + 2x_2 \leq 23 \end{cases}$$

The solution to this problem is $\mathbf{x} = [20 \quad 1.5]$, and the maximum value of the objective function is 157. Now suppose that, in addition to the above constraints, we require that x_1 and x_2 have integer values. With this additional constraint, the maximum value of the objective function is 156, and this value is attained at $\mathbf{y} = [0 \quad 2]$. The point in the feasible set with integer coordinates that is closest to $[20 \quad 1.5]$ is $[20 \quad 1]$, and the value of the objective function at the latter point is 118. Note that the points $\mathbf{x}$ and $\mathbf{y}$ are relatively far apart. In fact, it is relatively easy to modify this example to make the distance between $\mathbf{x}$ and $\mathbf{y}$ as close as you like to the maximum distance between two points in the feasible set for the original SMP (Exercise 10). ∎

The point of this example is, of course, that if you try to solve a linear programming problem with integer constraints by solving the problem with the same constraints except without the requirement that feasible points have integer coordinates, then the solution to the latter problem may not be at all close to the solution of the integer programming problem.

Exercises 5.4

1. Formulate a mathematical model for the situation discussed in this section. Be sure to identify your undefined terms, definitions, and axioms.

2. Find an algorithm for efficiently assigning five applicants to five jobs. Assume that the applicants are complete for the jobs. Estimate the efficiency of your method by counting the number of steps needed to make all the assignments, and use your method on the set of applicants given in Figure 5.7 (A_i is connected to J_k if and only if A_i is appropriate for J_k).

3. Establish the assertion in Case 2 of the induction proof of Theorem 5.2. That is, show that the set of applicants for the jobs in B is complete.

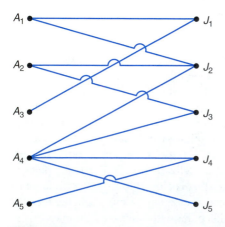

Figure 5.7

4. Discuss the situation of each applicant being unqualified, minimally qualified, or highly qualified for each job. Is there an analog to Theorem 5.2 in this case? What is the goal of an assignment, taking account of the qualifications of the applicant?

5. Using the setting of Exercise 2 (Figure 5.7), let $\{a_{ij}\}$ be a set of variables, $i = 1, \ldots, 5$, $j = 1, \ldots, 5$, that represents possible assignments of applicants to jobs. Thus $a_{ij} = 1$ if A_i is assigned to J_j, and $a_{ij} = 0$ if A_i is not assigned to J_j. Using the variables a_{ij}, form constraints to show that all jobs must be filled and no person can be assigned to more than one job.

6. Find an example of five applicants and five jobs, where the applicants are not complete for the jobs, but any subset of four jobs can be filled with appropriate applicants.

7. Formulate the job assignment problem as a linear programming problem with auxiliary constraints. Thus, find a matrix $\mathbf{A}$ and vectors $\mathbf{b}$ and $\mathbf{c}$ so that this assignment problem is the standard maximum problem $[\mathbf{A}, \mathbf{b}, \mathbf{c}]$ together with other constraints.

8. Develop a model for the consideration of a job assignment problem where each applicant is rated for each job on a scale from 0 to 5. As in Exercise 7, formulate this problem as a linear programming problem with auxiliary constraints, and exhibit the matrices and vectors used to write the problem as a standard maximum problem. Maximize the total rankings of people assigned to jobs.

9. Let the matrix $\mathbf{R} = [r_{ik}]$, which is given below, be the rating matrix representing the qualifications of four applicants for four jobs. Thus $r_{ik} = 3$ means that the applicant A_i is rated 3 for job J_k. Assume that higher ratings indicate more skill and experience relevant to the job. Find an assignment of applicants to job that maximizes the sum of the ratings of the applicants for the jobs they are assigned to fill. Is this assignment unique? Also, formulate this problem using the results of Exercise 8 and solve by using the simplex method.

$$
\mathbf{R} = \begin{array}{c} \\ A_1 \\ A_2 \\ A_3 \\ A_4 \end{array}
\begin{array}{cccc} J_1 & J_2 & J_3 & J_4 \end{array}
\left[\begin{array}{cccc}
0 & 1 & 0 & 1 \\
3 & 3 & 5 & 0 \\
2 & 1 & 1 & 1 \\
3 & 2 & 3 & 2
\end{array} \right]
$$

10. Suppose there are seven applicants for six jobs, and the goal is to fill as many jobs as possible with appropriate applicants. The matrix $\mathbf{A}$, given below, shows which applicants (rows) are appropriate for which jobs (columns). Thus an entry of 1 in the ij-th position of $\mathbf{A}$ means that applicant i is appropriate for job j. Fill as many jobs as possible with appropriate applicants. Is your solution unique?

$$
\mathbf{A} = \begin{array}{c} \\ 1 \\ 2 \\ 3 \\ 4 \\ 5 \\ 6 \\ 7 \end{array}
\begin{array}{cccccc} 1 & 2 & 3 & 4 & 5 & 6 \end{array}
\left[\begin{array}{cccccc}
0 & 1 & 1 & 0 & 0 & 0 \\
0 & 1 & 1 & 1 & 0 & 0 \\
1 & 1 & 1 & 1 & 1 & 0 \\
0 & 0 & 1 & 1 & 0 & 0 \\
0 & 0 & 0 & 1 & 0 & 0 \\
0 & 0 & 0 & 1 & 0 & 1 \\
0 & 0 & 0 & 1 & 0 & 1
\end{array} \right]
$$

11. Suppose that in the job assignment problem, one job can be filled by more than one applicant by using part-time work (for example, two people each working half-time). Formulate this new version of the problem as a linear programming problem.

12. Suppose that in a job assignment problem, one applicant can be assigned to more than one job by working part-time on each job. Formulate this version of the problem as a linear programming problem.

5.5 Network and Flows

In this section we shall consider a class of problems that can be viewed as linear programming problems with auxiliary constraints. We begin by recalling two problems from earlier in the chapter: the transportation problem and the job assignment problem.

In the transportation problem the goal is to find an optimal schedule for shipping mobile homes from manufacturing sites to retail outlets. This problem involves a constraint that was not explicitly taken into account in the general theory of linear programming. This constraint is the natural requirement that the coordinates of the solution vector be integers. Such a constraint is reasonable; it would be a strange decision to consider a schedule optimal if it required a mobile home to be divided. However, with this constraint, two questions immediately arise: First, do optimal schedules exist, and second, if they exist, how can they be found? It is clear, for example, that the simplex method discussed earlier is not generally applicable to problems that require a solution vector having integer coordinates. Indeed, the examples and the exercises provide illustrations of problems with integer data where the coordinates of the optimal vectors are not all integers. Actually, many integer programming problems, including the transportation problem, can be solved by the simplex method (such problems are called "natural" integer problems). However, even though the simplex method can be used for certain integer problems, it cannot be used for all of them, and in those cases where it *can* be applied, it is often not the best method. In this section, we shall introduce a new model (a network) that provides a second method for solving integer transportation problems. In addition to providing a model for the integer transportation problem, the concept of a network makes it possible to incorporate additional constraints, namely route capacities. Thus the network model provides a setting for problems that are more general than those considered before. Our development of this model will provide a computational scheme for determining optimal vectors, and this scheme is often more efficient than the simplex method when both methods are applicable.

The job assignment problem of Section 5.4 is also a linear programming problem with auxiliary constraints, and our new model will provide a means for solving this problem and generalizations of it. To make it clear how the job assignment problem has the framework of a linear programming problem, let $\mathcal{J} = \{J_1, J_2, \ldots, J_n\}$ be the set of available jobs, and let $\mathcal{A} = \{A_1, A_2, \ldots, A_m\}$ be the set of applicants. Each applicant has been interviewed, tested, and then rated as appropriate or not appropriate for each job. Recall (Section 5.4) that an applicant is appropriate for a job if that applicant is both qualified for the job and willing to accept it. The rating of the set of applicants is given by the $m \times n$ matrix $\mathbf{R} = [r_{ik}]$, where r_{ik} is 1 if A_1 is appropriate for J_k, and r_{ik} is 0 if A_1 is not appropriate for J_k. The first problem is to determine whether all jobs can be filled with appropriate applicants. The second problem is to find an actual assignment of applicants to jobs so that all the jobs are filled with appropriate applicants. In Section 5.4 we gave a condition for determining whether all jobs could be filled. However, the proof of that result was not constructive, and no algorithm was given for assigning applicants to jobs. In this section we shall provide a second proof of the condition of Section 5.4, and this proof will give a method for finding the desired assignment.

To phrase the job assignment problem in the form of a linear optimization problem, we use $m \times n$ matrices $\mathbf{S}$ of the form $\mathbf{S} = [s_{ik}]$, where

1. $s_{ik} = 0$ or 1, $i = 1, \ldots, m$, $k = 1, \ldots, n$

2. $\displaystyle\sum_{k=1}^{n} s_{ik} \leq 1$, $i = 1, \ldots, m$

3. $\displaystyle\sum_{i=1}^{m} s_{ik} \leq 1$, $k = 1, \ldots, n$

These matrices correspond in a one-to-one manner with assignments of applicants to jobs: If A_i is assigned to J_k, then $s_{ik} = 1$; otherwise, $s_{ik} = 0$. Condition 2 states that an applicant can be assigned to at most one job, and condition 3 states that at most one applicant is to be assigned to each job. Finally, we consider the quantity $\mathbf{R} * \mathbf{S} = \sum_{i=1}^{m} \sum_{k=1}^{n} r_{ik} s_{ik}$. Each term in this sum is either 0 or 1. If the term is a 1, then $s_{ik} = 1$ and $r_{ik} = 1$, and this means that the assignment determined by $\mathbf{S}$ fills the job J_k with an appropriate applicant A_i. Hence the quantity $\mathbf{R} * \mathbf{S}$ is a measure of the number of jobs filled by appropriate applicants under the assignment $\mathbf{S}$. The job assignment problem is then the problem of finding a matrix $\mathbf{S}$ that satisfies conditions 1–3 and that maximizes $\mathbf{R} * \mathbf{S}$ for all such matrices. In this form, the problem is similar to a linear programming problem (see Exercise 2); however, it has an additional constraint, condition 1. The model of this chapter explicitly includes this new condition. It is also useful in more general assignment problems.

■ Definitions and Notation

It is useful to discuss both the transportation problem and the assignment problem in a single abstract setting. The new setting is known as a network, and we obtain it by abstracting the essential idea from the transportation problem (TP from now on) in the following manner.

First, we replace each of the factories and each of the sales areas (stores) by an abstract concept called a vertex. A *vertex* can be thought of as a point in a plane or, more generally, as simply an element in a set. Next, in place of each of the routes between factories and sales area, we introduce the notion of an edge. Abstractly, an *edge* is simply an ordered pair of vertices, and it is used to indicate a directed connection between the vertices. For example, if v and w are vertices, then the edge (v, w) represents a directed connection from vertex v to vertex w, or, in terms of the TP, a route along which goods can flow from factory v to store w. Finally, it is useful to assign a nonnegative integer to each edge to reflect the *capacity* of the edge. For example, in a TP it may be that some stores can be provided with no more that a certain maximum amount of goods in each time period. This would be the case, for example, if the factories are in Japan, the stores are in the United States, and the goods must be transported in ships of a definite (known) size. The capacity of these ships would have to be considered in solving the TP, because it would be unrealistic to call a schedule optimal if it required shipments that exceeded the available capacity. Also, in the event that there is no direct route from one of the factories (denoted by vertex v) to one of the stores (denoted by vertex w), then we assign 0 capacity to the edge (v, w). We now make these ideas precise.

Definition 5.2 *A capacitated network* (or a graph with capacity) is a pair (V, k), where $V = \{v_i\}_{i=1}^n$ is a finite set of elements called *vertices* and k is a function defined on $V \times V$ with values in the set of nonnegative integers and satisfying $k(v, v) = 0$ for $v \in V$. The elements of $V \times V$ are called the *edges* of the network, and the quantity $k(v, w)$ is called the *capacity* of the edge $(v, w) \in V \times V$.

The next step in the process of abstraction deals with the notion of a flow. Recall that in the TP, one seeks a shipping schedule that is optimal among all the shipping schedules that satisfy certain constraints. In the setting of networks, the appropriate generalization of a shipping schedule is a flow.

Definition 5.3 A *flow*, in the network (V, k), is a function f, defined on $V \times V$, taking *only integer values* and satisfying the conditions

1. $f(v_i, v_j) = -f(v_j, v_i)$.
2. $f(v_i, v_j) \le k(v_i, v_j)$ for $i, j = 1, 2, \ldots, n$.

Note that a flow is constrained in three ways. The flow along an edge is *always an integer*, the flow can *never exceed the capacity* of that edge, and the flow in *one direction* along an edge *is always the negative of the flow in the other direction*.

Next, we need the idea of an originating point (the source of materials for the factories) and also that of a terminating point (the consumer who empties the stores). This leads to the following definition of two types of distinguished vertices.

Definition 5.4 A vertex $v_0 \in V = \{v_i\}_{i=1}^n$ is called an *originating point* for a flow f if $\sum_{i=1}^n f(v_0, v_i) > 0$. A vertex $v_t \in V$ is called a *terminating point* for a flow if $\sum_{i=1}^n f(v_i, v_t) > 0$. A flow with exactly one originating point, v_0, and exactly one terminating point, v_t, is called a *flow from v_0 to v_t*. The *value* of a flow from v_0 to v_t is the number $v(f) = \sum_{i=1}^n f(v_0, v_i)$. A zero flow (all $f(v_i, v_j) = 0$) is assigned value 0.

Remarks

1. The terms *source* and *sink* are often used in place of our terms *originating point* and *terminating point*.
2. The vertices v_0 and v_t are elements of the set $V = \{v_1, v_2, \ldots, v_n\}$. Because $k(v_i, v_i) = 0$ for every $v_i \in V$, we also have $f(v_i, v_i) = 0$ for every $v_i \in V$. In particular, $f(v_0, v_0) = f(v_t, v_t) = 0$.
3. The value of a nonzero flow is defined only for flows from a vertex v_0 to a vertex v_t. The zero flow is defined by the function that assigns zero to each edge.
4. If f is a flow from v_0 to v_t and $v_j \in V$, $v_j \ne v_0$, $v_j \ne v_t$, then $\sum_{i=1}^n f(v_i, v_j) = 0$.
5. If f is a flow from v_0 to v_t, then $v(f) = \sum_{i=1}^n f(v_i, v_t)$ (Exercise 10).

The central problem in our study of networks and flows is that of finding a flow with maximum value. One method of verifying the maximality of a flow is analogous to that used in the corresponding problem in linear programming. Recall that a vector may be

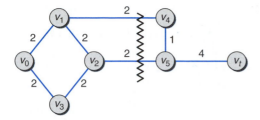

Figure 5.8

shown to be optimal for a linear programming problem by considering the dual problem and by verifying a certain equality. In the same way, a flow may be shown to be maximal by considering a second problem (a minimization problem) and by verifying a certain equality. This second problem may seem at first to be unrelated to the flow problem; however, there is a deep connection between the two problems, and it will be explored below (Theorem 5.3). To state the appropriate minimization problem, we first need the idea of a cut in a network.

Definition 5.5   A *cut* in the network (V, k), with respect to vertices v_0 and v_t, is any partition of the set V into two sets V_0 and V_t such that $v_0 \in V_0$ and $v_t \in V_t$. The *capacity* of the cut (V_0, V_t) is the number

$$k(V_0, V_t) = \sum_{\substack{u \in V_0 \\ v \in V_t}} k(u, v)$$

Remark   There is no flow involved in the definition of a cut. Hence the use of the symbols v_0 and v_t does not presuppose a flow from v_0 to v_t.

EXAMPLE 5.14   In the network shown in Figure 5.8, the numbers on the edges indicate capacities in both directions. Edges that are not shown have zero capacity.

A flow from v_0 to v_t is given by the function f:

$$f(v_0, v_1) = -f(v_1, v_0) = f(v_0, v_3) = -f(v_3, v_0) = f(v_1, v_2)$$
$$= -f(v_2, v_1) = f(v_3, v_2) = -f(v_2, v_3) = 1$$
$$f(v_2, v_5) = -f(v_5, v_2) = f(v_5, v_t) = -f(v_t, v_5) = 2$$
$$f(u, v) = 0 \quad \text{for all other edges } (u, v)$$

The value of f is $v(f) = f(v_0, v_1) + f(v_0, v_2) + f(v_0, v_3) + f(v_0, v_4) + f(v_0, v_5) + f(v_0, v_t) = 1 + 0 + 1 + 0 + 0 + 0 = 2$.

A cut with respect to v_0 and v_t is given by the pair of sets (V_0, V_t), where $V_0 = \{v_0, v_1, v_2, v_3\}$ and $V_t = \{v_4, v_5, v_t\}$. The cut is illustrated by the sawtooth line across the network. The capacity of this cut is given by $k(V_0, V_t) = k(v_1, v_4) + k(v_2, v_5) = 2 + 2 = 4$.   ■

The minimization problem of interest is the problem of finding a cut with minimal capacity. The intimate connection between this problem and the maximal flow problem is crucial to our work with networks.

■ Maximal Flows and Minimal Cuts

We begin with the following important theorem (note the similarity to Theorem 5.1).

THEOREM 5.3 (Max Flow–Min Cut) The value of any maximal flow from v_0 to v_t is equal to the capacity of any minimal cut with respect to these vertices.

Proof. The first step in the proof is to show that $v(f) \leq k(V_0, V_t)$ for every flow f from v_0 to v_t and every cut (V_0, V_t) with respect to these vertices. Once this fact has been established, then we know that if f and (V_0, V_t) are any flow and cut such that $v(f) = k(V_0, V_t)$, then f is a maximal flow and (V_0, V_t) is a minimal cut. Also, since every maximal flow has the same value and every minimal cut has the same capacity, it follows that if we can find *one* flow f and *one* cut (V_0, V_t) for which $v(f) = k(V_0, V_t)$, then this equality holds for every maximal flow and minimal cut. We have

$$v(f) = \sum_{k=1}^{n} f(v_0, v_k) = \sum_{\substack{v_i \in V_0 \\ v_k \in V_t}} f(v_i, v_k)$$

This follows from the definition of a flow from v_0 to v_t. Indeed, if $v_i \in V_0$, $v_i \neq v_0$, then $\sum_{j=1}^{n} f(v_i, v_j)$ must be equal to 0. Next, we have

$$v(f) = \sum_{\substack{v_i \in V_0 \\ v_j \in V_0}} f(v_i, v_j) + \sum_{\substack{v_i \in V_0 \\ v_j \in V_t}} f(v_i, v_j)$$

$$= 0 + \sum_{\substack{v_i \in V_0 \\ v_j \in V_t}} f(v_i, v_j)$$

(5.23) $$\leq \sum_{\substack{v_i \in V_0 \\ v_j \in V_t}} k(v_i, v_j) = k(V_0, V_t)$$

The proof is completed by exhibiting a flow f and a cut (V_0, V_t) for which equality holds. To this end, it is useful to examine inequality (5.23) to determine exactly when it is an equality. The inequality results from the condition that $f(v_i, v_k) \leq k(v_i, v_k)$ for every flow f. However, in (5.23) the pair of vertices (v_i, v_k) is such that $v_i \in V_0$ and $v_k \in V_t$. Hence, if the flow f and the cut (V_0, V_t) satisfy $f(v_i, v_k) = k(v_i, v_k)$ for all $v_i \in V_0$ and $v_k \in V_t$, then we have $v(f) = k(V_0, V_t)$.

Let f be any maximal flow from v_0 to v_t. Such flows always exist because there are only a finite number of possible flows (Exercise 2). We define a cut in the following way. An edge (v_i, v_j) is said to be *saturated* if $f(v_i, v_j) = k(v_i, v_j)$. Next we introduce the notion of a *path* from vertex v to vertex v' as an ordered set of edges of the form $\{(v, w_1), (w_1, w_2), \ldots, (w_p, v')\}$. A path from one vertex to another is called *unsaturated* if no edge in the path is saturated. We now define the set V_0 to be the vertex v_0 together with all vertices that can be reached by an unsaturated path from v_0. The set V_t consists of all other vertices. We claim that (V_0, V_t) is a cut with respect to the vertices v_0 and v_t. This will be true if $v_t \in V_t$. We proceed with a proof by contradiction and assume that $v_t \notin V_t$. Then $v_t \in V_0$, and by definition of V_0 there is an unsaturated path from v_0 to v_t. Let this path be

$P = \{(v_0, w_1), (w_1, w_2), \ldots, (w_p, v_t)\}$. Also, let $m = \min_{(w,w') \in P}[k(w, w') - f(w, w')]$. Since P is unsaturated, $m \geq 1$. We now define a flow f^* as follows:

$$f^*(w, w') = \begin{cases} f(w, w') + m, & (w, w') \in P \\ -[f(w', w) + m], & (w', w) \in P \\ f(w, w'), & \text{otherwise} \end{cases}$$

The flow f^* is again a flow from v_0 to v_t (Exercise 4), and because $m \geq 1$, we have $v(f^*) > v(f)$. But this is impossible because f is a maximal flow. Thus $v_t \in V_t$, and (V_0, V_t) is a cut with respect to v_0 and v_t.

The proof is now completed by the observation that $v(f) = k(V_0, V_t)$. This equality holds because for each $v_i \in V_0$ and $v_k \in V_t$, we must have $f(v_i, v_k) = k(v_i, v_k)$ or else v_k would not be in V_t. However, as noted above, if $f(v_i, v_k) = k(v_i, v_k)$ for all $v_i \in V_0$, $v_k \in V_t$, then $v(f) = k(V_0, V_t)$. ∎

The proof of Theorem 5.3 is constructive. It provides us with a means of finding maximal flows. The method is the following. One starts with any flow at all from v_0 to v_t. Then the sets V_0 and V_t are formed in the manner indicated in the proof. If the vertex v_t is in the set V_t, then the flow is already maximal. If $v_t \notin V_t$, then the flow is not maximal and a new flow must be constructed. The new flow is obtained in the manner used to obtain f^* from f in the proof of Theorem 5.3. The construction ensures that $v(f^*) > v(f)$. Either f^* is maximal and we are finished, or the process is to be continued, in which case we again form sets V_0 and V_t (this time for the new flow f^*) and again check the criteria: Is $v_t \in V_t$? This process can be continued until a maximal flow is obtained. Indeed, if the maximal flow has value $v(f)$, then it is clear that in at most $v(f)$ steps, this process yields a maximal flow. We illustrate the process with an example.

EXAMPLE 5.15 Find a maximum flow for the capacitated network shown in Figure 5.9. In the diagram, the numbers on the edges indicate capacities in both directions. Edges that are not shown have zero capacity. As an initial flow, we define f as follows: $f(v_0, v_1) = 1$, $f(v_0, v_3) = 1$, $f(v_1, v_2) = 1$, $f(v_1, v_4) = 0$, $f(v_3, v_2) = 1$, $f(v_2, v_5) = 2$, $f(v_4, v_5) = 0$, and $f(v_5, v_t) = 2$. The set V_0 is $\{v_1, v_2, v_3, v_4, v_5, v_0, v_t\}$ because each vertex can be reached from v_0 by an unsaturated path. Hence $v_t \in V_0$, and this choice of f is not maximal. The path $P = \{(v_0, v_1), (v_1, v_4), (v_4, v_5), (v_5, v_t)\}$ is unsaturated, and it connects v_0 to v_t. This path can hold only one more unit of flow since $f(v_0, v_1) = 1$ and $k(v_0, v_1) = 2$. Thus we define the flow f_1 to be the same as f with the following exception: $f_1(v_0, v_1) = 2 = -f_1(v_1, v_0)$, $f_1(v_1, v_4) = 1 = -f_1(v_4, v_1)$, $f_1(v_4, v_5) = 1 = -f_1(v_5, v_4)$, $f_1(v_5, v_t) =$

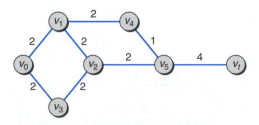

Figure 5.9

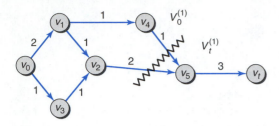

Figure 5.10

$3 = -f_1(v_t, v_5)$. Using the flow f_1, we form the sets $V_0^{(1)}$ and $V_t^{(1)}$ as in the proof of Theorem 5.3. We obtain

$$V_0^{(1)} = \{v_0, v_1, v_2, v_3, v_4\}$$
$$V_t^{(1)} = \{v_5, v_t\}$$

In this case, $v_t \in V_t^{(1)}$, and hence flow f_1 is a maximal flow and cut $(V_0^{(1)}, V_t^{(1)})$ is a minimal cut. The capacity of this cut is $k(V_0^{(1)}, V_t^{(1)}) = k(v_4, v_5) + k(v_2, v_5) = 2 + 1 = 3$.

The flow f_1 and the cut $(V_0^{(1)}, V_t^{(1)})$ are illustrated in Figure 5.10. Note that the arrows in Figure 5.10 indicate the positive direction of the flow. There is, of course, an equal flow in the direction opposite the arrow. ∎

It is not necessary to consider a very complicated example before it becomes clear that this method of finding an optimal flow is rather cumbersome. Fortunately, the algorithm can be carried out more efficiently by means of a tableau. We shall not provide the details of the tableau treatment. Many books in operations research offer numerous examples and discussions of the technique. We now turn to the use of Theorem 5.3 in the integer transportation problem and the job assignment problem.

■ Integer Transportation Problems

The integer transportation problem (ITP) can be phrased in the following way: Given sequences of nonnegative integers $\{s_i\}_{i=1}^m$ (supplies), $\{d_j\}_{j=1}^n$ (demands), $\{k_{ij}\}_{j=1,n}^{i=1,m}$ (route capacities), and real numbers $\{c_{ij}\}_{j=1,n}^{i=1,m}$ (costs), find a sequence of integers $F = \{f_{ij}\}_{j=1,n}^{i=1,m}$ (the shipping schedule) such that

$$(5.24) \qquad 0 \le f_{ij} \le k_{ij}, \qquad\qquad i = 1, \ldots, m, \; j = 1, \ldots, n$$

$$(5.25) \qquad \sum_{j=1}^n f_{ij} \le s_i, \qquad\qquad i = 1, \ldots, m$$

$$(5.26) \qquad \sum_{i=1}^m f_{ij} \ge d_j, \qquad\qquad j = 1, \ldots, n \qquad\qquad \text{and}$$

$$C(F) = \sum_{i=1}^m \sum_{j=1}^n c_{ij} f_{ij} \qquad \text{is a minimum over all such } F$$

As a first step in the solution of this problem, we note that it is sufficient to consider only the case in which $\sum_{i=1}^m s_i = \sum_{j=1}^n d_j$. The argument for this is as follows. First, if F satisfies

(5.25) and (5.26), then $\sum_{j=1}^{n} d_j \leq \sum_{j=1}^{n} \sum_{i=1}^{m} f_{ij} \leq \sum_{i=1}^{m} s_i$, and therefore, a necessary condition for the problem to have a solution is $\sum_{j=1}^{n} d_j \leq \sum_{i=1}^{m} s_i$ (supply must equal or exceed demand). Thus, if demand exceeds supply, then the problem stated above does not have a solution. However, there is a related problem involving rationing that does have a solution (see Exercise 11). Next, if $\sum_{j=1}^{n} d_j < \sum_{i=1}^{m} s_i$, then a new problem can be formed by introducing an additional store with demand $d_{n+1} = \sum_{i=1}^{m} s_i - \sum_{j=1}^{n} d_j$ and with the shipping costs to the new store being 0 from each factory. The new store is suggestively called a *dump*. This new problem satisfies the condition that supply is equal to demand, and solutions of the new problem are easily identified with solutions of the original problem (Exercise 5). Therefore, it is no restriction to assume that $\sum s_i = \sum d_j$, and we do so from now on.

We have already noted that a necessary condition for an ITP to possess a solution is for supply to equal or exceed demand. The condition that supply equal demand is not usually sufficient for the existence of a solution because of the limitations on shipping schedules that are imposed by the route capacities. A shipping schedule that satisfies (5.24), (5.25), and (5.26) will be called a *feasible schedule*. If there is a feasible schedule, then it follows that the demand at each store is less than the total capacity of the routes terminating at that store. Even this is not a sufficient condition because the supply may not be located correctly to utilize the available capacity over each route. With these preliminary comments in mind, we are now ready to establish a necessary and sufficient condition for the existence of a feasible shipping schedule. As a matter of notation, we let $\mathcal{W} = \{W_i\}_{i=1}^{m}$ and $\mathcal{S} = \{S_j\}_{j=1}^{n}$ be the factories and sales areas, respectively. Also, for subsets $\mathcal{W}' \subset \mathcal{W}$ and $\mathcal{S}' \subset \mathcal{S}$, we let $d(\mathcal{S}') = \sum_{S_i \in \mathcal{S}'} d_i$ and $s(\mathcal{W}') = \sum_{W_i \in \mathcal{W}'} s_i$ be the demand of the subset $\mathcal{S}'$ and the supply of the subset $\mathcal{W}'$, respectively. Also, we let $k(\mathcal{W}', \mathcal{S}') = \sum_{W_i \in \mathcal{W}, S_j \in \mathcal{S}} k_{ij}$ be the capacity between the subsets $\mathcal{W}'$ and $\mathcal{S}'$. The desired theorem has the following form.

THEOREM 5.4   A necessary and sufficient condition for the existence of at least one feasible shipping schedule for the ITP is that for each pair of subsets $\mathcal{W}' \subset \mathcal{W}$ and $\mathcal{S}' \subset \mathcal{S}$, the relation

$$(5.27) \qquad\qquad d(\mathcal{S}') - s(\mathcal{W}') \leq k(\mathcal{W} \setminus \mathcal{W}', \mathcal{S}')$$

is true. Note that $\mathcal{W} \setminus \mathcal{W}'$ is the set of elements in $\mathcal{W}$ but not in $\mathcal{W}'$.   ■

We point out that in this case, in contrast to the more general linear optimization problems from Section 5.2, the existence of a feasible vector guarantees the existence of an optimal vector (Exercise 6). It is not necessary to consider any sort of dual problem.

We do not provide a proof of Theorem 5.4, primarily because the proof does not aid us in finding an optimal shipping schedule.

We now have a necessary and sufficient condition for the existence of feasible shipping schedules for the ITP. In one sense this solves the ITP, because there are only a finite number of such schedules, and hence one can obtain an optimal schedule by trial and error. However, trial-and-error methods are not practical even for problems where n and m are only modestly large (say $m = n = 15$). Accordingly, considerable effort has been devoted to the development of efficient algorithms for obtaining optimal solutions. Many of these algorithms are based on the statement and proof of Theorem 5.3. They are often called Ford–Fulkerson methods, because these investigators first proved Theorem 5.3.

■ The Assignment Problem Revisited

In Section 5.3 we introduced the concept of a complete set of applicants in order to state a condition under which all the available jobs could be filled by appropriate applicants. In this section we shall give a second proof of that result. More important, we shall give a proof that leads to a method of actually finding an assignment of applicants to jobs. We begin by introducing some new notation. We use the setting of the job assignment problem as given in the introduction of this section. For any subset $\mathcal{J}_\alpha$ of the set of jobs $\mathcal{J}$, we let $\mathcal{A}(\mathcal{J}_\alpha)$ be the set of all applicants who are appropriate for at least one of the jobs in the set $\mathcal{J}_\alpha$. Also, we let $|\mathcal{J}_\alpha|$ and $|\mathcal{A}(\mathcal{J}_\alpha)|$ be the number of elements in the sets $\mathcal{J}_\alpha$ and $\mathcal{A}(\mathcal{J}_\alpha)$, respectively. Using this notation, we say that a set of applicants $\mathcal{A}$ is *complete* for a set of jobs $\mathcal{J}$ if $|\mathcal{A}(\mathcal{J}_\alpha)| \geq |\mathcal{J}_\alpha|$ for every subset $\mathcal{J}_\alpha \subset \mathcal{J}$. We are now ready to restate and re-prove a result on the existence of solutions for the job assignment problem.

THEOREM 5.2 The set of jobs $\mathcal{J} = \{J_1, \ldots, J_n\}$ can be filled from the set of applicants $\mathcal{A} = \{A_1, \ldots, A_m\}$ if and only if $\mathcal{A}$ is complete for $\mathcal{J}$.

Proof. If all the jobs can be filled by the applicants, then certainly $\mathcal{A}$ is complete for $\mathcal{J}$. In fact, for any subset $\mathcal{J}_\alpha$, the set of applicants who are assigned to fill these jobs is a subset of $\mathcal{A}(\mathcal{J}_\alpha)$, and hence $|\mathcal{J}_\alpha| \leq |\mathcal{A}(\mathcal{J}_\alpha)|$.

Next, suppose that $\mathcal{A}$ is complete for $\mathcal{J}$. It will be shown that all the jobs can be filled by considering a certain capacitated network. The vertices of this network consist of the elements of the set of applicants $\mathcal{A}$, the elements of the set of jobs $\mathcal{J}$, and two special vertices denoted by v_0 and v_t. Let $M = m + n$, and define the capacities of the edges of this network by the formulas

$$
\begin{aligned}
k(v_0, A_i) &= 1 & i &= 1, \ldots, m \\
k(J_k, v_t) &= 1 & k &= 1, \ldots, n \\
k(A_i, J_k) &= M, & &\text{if } A_i \text{ is appropriate for } J_k \\
k(A_i, J_k) &= 0, & &\text{if } A_i \text{ is not appropriate for } J_k \\
k(v, w) &= 0, & &\text{all other edges } (v, w)
\end{aligned}
$$

Pictorially, we represent such a network as shown in Figure 5.11. In Figure 5.11 we have indicated edges of capacity zero because they will play a role in our proof. In most cases, we omit such edges from the graph. The constant M assigned as the capacity of edges (A_i, J_k),

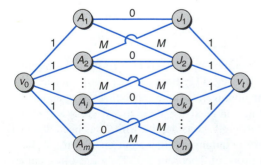

Figure 5.11

where A_i is appropriate for J_k, is chosen to be $m + n$ to be sure that this capacity is large enough to allow for the flow of multiple units (more than m or n) in either direction along such edges. Any such large number could be used in the proof. Of course, since the capacity into each A_i, along (v_0, A_i), and out of J_k, along (J_k, v_t) is set to 1, we will never end up with more than one applicant assigned to a single job or any job filled by more than one applicant.

Next, consider flows in this network, and let f be any maximal flow from v_0 to v_t. Such a maximal flow certainly exists. Indeed, the set of flows from v_0 to v_t is a finite set, and this set is nonempty since it contains the zero flow. It is useful to distinguish two cases for the value of the flow f. Case 1 is $v(f) = n$, and case 2 is $v(f) < n$. These cases cover all possibilities for $v(f)$ in view of the constraints on the capacities of the edges connected to v_t.

In case 1 ($v(f) = n$), the flow f provides an assignment of applicants to jobs that fills all available jobs. This assignment is given by the rule "assign applicant A_i to job J_k if and only if $f(A_i, J_k) = 1$." Since $v(f) = n$, it easily follows that each job is filled by exactly one appropriate applicant (Exercise 7).

The proof is concluded by showing that case 2 ($v(f) < n$) is impossible if $\mathcal{A}$ is complete for $\mathcal{J}$. To this end, note that if $v(f) < n$, then by the max flow–min cut theorem there is a minimal cut (V_0, V_t) with respect to v_0 and v_t with capacity less than n. Let $V_0 = \{v_0, A_1, \ldots, A_p, J_1, \ldots, J_q\}$. (There is no loss of generality in assuming this, since renumbering the jobs and applicants will always make it possible to express V_0 in this form. It is also possible, of course, that either all the A_i's or all the J_k's are missing from V_0.) The definition of the capacity of a cut states that $k(V_0, V_t)$ is the sum of capacities of all edges of the form (v, w), where $v \in V_0$ and $w \in V_t$. Such edges have one of the following forms:

1. $(v_0, A_i), p < i \le m$
2. $(A_i, J_r), i \le p, r > q$
3. $(A_i, J_r), i > p, r \le q$
4. $(J_r, v_t), r \le q$

Edges of types 1 and 4 have capacity 1, and edges of types 2 and 3 have either capacity 0 or $M = m + n$. Since $(V_0, V_t) < n$, no edge in the sum can have a capacity greater than n. Thus all edges of types 2 and 3 have capacity 0. Therefore, we have

$$n > k(V_0, V_t) = \sum_{i=p+1}^{m} k(v_0, A_i) + \sum_{r=1}^{q} k(J_r, v_t) = m - p + q$$

In an equivalent form, the above inequality states that $m - p < n - q$.

The proof is concluded by considering the set $\mathcal{J}_\alpha = \{J_{q+1}, \ldots, J_n\}$ and $\mathcal{A}(\mathcal{J}_n)$. The above argument has shown that $k(A_i, J_k) = 0$, for $i \le p$ and $k > q$. Therefore, $\mathcal{A}(\mathcal{J}) \subset \{A_{p+1}, \ldots, A_m\}$, and hence $|\mathcal{A}(\mathcal{J}_\alpha)| \le m - p$. But this in turn implies that

$$|\mathcal{A}(\mathcal{J}_\alpha)| \le m - p < n - q = |\mathcal{J}_\alpha|$$

Since this last inequality is a contradiction of the definition of completeness, it has been shown that the assumption $v(f) < n$ is incompatible with the assumption of completeness. This concludes the proof. ∎

This proof of Theorem 5.2 shows that the job assignment problem can be solved by finding a maximal flow in a certain capacitated network. As noted earlier, the proof

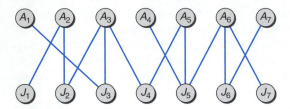

Figure 5.12

of Theorem 5.1 provides a constructive method for finding maximal flows in capacitated networks and hence a method of constructing solutions to the assignment problem.

EXAMPLE 5.16 Consider the sets of seven applicants and seven jobs shown in Figure 5.12, where a solid line between an applicant and a job means that the applicant is appropriate for the job.

Use the method described in the proof of the max flow–min cut theorem to assign the applicants to the jobs.

Solution. We begin by converting the information in Figure 5.12 into a capacitated network with an originating point, S, and a terminating point, T. We use the capacities given in Figure 5.11, where $M = 7 + 7 = 14$. This gives the network shown in Figure 5.13, but this time we have not included the edges with capacity zero.

We begin assigning applicants to jobs by using a flow from S to T. Our initial flow is the simple one wherein we assign any applicant A_i to the job J_i, provided that there is a positive capacity from A_i to J_i. This gives the assignments $A_2 \rightarrow J_2$, $A_3 \rightarrow J_3$, $A_5 \rightarrow J_5$, and $A_6 \rightarrow J_6$, as well as the associated flows of 1 along (S, A_2), (S, A_3), (S, A_5), and (S, A_6)

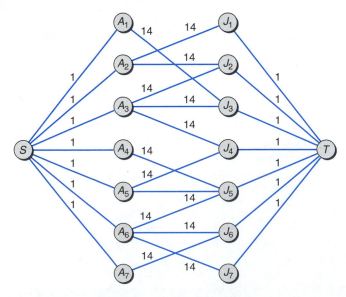

Figure 5.13

and along (J_2, T), (J_3, T), (J_5, T), and (J_6, T). If we call the associated flow f_1, then $v(f_1) = 4$ since the flow out of S and into T has four units. We now follow the proof of the max flow–min cut theorem (5.2) and look for an unsaturated path from S to T. Such a path is given by $S \rightarrow A_7 \rightarrow J_6 \rightarrow A_6 \rightarrow J_7 \rightarrow T$. If we send one unit along this path, we obtain flow f_2 with the assignments $S \rightarrow A_7$, $A_2 \rightarrow J_2$, $A_3 \rightarrow J_3$, $A_5 \rightarrow J_5$, $A_6 \rightarrow J_7$, $A_7 \rightarrow J_6$, and $J_7 \rightarrow T$. Since we now have five jobs filled with appropriate applicants, $v(f_2) = 5$. We now repeat the process, first finding an unsaturated path from S to T and then sending one unit along that path. We obtain the unsaturated path $S \rightarrow A_4 \rightarrow J_5 \rightarrow A_5 \rightarrow J_4 \rightarrow T$ and the new flow f_3 with the assignment $A_2 \rightarrow J_2$, $A_3 \rightarrow J_3$, $A_4 \rightarrow J_5$, $A_5 \rightarrow J_4$, $A_6 \rightarrow J_7$, and $A_7 \rightarrow J_6$. We have $v(f_3) = 6$. Once again we look for an unsaturated path from S to T and then fill the path by sending one unit along the path. This gives flow f_4. The path is $S \rightarrow A_1 \rightarrow J_3 \rightarrow A_3 \rightarrow J_2 \rightarrow A_2 \rightarrow J_1 \rightarrow T$, and the related assignment is $A_1 \rightarrow J_3$, $A_2 \rightarrow J_1$, $A_3 \rightarrow J_2$, $A_4 \rightarrow J_5$, $A_5 \rightarrow J_4$, $A_6 \rightarrow J_7$, and $A_7 \rightarrow J_6$. We now have $v(f_4) = 7$, and since this is the maximal possible value (all edges from S are full), we have a maximal flow and all jobs are filled with appropriate applicants.   ∎

Remark   The example above is small enough (7 jobs and 7 applicants) that we could have filled the jobs quickly by simple trial and error on Figure 5.12. However, once Figure 5.12 gets much larger, say 25 jobs and 25 applicants, it is very difficult to use an "eyeball" method. One needs to use a computerized method, and one such method is based on the proof of Theorem 5.2, as in the example. Also, we should note that it may not always be possible to fill all jobs, because we might not have a complete set of applicants. In such a case, the maximal flow will fill as many jobs as possible.

Exercises 5.5

1. Consider the linear programming problem

$$\text{maximize} \quad x + 48y$$

$$\text{subject to} \quad \begin{cases} x \geq 0, \ y \geq 0 \\ x + 50y \leq 202 \\ 15x + y \leq 300 \end{cases}$$

   (a) Use the simplex algorithm to find the solution to this problem. Express the values of x^*, y^* and the objective function correct to three decimal places.
   (b) Solve this problem with the additional constraint that x and y must be integers.
   (c) Find the point with integer coordinates that is closest to the solution of the original problem. Compare the value of the objective function at this point and the value of the objective function found in part (b).
   (d) Show that the point with integer coordinates that is closest to the solution of the original problem is not in the feasible set for either the original problem or the problem of part (b).

2. Refer to the problem of Exercise 1. Define the *diameter* of a bounded set as the maximum distance (the ordinary Euclidean distance) between any two points in the set.

   (a) Show that the distance between the point (x^*, y^*) of Exercise 1, part (a), and the solution of the problem in part (b) is more than 97% of the diameter of the feasible set.

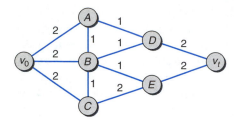

Figure 5.14

(b) Modify the problem of Exercise 1 so that the distance between the solution of the original problem and the solution of the same problem with integer constraints is at least 99% of the diameter of the feasible set.

(c) Find a new objective function for the problem of Exercise 1 with the following characteristics: Let (x^*, y^*) be a solution of the new problem (without integer constraints), and let (x', y') be the point in the feasible set with integer coordinates that is closest to (x^*, y^*). The solution of the new problem with integer constraints is at least 30% larger that the value of the new objective function at the point (x', y').

3. Show that there are only a finite number of possible flows in a capacitated network.

4. Show that the function f^* of the proof of Theorem 5.3 actually defines a flow from v_0 to v_t. Note that, in particular, it must be shown that f^* is integral-valued and that f^* has a unique originating point and a unique terminating point.

5. As we noted, an ITP in which supply exceeds demand can be converted into an ITP in which supply equals demand. Show that there is a one-to-one correspondence between the solutions of these two problems.

6. Show that the existence of a feasible solution for the ITP implies the existence of an optimal vector.

7. Show that the assignment rule of the proof of Theorem 5.2 actually works. That is, show that each available job is filled by one and only one appropriate applicant.

8. In the network shown in Figure 5.14, the numbers represent capacities in both directions. All edges not given numbers have zero capacity. Find a maximal flow and minimal cut with respect to v_0 and v_t. Take as your initial flow one with $f(v_0, v_i) = 1$ for all vertices v_i with $k(v_0, v_i) \neq 0$.

9. Same as Exercise 8 for the network shown in Figure 5.15.

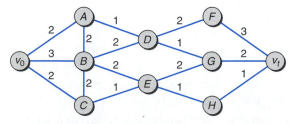

Figure 5.15

10. Show that if f is a flow from v_0 to v_t and that if $V = \{v_i\}_{i=1}^n$ is the set of vertices of the network, then $v(f) = \sum_{i=1}^n f(v_i, v_t)$.

11. Consider an integer programming problem in which supply is less than demand. Formulate an equivalent problem in which supply equals demand by introducing a dummy factory. Discuss how the solution of this new problem solves the rationing problem that results when demand exceeds supply.

12. In Figure 5.6, we have a set of five applicants and five jobs, and the applicants are complete for the jobs. Formulate this assignment problem as a linear programming problem, and solve it using a computer package such as Maple or Excel.

13. Repeat Exercise 12, but use the applicants and jobs shown in Figure 5.12. Note that the matrix A is now 49 by 14.

14. Suppose that there are four applicants for six jobs and that the matrix giving the appropriate applicants for each job is shown below as matrix $\mathbf{A}$: applicants are rows, jobs are columns. Training costs for the applicant/job pairs are shown in matrix $\mathbf{B}$. The task is, first, to fill as many jobs as possible and, second, to do so with minimal training costs.

$$\mathbf{A} = \begin{bmatrix} 1 & 1 & 1 & 0 & 0 & 0 \\ 0 & 1 & 0 & 1 & 0 & 0 \\ 0 & 0 & 0 & 1 & 1 & 0 \\ 0 & 0 & 0 & 1 & 1 & 1 \end{bmatrix} \qquad \mathbf{B} = \begin{bmatrix} 100 & 200 & 600 & 0 & 0 & 0 \\ 0 & 500 & 0 & 500 & 0 & 0 \\ 0 & 0 & 0 & 300 & 400 & 0 \\ 0 & 0 & 0 & 600 & 700 & 900 \end{bmatrix}$$

    (a) Formulate a mathematical problem whose solution achieves the goals.
    (b) Solve the problem posed in part (a).

15. A capacitated network is given by $V = \{v_0, A, B, C, D, E, F, G, H, v_t\}$, and k is

$$\begin{array}{llll} k(v_0, A) = 8 & k(v_0, B) = 5 & k(A, D) = 2 & k(A, C) = 6 \\ k(B, D) = 3 & k(B, E) = 3 & k(D, G) = 3 & k(C, F) = 5 \\ k(C, G) = 2 & k(E, G) = 2 & k(E, H) = 2 & k(G, H) = 1 \\ k(C, v_t) = 2 & k(F, v_t) = 4 & k(G, v_t) = 4 & k(H, v_t) = 2 \end{array}$$

Find a maximal flow using the algorithm of Section 5.5.

Addendum for Students and Teachers on Projects and Presentations

A.0   Introduction

In Chapter 1, we discussed the cycle of model building and noted the important final step of connecting your results with the real world. Of course, in order for you to make this connection, the situation you are studying must be one that actually arises in the real world and requires some modeling and analysis. We begin this appendix by describing the important role of projects and project activity in learning mathematical modeling. We discuss several different types of projects and consider how appropriate topics for projects can be identified. Finally, we address the importance of communicating results and several ways of evaluating projects and project activity. The goal of this discussion is to help students, teachers, and researchers improve their mathematical modeling skills.

A.1   The Role of Projects and the Types Useful in Learning Model Building

How does one go about practicing the cycle of model building described in Chapter 1? Of course, there is no substitute for working on actual problems that arise in nature or society, but such problems are typically quite complex and may be beyond the reach of most beginners. Thus, as with most activities, it is best to start with problems that are relatively straightforward and work up to serious research problems. In selecting and working on examples of the modeling process, it is important to be aware that modeling projects come in many sizes and shapes. Also, as we noted in Chapter 1, mathematical modeling may be undertaken for several reasons. For example, models may be used by a company for decision making, by a governmental unit to form policies, or by a team of scientists to help them understand certain natural phenomena. In working on modeling projects, it is important to be certain of the purpose of the activity, because that purpose will play a major role in decisions about the type of model and the sort of information to be produced. In general, determining the purpose includes determining the goal of the study and the audience for the report. A precise specification of the goal of a project is an essential first step. This includes a clear understanding of the meaning of any terms used in describing the problem and the objectives of the study—and agreement on this by all those involved in the project. More than one modeling project has failed to meet its objectives because there was misunderstanding at this point and, as a consequence, the project results had little to do with the interests of those who posed the original problem. It is also essential to understand the audience for whom the work is intended. If the task is to produce a report that can be used to implement a recommendation, then technical detail may be critical. On the other hand, if the task is to produce a report that surveys the options for management decision making, then including technical detail may be unnecessary or even counterproductive.

To organize our discussion of modeling project activity, it is helpful to classify projects in two ways: first by the degree of structure in the specifications of the project and second by the origin of the project. Projects can be very structured, very unstructured, or somewhere in between, and the source of the project may be the instructor of the course, a team of students in the course, or a client outside the setting of the course. We use the terms *structured, unstructured,* and *client-driven* to indicate characteristics of projects, and a specific project may have more than one characteristic. Also, different projects may have a characteristic in different degrees. For instance, one project in a group may be completely client-driven, and another may have originated as client-driven but also have aspects that go beyond the original client's interests. In addition, for example, a project may have one part that is highly structured and another part that is relatively unstructured. The terminology is useful, but it must be used in a flexible way.

Structured projects are the usual starting point for someone who is learning the modeling process. These are projects for which someone, perhaps an instructor in a modeling course or a group leader in a corporate setting, specifies the goals of the study, the basic setting (the details of the aspects of the real world to be studied), the general nature of the model to be used, and the sort of results to be obtained. For instance, the goal might be to determine how best to use a rectangular field with specified size for a parking lot for regular passenger cars. The term *best* might be specified to mean that the largest number of cars can be safely parked. In a structured project, criteria for safe parking and the appropriate entrance/exit constraints would be given as part of the background information for the problem. Several examples of structured projects are given in Section A.2.

Unstructured projects are those in which some of the key assumptions are only partially specified or are not specified at all, and some real-world aspects of the situation must be studied and quantified as part of the modeling task. For instance, in the parking lot situation above, a part of the project might be researching standards for safe parking and the entrance/exit constraints. Or it might be that the size of the field, its dimensions, or the nature of the vehicles to be accommodated are unspecified. In such cases, the modeling team might need to collect information on the possible locations for a lot, the types of vehicles that would use it, the cost implications of various alternatives, and so on. It is likely that a useful report would provide information on several scenarios.

Client-driven projects are usually assigned to a modeling team by a company or a governmental unit. Such a project is presented as a problem that must be solved to enable the management of the company to make a decision or plan a course of action. The modeling team must work with the client to understand the setting and to determine which aspects of the problem are important and which can be ignored or simplified in the course of solving the problem. For example, the problem might be one in routing and scheduling delivery trucks. Key facts might be that the company owns four trucks and that the maximum number of hours that drivers are permitted to drive per day is ten. On the other hand, it might not be important that two of the trucks have Cummins engines and two have Ford engines. In such a situation, the modeling team, serving as consultants to the company, must work with company management to learn the key goals of the project. Is the goal simply to minimize delivery costs, or are the goals more complicated, perhaps involving customer preferences and company growth plans?

In all of these types of projects, and especially in client-driven projects, presentation of the results is a critical step. It is not enough to formulate a model, carry out a careful study of the model, and reach conclusions. The modeling team must be able to explain and defend its work. In many cases, the explanation will be presented to individuals who have not studied the problem in the same detail as the members of the team. The team must prepare both oral and written presentations, and the value of the work is often largely determined by the strength of the presentation. In Section A.3, we discuss the goals and essential items needed to make a good presentation. First, however, in Section A.2, we describe a number of modeling projects that have been used to enable students (and others) to gain experience in the modeling process. These are samples of the projects that we assigned to students to practice modeling.

Many of the examples in Chapters 2 through 5 provide siuations (or problems) that could be used as projects or are simplifications of such situations. For example, there are many articles in recent literature in biology and ecology wherein models similar to those discussed in Chapters 2, 3, and 4 are used to understand phenomena in those fields, and many of the topics considered in those articles can be used as a basis for projects. That is, the setting and the data provide the raw material for worthwile project activity. Similarly, the literature in psychology and business provides useful project settings in which the models developed in this book can be used. Of course, the models discussed here are simply examples of useful mathematical models, and in many situations in which the modeling process is useful, the resulting models involve mathematical ideas not pursued explicitly in this book. Some of the examples of project activity included in this appendix can be approached using ideas discussed in the book, but several have natural models that involve ideas not systematically developed here. However, all can be approached using the basic process developed in Chapter 1.

A.2   Examples of Projects

Each of the examples described below was given to a team of students in one of our classes. In some cases the students worked in class for 75 minutes. In other cases they worked over a weekend, picking up the project on Friday and turning in the report on Monday. In a few cases, projects lasted several weeks.

Typically, we give the following directions to teams preparing student presentations and project reports:

1. Specify the goal of the project and include a clear statement of the meaning of terms.
2. Carefully state the assumptions you make in your approach to solving the problem. Provide a rationale for your assumptions when appropriate.
3. Carry out an analysis that supports your conclusions or recommendations.
4. If appropriate, provide error estimates that give bounds for your numerical results. That is, give numbers that you are confident are below and above your answer.

5. Discuss the strengths and weaknesses of your approach to the problem.
6. Include any pertinent computer files as well as a list of references (print and Web).

In each of the following projects, we indicate the time period provided for completion (one class period, one weekend, or long-term), the setting for the project, and the task to be completed. In a few cases we also provide a list of activities that should be carried out to complete the assigned task. Normally we would not provide this list to the team assigned to carry out the project. Part of the project would be to create this list.

■ Locating a Community College Campus (Weekend Project, 72 hours)

The Setting. The board of regents of a community college system plans to locate a new campus in Metroburg. The board believes that the community college audience will come primarily from the area of Metroburg shown on the map in Figure A.1. The area is divided into three subareas (labeled A, B, and C on Figure A.1), each of which has its own characteristics and a business core with relatively few residents. Subarea A is in the western part of Metroburg, subarea B is in the northeast, and subarea C is in the southeast. The area adjacent to the river (shaded on the map) is used primarily for business/industry and parks and also has relatively few residents.

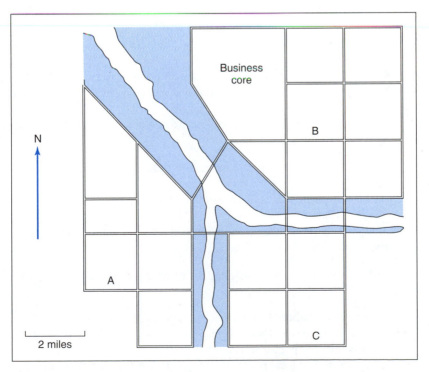

Figure A.1

Table A.1

District	Population	Number of High School Graduates per Year	Participation Rate among Nontraditional Students (students/1000 residents)
A	12,000	250	10
B	24,000	325	13
C	16,000	225	18

Because the college has no residence halls, most of the students commute from their homes, and commuting time is known to be an important factor to such students. The arterial streets used by most commuters are shown on the map, and on these streets the speed limit is 45 mph. In addition to the arterial streets shown, most of the rest of the city is served by a typical east–west/north–south grid of residential/small business streets with a speed limit of 30 mph. There are four bridges over the east and west forks of the Mulberry River, and each time a bridge is used, the expected commuting time on the segment containing the bridge is approximately double that anticipated on the basis of distance alone.

Table A.1 includes data on population size at the most recent census, numbers of high school graduates, and postsecondary participation among nontraditional students for the most recent year. These data are given for each of the three subareas.

The Task.   The task is to recommend a location for the community college campus. The recommended location should respond to the concern of reducing average commuting time as much as possible. The availability or cost of land for the campus need not be considered at this stage.

Task Activity Plan.   It is natural to consider the basic assumptions as grouped into two classes: (1) assumptions having to do with the population, and (2) assumptions having to do with traffic flow.

Using only the information provided in the setting, we can take the following demographic assumptions as reasonable:

- High school graduates who will attend the community college are uniformly distributed within each district.
- Individuals who will be nontraditional students at the community college are uniformly distributed within each district.
- The percentage of high school graduates who will attend the community college is the same in each district.

Using only the information provided in the setting, we make the following assumptions about traffic flow:

- Differential speeds on local and arterial streets are the same in all districts.
- The relation between actual speeds and speed limits is the same in all districts.
- Actual speeds are independent of population density.
- All arterial streets are shown, and local streets run only east–west and north–south.
- Traffic conditions are similar in all districts, and travel time is proportional to distance on local streets, on arterial streets, and on bridges.

The task of "determining a recommended location" should be simplified. For example, we could phrase it as follows: Identify a finite set S of possible locations—the centers or the corners of the blocks are natural candidates—and provide an argument to justify considering only these locations. Such an argument might show, for instance, that for any possible location P, there is a location P' in S such that the average commuting time to P is no less than the average commuting time to P'.

An evaluation of the results should include a comparison of the average commuting times for the locations in S. Is there a subset of S such that the average commuting times to all locations in that subset are about the same and smaller than the average commuting time to the locations not in the subset? If so, what are the implications?

■ Construction of an Earthen Dam (Weekend Project, 72 hours)

The Setting.   A landowner plans to construct an earthen dam across the mouth of a small hollow to create a small lake. The site of the proposed dam is shown in the upper left corner of Figure A.2, where the axis of the dam is a heavy straight line. The contour intervals in

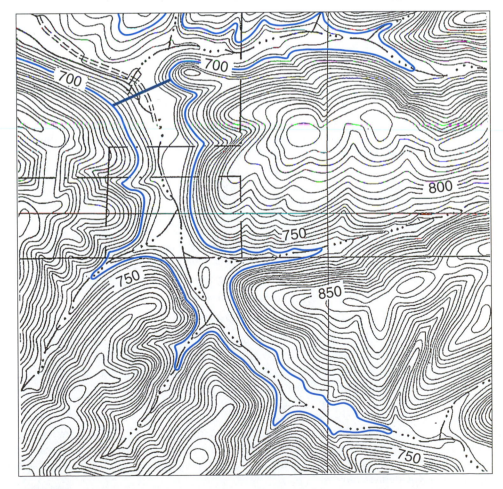

Figure A.2

Figure A.2 are 10 feet, and the contours at multiples of 50 feet are heavier lines and are labeled. The scale on the map in Figure A.2 is 1:15000.

The Task.   The landowner plans a lake in which the pool height is 700 feet, and she is interested in estimates on the amount of earth necessary to construct the dam and the volume of water impounded when the lake is at normal pool height. The estimates should be accompanied by a discussion of their accuracy.

Task Activity Plan.   To estimate the earth required for the dam, you will need information on the cross section of the dam and the height of the top of the dam in comparison to the planned pool height. You will need to consider what is involved in maintaining the height of the lake at the pool level and the implications of this for the amount of earth required for the dam.

Estimating the volume of water impounded when the lake is at normal pool height requires deciding how to use the contour information provided in Figure A.2.

Providing error estimates for your answer requires paying attention to the details of your approach—keeping track of how the various parts of your data are used and then determining how much the actual values of the data could vary from the values used in your estimates.

Earthen dams are common, and it is likely that there are examples in your area. Details about dam construction can be found in the literature and on the Web.

■ Allocating Teachers (90-minute, In-Class Project)

The Setting.   A new wing has been added to the Magnet School for Math, Science, and Technology so that the student population can be increased by 142 students, from 480 to 622. The new sophomore class (for the 2005–2006 school year) will have 142 more students than the graduating senior class (the class of 2005). To accommodate the increase, the size of the faculty is to be increased by 7. There has been an extended discussion about which departments should be allotted the additional teachers. At the time of the enrollment increase, there were

6 math instructors	5 English instructors
3 chemistry instructors	3 foreign language instructors:
3 physics instructors	1 Spanish, 1 German, 1 French
4 biology instructors	1 music instructor
4 social studies instructors	1 art instructor

Several questions have been discussed, including the following: Should every department (English, Social Studies, Math, Physics, Biology, Chemistry, Foreign Language, Music, and Art) each receive one teacher, or does the demand on some departments argue for their receiving two new teachers while other departments get none? Should any department get three new members?

The Task.   Given the present number of students taking classes in each department (see Table A.2), how would you apportion the seven new faculty members to departments *most fairly?* Please describe your method in detail. Also, explain why you think your method is fair and how you define "fair."

Table A.2   Course Enrollment Totals, September 1994

Subject Area	10th	11th	12th	Total
Art	21	33	15	69
Biology	198	95	26	319
Chemistry	59	126	109	294
English	183	155	151	489
French	41	32	49	122
German	19	22	10	51
Mathematics	184	201	262	647
Music	20	26	49	95
Physics	50	58	183	291
Social Studies	183	131	59	373
Spanish	51	26	33	110

■ A Pipeline Flow Problem (75-minute, In-Class Project)

The Setting.   The Alaska pipeline consists of a network of pipes with various flow capacities (gallons/second). Flow through the network must satisfy three rules:

1. Flow is allowed only in the direction indicated on each pipe.
2. The actual flow in any pipe cannot exceed the pipe's capacity.
3. The amount of flow arriving at a node must equal the amount of flow leaving the node.

The Task.   Given the graph theory model of the pipeline shown in Figure A.3 (the edge weights represent flow capacities, and the arrows indicate direction of flow), we are interested in the maximum possible flow from Prudhoe Bay to Seward.

1. Find the maximum possible flow. Provide a proof that your solution is indeed the largest possible flow and/or a general algorithm (method) for finding the maximum flow in a network.
2. If the maximum flow had to be increased by at least 3 units of flow, but for as little cost as possible, what changes would you propose? (Make some reasonable assumptions about relative costs.)

■ An Irrigation Problem (90-minute, In-Class Project)

The Setting.   A field to be irrigated is a rectangular area 200 feet wide and 500 feet long. You have a mobile irrigation system consisting of a pipe mounted on wheels. The assembly is set across the 200-foot width of the field and moves from one end to the other at constant speed. Each nozzle sprays water over a circular area with a 10-foot radius centered at the nozzle. Water sprays from each nozzle at the rate of 2 cubic feet per minute.

Nozzles can be located along the pipe at places you choose, subject to the constraint that they must be at least 10 feet apart. You can select the locations of the nozzles and the speed at which the device moves across the field, subject to the constraint that the speed cannot exceed 3 feet per minute.

Each part of the field must receive at least 0.5 inch of water.

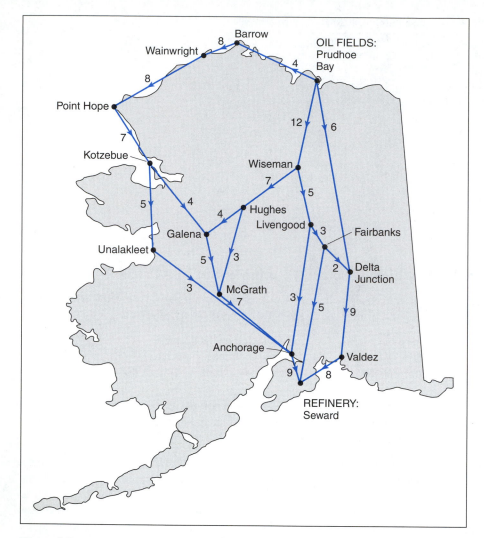

Figure A.3

The Task. The primary goal is to design a field irrigation system that provides the water required with the equipment available. A secondary goal is to accomplish this while using the least possible amount of water.

The primary task is to select locations of the nozzles and a speed for the device so that you provide the water required. Do so, and, for your selection of nozzle locations and speed, determine how much water is used and how long it takes to irrigate the field. Also consider the water that exceeds the requirement—that is, areas of the field that receive more than 0.5 inch.

The next task is to refine your choices to reduce the amount of water used as much as possible, while continuing to meet the requirements.

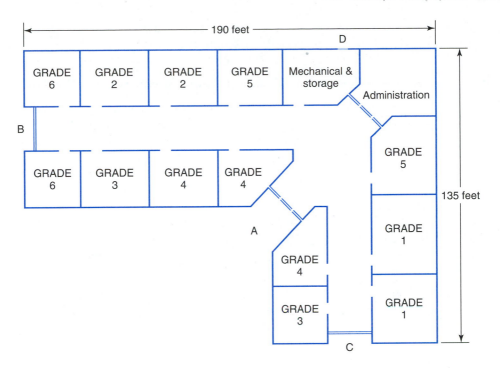

Figure A.4

■ An Evacuation Plan for an Elementary School (75-minute, In-Class Project)

The Setting. The superintendent of the Mulberry Consolidated School Corporation (MCSC) has asked the principal of the Topnotch Elementary School to update the emergency evacuation plan for the school.

A floor plan for the school, approximately to scale, is shown in Figure A.4. Classrooms are currently assigned to the six grade levels as shown on the plan. There are two double doors at the main entrance (the entrance labeled A on the plan), and there is one double door at each of the other entrances (B and C on the plan). There is a service door in the maintenance room (shown as D on the plan), but that door does not meet codes for student use.

The Task. The task is to formulate an emergency evacuation plan for the school. Part of the task is to decide what constitutes an evacuation plan, how different plans should be compared, and what makes an evacuation plan a good one.

Could a better evacuation plan be devised if the rooms were assigned to grade levels in a different way?

■ A Vaccination Problem (Long-Term, Several Weeks)

The Setting. Health officials in Metropolis are concerned that they may need to vaccinate a large percent of the local population for small pox in a short time, and yet they have not made the required preparations.

The Task. Design a plan to carry out an emergency inoculation from one site in a modest-sized city (of about 100,000 people) in a period of five days.

■ A Credit Union Scheduling Problem (Long-Term, Several Weeks)

The Setting. The manager of the local credit union branch has asked for help in scheduling part-time tellers. The problem is that demand for tellers varies greatly depending on the time of the day, the day of the week, the day of the month, and the month of the year. If there are too many tellers, then some are idle for long periods of time (at a cost to the credit union), and if there are too few, then customers wait a long time in a queue and are unhappy (at a cost to the credit union).

The Task. Work with the manager of the credit union to develop a satisfactory method of scheduling part-time tellers so that goals of the manager are achieved.

Task Activities.

- Determine the current pattern of arrivals at the credit union, and develop a method to forecast arrivals by day and time of day. It is very likely that the credit union has transaction records that will help greatly in developing the forecasting methods.
- Collect data on service times so that a queuing model can be developed. This may require that team members visit the branch and record service times.
- Develop a queuing model that relates the number of tellers in service to the average length of the queue and the average time a customer spends in the branch.
- Work with the manager to specify goals for the time an average customer should spend in line.
- Develop a formula for the number of tellers needed to achieve the manager's goal for the average time a customer should spend in line.
- Develop a scheduling algorithm for part-time tellers that accommodates the requirements for part-time work (for example, a shift may need to be a least 3 hours long) and achieves the goals of the manager at a minimum cost.

A.3   Reports and Presentations

As we noted earlier, it is not enough for a modeling team to do a good job of studying the problem presented to them. Team members also need to communicate the results of their work effectively, convince others that their approach and results are worthwhile, and demonstrate that, in some sense, the task given to them has been accomplished. This usually involves writing a report that describes and defends the work done, and it may also involve a presentation to a client or other audience.

An essential goal is to communicate effectively. The likelihood of effective communication is enhanced if the report or presentation is well organized and clearly presented. The audience (for the report and for the presentation) may well include people who have not been involved in earlier discussions and who know much less about the situation than you. In order for your work to be accessible to such an audience, it is essential that the terminology and notation be clearly defined and carefully used. If part of your audience does not have a technical background—for instance, if people with general management responsibilities are included—this must be considered in designing the presentation. After a report or presentation has been prepared, it should be reviewed to make sure that the results are clearly stated and the most important results are given the greatest emphasis.

Some of the main points that should be considered in preparing the report or presentation follow.

1. What is the problem or area of study? This may seem obvious to the team studying the problem, but it may not be obvious to someone who is just now presented with the results of the study. It is important to set the scene carefully so that the results of the study will make sense to anyone interested who reads or listens to a presentation about it. Many times, a problem is originally posed in a vague or imprecise way, and part of the modeling task is to clarify what earlier had only been understood or stated approximately. An important step in a report or presentation is to convince the reader (or listener) that the problem studied by the modeling team is an appropriate one for the goals of the study originally proposed to the team.

2. How was the study carried out? Were data collected by the team or provided by someone else? Was statistical sampling used, and if so, how was the sample selected? How were the data used? Was the study an analytical one or was it a simulation? Or were some parts of each involved?

3. What are the results of the study? The team needs to describe what its members have learned about the area of study and their solution to any problems posed to them. This is a critical step in the process, of course, and more than one method of presentation may be needed. For example, graphs and charts may be prepared to display results, and it may be useful to compare the results obtained in the current study with those obtained in earlier studies or by different methods in the same study.

4. What are the conclusions of the study and how valid are they? This part of the report or presentation is the one where the results of the study are compared to the real world. They are interpreted in light of the problem posed at the start of the study. As part of this comparison, the team needs to look critically at its work and note the strengths and limitations (if any) of the work. Frequently, it is helpful to note how the assumptions are related to the conclusions. If you know that an assumption is *not* completely valid, how does this affect the results of the study? This is also the place to note the options for additional study of this or related problems that may be of interest to the clients.

A.4   **Evaluating Project Reports**

Project reports cannot be evaluated in the same way as one would evaluate a traditional mathematics exam. In general, there is no single correct answer. Even the successful methods used on a specific project can vary greatly from one team to another. There are two complementary methods that are often used to evaluate a project report. The first method is to create and use a rubric. In this method, one would list all the items needed for successful completion of the project, assign to each item a number of points corresponding to its importance, and then look for each item in the report and assign it a full or partial credit based on the results and the report. Finally, the points for all items are added up, and the project receives a score (or grade). One such rubric is shown in Table A.3.

Table A.3

Subscores	
Executive Summary	0–10
Assumptions/Rationale	0–10
Model Design and Development	0–40
Evaluations/Sensitivity/Strengths/Weaknesses	0–15
Results/Conclusions/Recommendations	0–15
Style/Format/Clarity	0–10

An alternative to using a rubric is to grade projects holistically. In this method, the project report is considered in its entirety, and a summative evaluation (a score or grade) is given to the entire report without assigning specific scores (or grades) to the individual sections or pieces. If one is grading the work of several teams that all worked on the same project, one might try to rank-order the reports holistically before attempting to assign a final score or grade.

Of course, the methods of evaluating by a rubric and evaluating holistically are not mutually exclusive. One could use a rubric to obtain an initial score and then use a holistic method to refine that score and assign a final evaluation. In the end, there is always a goal(s) for the project, and any evaluation of a project report should measure the progress made toward achieving that goal.

A.5 Sources of Projects

Working on projects is an extremely important part of learning how to do mathematical modeling, and it is important to have creative ideas about how to generate these projects. We comment both on client-driven projects and on self-initiated or instructor-initiated projects. Because client-driven projects usually require significant contacts outside of the classroom and the university, and because they call for efforts that may not be a part of the past experiences of students (or instructors), we comment on such projects first.

■ Client-Driven Projects

We begin with a comment on methods that may appear to be a natural first step but that we have found *not* to be particularly useful. The first such method is writing to executives of corporations or government agencies and asking about possible projects. In the absence of a previous connection with the executive, such "cold call" letters are rarely effective. The same approach conducted by phone—that is, a call asking for contacts, to be followed by calls to the suggested people—also rarely works.

Corporate managers and executives in government agencies are much more likely to respond if you have been introduced or recommended by someone who is already known to them. With this in mind, it tends to be more fruitful to contact people you know (neighbors, friends, relatives, and the like), discuss the project concept with them, and ask whether they know of opportunities for team projects or of other people who might know of opportunities. It may be useful for you to have specific firms, or types of firms or agencies, in mind when you broach the subject. Once you have a contact at a company or agency, it is much easier to find the right person to talk with or to gain an introduction to that person. When you contact an individual who may be interested (or may know of someone who might be), suggesting types of projects and having examples at hand will frequently spark immediate interest. In today's market, projects that involve forecasting, scheduling, quality control, or inventory control often generate a positive corporate response.

In our experience, especially good sources of projects are university offices and city or state agencies that work with small businesses. Such agencies can often suggest companies that may be interested. And frequently, they will make the initial introductions.

The likelihood of interesting a client in working with you will be greatly enhanced if you have good examples of past work, either by you or by other teams with similar backgrounds. It is unnecessary to have a detailed description of your work—and perhaps it is even undesirable to offer such a description unless it is requested—but being prepared to describe, in a few sentences, some past projects and what the modeling team was able to do is frequently valuable in increasing a client's interest. On the other hand, it is important to be realistic in describing what you can (and what you cannot) accomplish. It is in neither your nor the client's interest to create false expectations. Throughout discussions with potential clients, the primary concern must be the educational value of the experience for students. For that reason, projects that are primarily programming, or data collection and entry, while perhaps of high interest to a client, are rarely appropriate vehicles for students to learn mathematical modeling.

■ Self-Initiated or Instructor-Initiated Projects

Client-driven projects are usually major projects that require extended time for completion—a semester or even an academic year. Most projects in modeling courses are not client-driven and are carried out in shorter periods, frequently a few weeks. They are interesting and challenging situations that someone—an instructor, a person in another field, or a student—has identified as a modeling project. Some such projects require direct data collection. For example, a team studying a city or university bus system or a team studying the allocation of shared printers to student workstations may need to collect data. However, many projects can use data that were collected by others. Such data can be found in books, in periodical literature, in technical reports produced by research groups in social science or life science departments, or on the Web.

Many journals in the social and life sciences contain articles whose goal is to use mathematical concepts and techniques to understand situations in those fields. These articles are frequently written from the perspective of the discipline, and the mathematical modeling activity is in the background, as is appropriate for the intended audience. Such articles, or several such related articles, may form the basis of a modeling project. The task would be for the team carefully to identify the assumptions, work through the details of model construction, confirm the mathematical arguments, and perhaps look at alternative models. It is sometimes worthwhile to identify just where the assumptions are used in development of the model and to consider whether there are additional assumptions that may not be explicitly stated.

As students and instructors explore options for such projects, several questions need to be considered. First, does the team have enough knowledge of the real-world situation to make informed choices when building a model? If the team does not have such knowledge, can access to it be arranged? Discussing the situation with someone knowledgeable or reading an especially good survey article on the area will often provide the necessary background information.

Second, once the situation is identified and enough is known about it to begin to consider model construction, does the team have the technical skills needed to conduct the study? Such skills may be conceptual, mathematical, or computational. These questions are difficult to answer in advance, because the problems that arise in modeling frequently involve questions that do not fit neatly into standard subject matter areas. An experienced instructor can provide valuable advice.

■ The Mathematical Contest in Modeling

Since 1984, an international Mathematical Contest in Modeling (MCM) has been administered under the auspices of COMAP, Inc. Each year the contest takes place in February, the papers are then graded, and the results are announced about mid-March. Each team selects a problem from two options, and devotes a weekend to solving the problem and writing a report. The entries are graded, and the best papers are recognized. Each year some of the best papers are published in *The UMAP Journal*, and selected papers from the first ten years, 1984–1994, were published in a special edition, *The UMAP Journal: Tools for Teaching 1994*. Other problems posed for the MCM during the period 1984–1994 are also described in that volume. A similar volume covering the period 1995–2004 was scheduled for publication in 2005. The contest problems are usually quite challenging, but simplified versions may be appropriate for projects, or they may generate ideas about related problems.

Index